Methods in Molecular Biology

Series editor
John M. Walker
School of Life and Medical Sciences
University of Hertfordshire
Hatfield, Herts., United Kingdom

For further volumes:
http://www.springer.com/series/7651

Chikungunya Virus

Methods and Protocols

Edited by

Justin Jang Hann Chu

Department of Microbiology and Immunology, National University of Singapore, Singapore

Swee Kim Ang

Department of Microbiology and Immunology, National University of Singapore, Singapore

Humana Press

Editors
Justin Jang Hann Chu
Department of Microbiology and Immunology
National University of Singapore
Singapore

Swee Kim Ang
Department of Microbiology
and Immunology
National University of Singapore
Singapore

ISSN 1064-3745 ISSN 1940-6029 (electronic)
Methods in Molecular Biology
ISBN 978-1-4939-8096-3 ISBN 978-1-4939-3618-2 (eBook)
DOI 10.1007/978-1-4939-3618-2

Printed on acid-free paper

This Humana Press imprint is published by Springer Nature
The registered company is Springer Science+Business Media LLC New York

Preface

Chikungunya virus (CHIKV) is a mosquito-borne, single-stranded, positive-sense RNA virus belonging to the *alphavirus* genus of the family *Togaviridae*. CHIKV infection in humans has re-emerged as a significant infectious disease which is no longer confined to developing countries in the tropics due to recent outbreaks in various geographical regions worldwide including America and Europe. *Chikungunya: Methods and Protocols* provides the increasing number of CHIKV researchers a useful handbook covering multidisciplinary approaches on various aspects of CHIKV research brought together by leading laboratories across the globe. Topics covered include techniques in clinical and diagnostic virology, basic protocols in cell and virus culture, and bioinformatics and proteomics approaches in cellular response studies. In addition, methods in immunology and animal model studies, as well as different strategies of antivirals and vaccine development for therapeutics against CHIKV infection are also covered in this comprehensive handbook.

Singapore *Justin Jang Hann Chu*
Singapore *Swee Kim Ang*

Contents

Preface v
Contributors . . . . . . *xi*

PART I CLINICAL AND DIAGNOSTIC VIROLOGY

1 Evolution and Epidemiology of Chikungunya Virus 3
Giada Rossini, Maria Paola Landini, and Vittorio Sambri

2 Molecular Epidemiology of Chikungunya Virus by Sequencing 11
Ravindran Thayan, Mohd Apandi Yusof, Zainah Saat, Shamala Devi Sekaran, and Seok Mui Wang

3 Advanced Genetic Methodologies in Tracking Evolution and Spread of Chikungunya Virus 21
Hapuarachchige Chanditha Hapuarachchi and Kim-Sung Lee

4 Synthetic Peptide-Based Antibody Detection for Diagnosis of Chikungunya Infection with and without Neurological Complications 39
Rajpal S. Kashyap, Shradha S. Bhullar, Nitin H. Chandak, and Girdhar M. Taori

5 Expression and Purification of E2 Glycoprotein from Insect Cells (Sf9) for Use in Serology 51
Chong Long Chua, I-Ching Sam, and Yoke Fun Chan

6 Diagnostic Methods for CHIKV Based on Serological Tools 63
Paolo Gaibani, Maria Poala Landini, and Vittorio Sambri

7 Utilization and Assessment of Throat Swab and Urine Specimens for Diagnosis of Chikungunya Virus Infection 75
Chandrashekhar G. Raut, H. Hanumaiah, and Wrunda C. Raut

PART II CELL CULTURE, VIRUS REPLICATION, AND CELLULAR RESPONSES

8 Propagation of Chikungunya Virus Using Mosquito Cells 87
Swee Kim Ang, Shirley Lam, and Justin Jang Hann Chu

9 Infectious Viral Quantification of Chikungunya Virus—Virus Plaque Assay . . . 93
Parveen Kaur, Regina Ching Hua Lee, and Justin Jang Hann Chu

10 Detection and Quantification of Chikungunya Virus by Real-Time RT-PCR Assay 105
Seok Mui Wang, Ummul Haninah Ali, Shamala Devi Sekaran, and Ravindran Thayan

11 Chikungunya Virus Infection of *Aedes* Mosquitoes 119
Hui Vern Wong, Yoke Fun Chan, I-Ching Sam, Wan Yusof Wan Sulaiman, and Indra Vythilingam

12 Analysis of CHIKV in Mosquitoes Infected via Artificial Blood Meal 129
Jeremy P. Ledermann and Ann M. Powers

13 Chikungunya Virus Growth and Fluorescent Labeling: Detection of Chikungunya Virus by Immunofluorescence Assay 143
Meng Ling Moi and Tomohiko Takasaki

14 Virus Isolation and Preparation of Sucrose-Banded Chikungunya Virus Samples for Transmission Electron Microscopy 153
Chang-Kweng Lim

15 Viral–Host Protein Interaction Studies Using Yeast Two-Hybrid Screening Method 163
Namrata Dudha and Sanjay Gupta

16 Application of GelC-MS/MS to Proteomic Profiling of Chikungunya Virus Infection: Preparation of Peptides for Analysis 179
Atchara Paemanee, Nitwara Wikan, Sittiruk Roytrakul, and Duncan R. Smith

17 Bioinformatics Based Approaches to Study Virus–Host Interactions During Chikungunya Virus Infection 195
Sreejith Rajasekharan and Sanjay Gupta

18 T-Cell Epitope Prediction of Chikungunya Virus 201
Christine Loan Ping Eng, Tin Wee Tan, and Joo Chuan Tong

Part III Immunology and Animal Model Studies

19 Mouse Models of Chikungunya Virus 211
Lara J. Herrero, Penny A. Rudd, Xiang Liu, Stefan Wolf, and Suresh Mahalingam

20 Generation of Mouse Monoclonal Antibodies Specific to Chikungunya Virus Using ClonaCell-HY Hybridoma Cloning Kit 225
Chow Wenn Yew and Yee Joo Tan

21 Immunohistochemical Detection of Chikungunya Virus Antigens in Formalin-Fixed and Paraffin-Embedded Tissues 235
Chun Wei Chiam, I-Ching Sam, Yoke Fun Chan, Kum Thong Wong, and Kien Chai Ong

Part IV Antivirals and Vaccines

22 Antiviral Strategies Against Chikungunya Virus 243
Rana Abdelnabi, Johan Neyts, and Leen Delang

23 A Real-Time Cell Analyzing Assay for Identification of Novel Antiviral Compounds against Chikungunya Virus 255
Keivan Zandi

24 Using Bicistronic Baculovirus Expression Vector System to Screen the Compounds That Interfere with the Infection of Chikungunya Virus 263
Szu-Cheng Kuo, Chao-Yi Teng, Yi-Jung Ho, Ying-Ju Chen, and Tzong-Yuan Wu

25 Neutralization Assay for Chikungunya Virus Infection: Plaque Reduction Neutralization Test 273
Nor Azila Muhammad Azami, Meng Ling Moi, and Tomohiko Takasaki

26 Reverse Genetics Approaches for Chikungunya Virus 283
Patchara Phuektes and Justin Jang Hann Chu

27 Production of Chikungunya Virus-Like Particles and Subunit Vaccines in Insect Cells 297
Stefan W. Metz and Gorben P. Pijlman

28 Protocols for Developing Novel Chikungunya Virus DNA Vaccines 311
Christopher Chung, Kenneth E. Ugen, Niranjan Y. Sardesai, David B. Weiner, and Kar Muthumani

Index *333*

Contributors

RANA ABDELNABI • *Laboratory of Virology and Chemotherapy, Department of Microbiology and Immunology, Rega Institute for Medical Research, KU Leuven – University of Leuven, Leuven, Belgium*

UMMUL HANINAH ALI • *Department of Medical Microbiology, Faculty of Medicine, University of Malaysia, Jalan Universiti, Kuala Lumpur, Malaysia; Virology Unit, Infectious Diseases Research Center, Institute for Medical Research, Kuala Lumpur, Malaysia*

SWEE KIM ANG • *Department of Microbiology and Immunology, Yong Loo Lin School of Medicine, National University Health System, National University of Singapore, Singapore, Singapore*

NOR AZILA MUHAMMAD AZAMI • *Department of Virology I, National Institute of Infectious Diseases, Tokyo, Japan; Graduate School for Comprehensive Human Sciences, University of Tsukuba, Tsukuba, Japan*

SHRADHA S. BHULLAR • *Biochemistry Research Laboratory, Central India Institute of Medical Sciences, Nagpur, India*

YOKE FUN CHAN • *Department of Microbiology, Faculty of Medicine, University of Malaya, Kuala Lumpur, Malaysia*

NITIN H. CHANDAK • *Biochemistry Research Laboratory, Central India Institute of Medical Sciences, Nagpur, India*

YING-JU CHEN • *Bioengineering Group, Institute of Biologics, Development Center for Biotechnology, New Taipei, Taiwan*

CHUN WEI CHIAM • *Department of Medical Microbiology, Faculty of Medicine, University of Malaya, Kuala Lumpur, Malaysia*

JUSTIN JANG HANN CHU • *Department of Microbiology and Immunology, Yong Loo Lin School of Medicine, National University Health System, National University of Singapore, Singapore, Singapore*

CHONG LONG CHUA • *Department of Medical Microbiology, Faculty of Medicine, University of Malaya, Kuala Lumpur, Malaysia*

CHRISTOPHER CHUNG • *Department of Pathology and Laboratory Medicine, University of Pennsylvania School of Medicine, Philadelphia, PA, USA*

LEEN DELANG • *Laboratory of Virology and Chemotherapy, Department of Microbiology and Immunology, Rega Institute for Medical Research, KU Leuven – University of Leuven, Leuven, Belgium*

NAMRATA DUDHA • *Center for Emerging Diseases, Department of Biotechnology, Jaypee Institute of Information Technology, Noida, Uttar Pradesh, India*

CHRISTINE LOAN PING ENG • *Department of Biochemistry, Yong Loo School of Medicine, National University of Singapore, Singapore, Singapore*

PAOLO GAIBANI • *Regional Reference Centre for Microbiological Emergencies (CRREM), Unit of Microbiology, S. Orsola-Malpighi Hospital, Bologna, Italy*

SANJAY GUPTA • *Center for Emerging Diseases, Department of Biotechnology, Jaypee Institute of Information Technology, Noida, Uttar Pradesh, India*

H. HANUMAIAH • *National Institute of Virology, Bangalore Unit, Bangalore, India*

Hapuarachchige Chanditha Hapuarachchi • *Environmental Health Institute, National Environment Agency, Singapore, Singapore*
Lara J. Herrero • *Emerging Viruses and Inflammation Research Group, Institute for Glycomics, Griffith University–Gold Coast Campus, Southport, QLD, Australia*
Yi-Jung Ho • *Institute of Prevention Medicine, National Defense Medical Center, Taipei, Taiwan*
Rajpal S. Kashyap • *Biochemistry Research Laboratory, Central India Institute of Medical Sciences, Nagpur, India*
Parveen Kaur • *Department of Microbiology and Immunology, Yong Loo Lin School of Medicine, National University Health System, National University of Singapore, Singapore, Singapore*
Szu-Cheng Kuo • *Institute of Prevention Medicine, National Defense Medical Center, Taipei, Taiwan*
Shirley Lam • *Department of Microbiology and Immunology, Yong Loo Lin School of Medicine, National University Health System, National University of Singapore, Singapore, Singapore*
Maria Poala Landini • *Unit of Microbiology, Regional Reference Centre for Microbiological Emergencies (CRREM), St. Orsola Malpighi Hospital, Bologna, Italy; DIMES, University of Bologna, Bologna, Italy*
Jeremy P. Ledermann • *Centers for Disease Control and Prevention, Fort Collins, CO, USA*
Kim-Sung Lee • *School of Life Sciences and Chemical Technology, Ngee Ann Polytechnic, Singapore, Singapore*
Regina Ching Hua Lee • *Department of Microbiology and Immunology, Yong Loo Lin School of Medicine, National University Health System, National University of Singapore, Singapore, Singapore*
Chang-Kweng Lim • *Laboratory of Neurovirology, Department of Virology 1, National Institute of Infectious Diseases, Tokyo, Japan*
Xiang Liu • *Emerging Viruses and Inflammation Research Group, Institute for Glycomics, Griffith University–Gold Coast Campus, Southport, QLD, Australia*
Suresh Mahalingam • *Emerging Viruses and Inflammation Research Group, Institute for Glycomics, Griffith University–Gold Coast Campus, Southport, QLD, Australia*
Stefan W. Metz • *Department of Microbiology and Immunology, University of North Carolina, Chapel Hill, NC, USA; Laboratory of Virology, Wageningen University, Wageningen, The Netherlands*
Meng Ling Moi • *Department of Virology 1, National Institute of Infectious Diseases, Tokyo, Japan; Institute of Tropical Medicine, Nagasaki University, Nagasaki, Japan*
Kar Muthumani • *Department of Pathology and Laboratory Medicine, University of Pennsylvania School of Medicine, Philadelphia, PA, USA*
Johan Neyts • *Laboratory of Virology and Chemotherapy, Department of Microbiology and Immunology, Rega Institute for Medical Research, KU Leuven – University of Leuven, Leuven, Belgium*
Kien Chai Ong • *Department of Biomedical Science, Faculty of Medicine, University of Malaya, Kuala Lumpur, Malaysia*

ATCHARA PAEMANEE • *Institute of Molecular Biosciences, Mahidol University, Nakorn Pathom, Thailand; National Center for Genetic Engineering and Biotechnology (BIOTEC), National Science and Technology Development Agency, Pathum Thani, Thailand*

PATCHARA PHUEKTES • *Department of Pathobiology, Faculty of Veterinary Medicine, Khon Kaen University, Khon Kaen, Thailand*

GORBEN P. PIJLMAN • *Laboratory of Virology, Wageningen University, Wageningen, The Netherlands*

ANN M. POWERS • *Centers for Disease Control and Prevention, Fort Collins, CO, USA*

SREEJITH RAJASEKHARAN • *Center for Emerging Diseases, Department of Biotechnology, Jaypee Institute of Information Technology, Noida, Uttar Pradesh, India*

CHANDRASHEKHAR G. RAUT • *National Institute of Virology, Bangalore Unit, Bangalore, India*

WRUNDA C. RAUT • *Dr. D. Y. Patil Medical College, Hospital and Research Center, Pune, India*

GIADA ROSSINI • *Unit of Microbiology, Regional Reference Centre for Microbiological Emergencies (CRREM), St. Orsola Malpighi Hospital, Bologna, Italy*

SITTIRUK ROYTRAKUL • *National Center for Genetic Engineering and Biotechnology (BIOTEC), National Science and Technology Development Agency, Pathum Thani, Thailand*

PENNY A. RUDD • *Emerging Viruses and Inflammation Research Group, Institute for Glycomics, Griffith University–Gold Coast Campus, Southport, QLD, Australia*

ZAINAH SAAT • *Virology Unit, Infectious Diseases Research Centre, Institute for Medical Research, Kuala Lumpur, Malaysia*

I-CHING SAM • *Department of Medical Microbiology, Faculty of Medicine, University of Malaya, Kuala Lumpur, Malaysia*

VITTORIO SAMBRI • *DIMES, University of Bologna, Bologna, Italy; Unit of Microbiology, The Greater Romagna Area Hub Laboratory, Pievesestina, Italy*

NIRANJAN Y. SARDESAI • *Inovio Pharmaceuticals, Plymouth Meeting, PA, USA*

SHAMALA DEVI SEKARAN • *Department of Medical Microbiology, Faculty of Medicine, University of Malaysia, Jalan Universiti, Kuala Lumpur, Malaysia*

DUNCAN R. SMITH • *Institute of Molecular Biosciences, Mahidol University, Nakorn Pathom, Thailand; Center for Emerging and Neglected Infectious Diseases, Mahidol University, Bangkok, Thailand*

WAN YUSOF WAN SULAIMAN • *Department of Parasitology, Faculty of Medicine, University of Malaya, Kuala Lumpur, Malaysia*

TOMOHIKO TAKASAKI • *Department of Virology 1, National Institute of Infectious Diseases, Tokyo, Japan*

YEE JOO TAN • *Monoclonal Antibody Unit, Institute of Molecular and Cell Biology (IMCB), Agency for Science, Technology and Research, Singapore, Singapore; Department of Microbiology, Yong Loo Lin School of Medicine, National University Health System (NUHS), National University of Singapore, Singapore, Singapore*

TIN WEE TAN • *Department of Biochemistry, Yong Loo School of Medicine, National University of Singapore, Singapore, Singapore*

GIRDHAR M. TAORI • *Biochemistry Research Laboratory, Central India Institute of Medical Sciences, Nagpur, India*

CHAO-YI TENG • *Bioengineering Group, Institute of Biologics, Development Center for Biotechnology, New Taipei, Taiwan*
RAVINDRAN THAYAN • *Virology Unit, Infectious Diseases Research Centre, Institute for Medical Research, Kuala Lumpur, Malaysia*
JOO CHUAN TONG • *Department of Biochemistry, Yong Loo School of Medicine, National University of Singapore, Singapore, Singapore; Institute of High Performance Computing, Agency for Science, Technology and Research, Singapore, Singapore*
KENNETH E. UGEN • *Department of Molecular Medicine, University of South Florida Morsani College of Medicine, Tampa, FL, USA*
INDRA VYTHILINGAM • *Department of Parasitology, Faculty of Medicine, University of Malaya, Kuala Lumpur, Malaysia*
SEOK MUI WANG • *Faculty of Medicine, Jalan Hospital, Universiti Teknologi MARA (UiTM), Selangor, Malaysia*
DAVID B. WEINER • *Department of Pathology and Laboratory Medicine, University of Pennsylvania School of Medicine, Philadelphia, PA, USA*
NITWARA WIKAN • *Institute of Molecular Biosciences, Mahidol University, Nakorn Pathom, Thailand*
STEFAN WOLF • *Emerging Viruses and Inflammation Research Group, Institute for Glycomics, Griffith University–Gold Coast Campus, Southport, QLD, Australia*
KUM THONG WONG • *Department of Pathology, Faculty of Medicine, University of Malaya, Kuala Lumpur, Malaysia*
HUI VERN WONG • *Department of Microbiology Faculty of Medicine, University of Malaya, Kuala Lumpur, Malaysia*
TZONG-YUAN WU • *Department of Bioscience Technology, Chung Yuan Christian University, Chung-Li, Taiwan*
CHOW WENN YEW • *Monoclonal Antibody Unit, Institute of Molecular and Cell Biology (IMCB), Agency for Science, Technology and Research, Singapore, Singapore*
MOHD APANDI YUSOF • *Virology Unit, Infectious Diseases Research Centre, Institute of Medical Research, Kuala Lumpur, Malaysia*
KEIVAN ZANDI • *Tropical Infectious Disease Research and Education Center (TIDREC), Department of Medical Microbiology, Faculty of Medicine, University of Malaya, Kuala Lumpur, Malaysia*

Part I

Clinical and Diagnostic Virology

Chapter 1

Evolution and Epidemiology of Chikungunya Virus

Giada Rossini, Maria Paola Landini, and Vittorio Sambri

Abstract

Chikungunya is a mosquito-borne Alphavirus that is spreading worldwide in the tropical areas and that has a 11.8 kb RNA genome. The most relevant vectors belong to the genus *Aedes* and contribute to the diffusion of the three different genotypes of the virus from the original site of first identification in East Africa. Recently, an additional site of origin has been identified in Asia. The epidemiology of Chikungunya has been extensively evaluated from 2004 when the virus initiated its travel eastbound from the coast of Africa to the Indian Ocean. It is noteworthy that this diffusion has been mainly sustained by *Ae. albopictus*, a new vector to which the virus become adapted due to the mutation E1-Ala226Val. This mutation was also identified during the first, even small, outbreaks of Chikungunya-related disease outside the tropics that occurred in Northern Italy in 2007 and in Southern France in 2010. Three years later the virus appeared for the first time in the Western hemisphere and since then, in less than 24 months spread to North and South America.

Key words Chikungunya virus, Arbovirus, Mosquitoes, Evolution

1 Introduction

Chikungunya virus (CHIKV) is a mosquito-borne virus that, since 2004, has caused millions of human cases. CHIKV, which previously had a geographic range restricted to sub-Saharan Africa, the Indian subcontinent, and South East Asia, has recently moved to subtropical latitudes and to the western hemisphere. The mosquito vectors for this virus are globally distributed in tropical and temperate zones, providing the opportunity for CHIKV to continue to expand into new geographic regions [1, 2].

2 Virus, Transmission Cycle and Vectors

Chikungunya virus (CHIKV) belongs to the family *Togaviridae*, genus Alphavirus. CHIKV grouped within the Semliki Forest virus antigenic complex, which also consists of O'nyong nyong virus, Mayaro virus, and Ross River virus. The 11.8 kb genome of

Justin Jang Hann Chu and Swee Kim Ang (eds.), *Chikungunya Virus: Methods and Protocols*, Methods in Molecular Biology, vol. 1426,DOI 10.1007/978-1-4939-3618-2_1,

CHIKV consists of a linear, single-stranded positive sense RNA. CHIKV infection causes a generally self-limiting disease characterized by an abrupt onset of high fever, severe arthralgia, and skin rush. Sporadically, CHIKV infection can be associated with neurologic and cardiac complications. The illness is often associated with prolonged and incapacitating arthritis; for this reason, large CHIKV epidemics have severe economic impacts, highlighting the significant public health threat posed by CHIKV infection.

Phylogenetic analysis of CHIKV [3] identified three geographically associated genotypes: the West African, the East/Central/South African (ECSA), and the Asian genotype. The CHIKV strains belonging to different geographic lineages exhibit differences in their transmission cycles [4] in Asia. CHIKV circulates primarily in an urban transmission cycle involving the peridomestic mosquitoes (*Aedes aegypti* and *Aedes albopictus*) and humans. In contrast, in Africa the virus is maintained in a zoonotic cycle between forest dwelling *Aedes* spp. mosquitoes (e.g., *Ae. furcifer*, *Ae. africanus*) and nonhuman primates, and epidemic emergence involves the transition from the sylvatic cycle into an urban cycle where peridomestic mosquito vectors transmit among humans. Epidemics in rural Africa usually occur on a much smaller scale than in Asia, likely a result of the lower human population densities, and possibly more stable herd immunity. Although there are at least three genotypes, CHIKV consists of only one serotype. Consequently, an outbreak in a population generates herd immunity conferring lifelong protection. This may explain why massive outbreaks of CHIKV occur only when significant naïve population is developed in endemic areas, which usually takes decades [5].

3 History and Origin of the Virus

The virus is believed to have originated in Africa, where it still circulates enzootically among nonhuman primates and is transmitted by arboreal *Aedes* mosquitoes [3, 4]. Historic accounts suggest that febrile illness with rush and arthralgia, probably caused by CHIKV infection, has occurred in Africa for centuries or longer, during the eighteenth and nineteenth centuries [6] and probably originally described as dengue-associated disease.

The first confirmed CHIKV fever outbreak was reported in 1952 in Makonde Plateau in Tanzania, East Africa, where the virus was first isolated [7]. The first evidence of sylvatic enzootic CHIKV cycle was in Uganda and subsequently, CHIKV was discovered in many parts of sub-Saharan Africa, with transmission mainly by arboreal mosquitoes [8]. Phylogenetic studies placed representative CHIKV strains from all of these sub-Saharan African locations into one clade termed the East/Central/South American (ECSA) lineage [3, 4]. In the twentieth century, enzootic CHIKV cycles were recognized in many parts of sub-Saharan Africa and sporadic outbreaks in Africa and Asia. In Asia, the virus was first identified in 1958 in Bangkok [9] during an outbreak sustained by *Ae. aegypti*. In 1963, India reported the first CHIKV outbreaks in Calcutta and Madras [10–12].

Initially, the Asian outbreak in the 1950s–1960s has had attributed to a spread of the virus from Africa, but sequencing and phylogenetic studies [3, 4] demonstrated that CHIKV strains isolated during the Asian outbreaks from 1958 to 1973 comprise a monophyletic group in the East, Central, South Africa (ECSA) clade; this lineage was then termed the Asian epidemic lineage. The Asian CHIKV lineage became extinct in India after 1973 but continued to circulate in Southeast Asia, occasionally detected during small-medium sized epidemics.

4 Epidemiology and Spread of CHIKV Infection from 2004

Prior to 2004, few sites of CHIKV epidemics outside the enzootic distribution were known [1]: large documented outbreaks in Thailand in 1962–1964 and in India in 1963–1964 and 1974 [13, 14]; after that no large-scale epidemic activity was reported for over 30 years, most probably for establishment of herd immunity in the population [4].

The epidemiology and spread of CHIKV infection took a turn in 2004 when a documented CHIKV fever outbreaks occurred in costal Kenya sustained [15]. The outbreaks spread into islands in the Indian Ocean and to Indian subcontinent in 2005, presumably via infected air travelers, where it caused explosive epidemics involving millions of people [16–19]. The causative virus strain was a new lineage that originated from the ECSA lineage and named Indian Ocean Lineage (IOL; Fig. 1; Table 1; [20–22]). Two distinct sublineages were identified for the Indian Ocean and Indian subcontinent strains indicating independent emergences from East Africa into the Indian Ocean and the Indian subcontinent [4]. The best studied is the outbreak on the French Island of La Rèunion which involving about 300, 000 cases [17] and the principal vector was *Ae. albopictus*. The great magnitude of these outbreaks was influenced by several factors such as increased air travels, the lack of previous protective immunity of human population in the Indian Ocean basin, the high density of the vector *Ae. albopictus*, and also a series of adaptive mutations in the virus strains of the new epidemic IOL lineage, which mediated enhanced virus transmission by *Ae. albopictus*. In the Indian Ocean basin, for the first time, *Ae. albopictus* was implicated as a major vector in a CHIKV epidemic. During the La Rèunion outbreak, genomic sequencing of CHIKV isolates revealed an amino acid mutation (Ala 226 Val) in the E1 envelope glycoprotein [21] that was

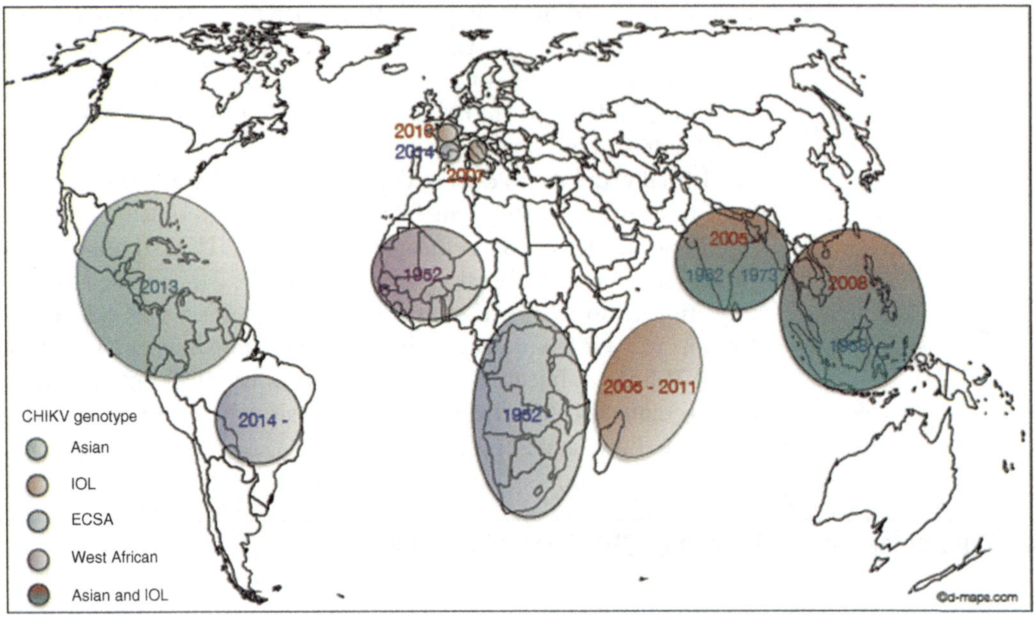

Fig. 1 CHIKV distribution in the world. The map shows the spread of Asian (*green circle*), ECSA (*blue circle*), Indian Ocean Lineage (IOL) (*red circle*), and West African (*purple circle*) lineages of CHIKV [37–39]. The map template was from http://d-maps.com

hypothesized to be involved in the use of *Ae. albopictus* as vector for transmission. Experimental infections also supported the hypothesis that E1-Ala226Val substitution confers adaptation of the virus to *Ae. albopictus*, enabling enhanced transmissibility of the virus to this vector species [23, 24]. This has likely resulted in explosive outbreaks in areas where *Ae. albopictus* is particularly abundant. The improved adaptation of E1-Ala226Val variant to *Ae. albopictus* had a significant impact on the epidemiology of CHIKV globally due to the extensive geographic distribution of *Ae. albopictus*.

During the peak of CHIKV epidemics involving IOL strains in the Indian Ocean Basin and Asia, thousands of infected travelers imported CHIKV into all regions of the world. These resulted in the initiation of small outbreaks in northern Italy in 2007, via a viremic traveler from India [25, 26] and in southern France in 2010 [27], both sustained by *Ae. albopictus* transmission (Fig. 1; Table 1). These two European outbreaks underscored the risk for temperate regions, to be susceptible to CHIKV circulation due to its ability to use *Ae. albopictus* as a vector. The epidemic continued to spread further in 2008 and 2009 with a large outbreak of almost 50 000 people in southern Thailand [28] (Table 1), associated with the CHIKV IOL suggesting that this lineage was displacing the older Asian lineage, which had remained there since the 1950s. However, in Southeast Asia there has been for years the co-circulation of Asian and the ECSA/IOL lineages [29, 30].

Table 1
Selected CHIKV outbreaks

Year	Country/territory	CHIKV lineage	Vector	Reference
1952	Tanzania	ECSA	*Ae. aegypti*	[7]
1958	Thailand (Bangkok)	Asian	*Ae. aegypti*	[9]
1962	India (Calcutta, Madras)	Asian	*Ae. aegypti*	[10–12]
2004	Costal Kenya	IOL	*Ae. aegypti*	[15]
2005–2011	Comoros, Mauritius, La Rèunion	IOL	*Ae. albopictus*	[16, 17, 21]
2005–2008	India, Sri Lanka	IOL	*Ae. albopictus/Ae. aegypti*	[18, 19]
2006	Malaysia	Asian		[29]
2007	Italy	IOL	*Ae. albopictus*	[25, 26]
2008	Thailand	IOL	*Ae. albopictus*	[28, 30]
2008	Singapore	IOL	*Ae. aegypti*	[40]
2010	France	IOL	*Ae. albopictus*	[27]
2013 -	Caribbean islands	Asian	*Ae. aegypti*	[31]
2014	Brazil	Asian/ECSA	*Ae. albopictus/ Ae. aegypti*	[34, 35]
2014	France	ECSA	*Ae. albopictus*	[36]

5 CHIKV Infection in Western Hemisphere

The first known autochthonous CHIKV cases in the Western Hemisphere occurred in October 2013 on the island of Saint Martin (Fig. 1; Table 1; [31]); thereafter, autochthonous cases were reported from numerous other Caribbean islands. Infected travelers have carried the virus from the island countries around the region, leading to autochthonous CHIKV cases occurring on the mainland of South America. Sequence analysis demonstrated that the current outbreak in the Caribbean was caused by an Asian genotype of CHIKV, possibly imported from Southeast Asia or Oceania [32]. CHIKV strains circulating in the Caribbean were closely related to strains isolated in China during 2012 and from Yap, Federated States of Micronesia, during 2013–2014 and very similar to strains circulating in British Virgin Islands in January 2014 [32], demonstrating the movement of the Asian genotype from East Asia to the western Pacific. This Asian CHIKV has spread into Central American countries, most of South America, northern Mexico and Florida; by August 2014, the Pan American Health Organization/World Health Organization (PAHO/WHO) reported more than 650,000 autochthonous cases.

To date, the outbreaks in the Caribbean region have been primarily transmitted by *Ae. aegypti*. The Asian genotype CHIKV strains are known to be constrained in their ability to adapt to *Ae. albopictus* mosquitoes [33], suggesting that this may limit the spread of Asian genotype CHIKV strains into temperate regions, including the southern United States, where *Ae. albopictus* mosquitoes are more commonly found than are *Ae. aegypti* mosquitoes.

In Brazil, autochthonous transmission was first detected in September 2014 in the city of Oiapoque, Amapá State and the Asian lineage from ongoing epidemics in the Caribbean and South America was identified; but surprisingly a separate CHIKV introduction, which comprises ECSA genotype, was discovered and confirmed in Feira de Santana during 2014. Genetic and epidemiologic data suggest that the ECSA genotype was introduced from Angola via a viremic traveler (Fig. 1; Table 1; [34, 35]). In the same period, in Europe an outbreak of CHIKV involving 12 autochthonous cases took place in France in the city Montpellier, located on the Mediterranean coast, from September to October 2014. The index case was a viremic traveler returning from Cameroon, infected by a strain belonging to the ECSA genotype with the E1-Ala 226 Val adaptive mutation (Fig. 1; Table 1; [36]).

6 Concluding Remarks

The capacity of Chikungunya virus to geographically spread worldwide is basically linked to the appearance of spontaneous mutations in the viral genome that can have as final consequence the adaptation of the virus to different vector species. It is so of outmost relevance to implement a worldwide surveillance system that can monitor the diffusion of the virus and the appearance of new genetic variants. This can be achieved only by integration of an efficient public health system with sensitive and specific diagnostic tools that laboratories worldwide can put in place to achieve a fast identification of infected patients and virus variants.

References

1. Powers AM (2015) Risks to the Americas associated with the continued expansion of chikungunya virus. J Gen Virol 96(Pt 1):1–5. doi:10.1099/vir.0.070136-0
2. Morrison TE (2014) Reemergence of chikungunya virus. J Virol 88(20):11644–11647. doi:10.1128/JVI.01432-14
3. Powers AM, Brault AC, Tesh RB, Weaver SC (2000) Re-emergence of Chikungunya and O'nyong-nyong viruses: evidence for distinct geographical lineages and distant evolutionary relationships. J Gen Virol 81(Pt 2):471–479
4. Volk SM, Chen R, Tsetsarkin KA, Adams AP, Garcia TI, Sall AA, Nasar F, Schuh AJ, Holmes EC, Higgs S, Maharaj PD, Brault AC, Weaver SC (2010) Genome-scale phylogenetic analyses of chikungunya virus reveal independent emergences of recent epidemics and various evolutionary rates. J Virol 84(13):6497–6504. doi:10.1128/JVI.01603-09

5. Ng LC, Hapuarachchi HC (2010) Tracing the path of Chikungunya virus—evolution and adaptation. Infection, genetics and evolution. Infect Genet Evol 10(7):876–885. doi:10.1016/j.meegid.2010.07.012
6. Carey DE (1971) Chikungunya and dengue: a case of mistaken identity? J Hist Med Allied Sci 26(3):243–262
7. Lumsden WH (1955) An epidemic of virus disease in Southern Province, Tanganyika Territory, in 1952-53. II General description and epidemiology. Trans R Soc Trop Med Hyg 49(1):33–57
8. Coffey LL, Failloux AB, Weaver SC (2014) Chikungunya virus-vector interactions. Viruses 6(11):4628–4663. doi:10.3390/v6114628
9. Hammon WM, Sather GE (1964) Virological findings in the 1960 hemorrhagic fever epidemic (dengue) in Thailand. Am J Trop Med Hyg 13:629–641
10. Myers RM, Carey DE, Reuben R, Jesudass ES, De Ranitz C, Jadhav M (1965) The 1964 epidemic of dengue-like fever in South India: isolation of chikungunya virus from human sera and from mosquitoes. Indian J Med Res 53(8):694–701
11. Shah KV, Gibbs CJ Jr, Banerjee G (1964) Virological investigation of the epidemic of haemorrhagic fever in Calcutta: isolation of three strains of chikungunya virus. Indian J Med Res 52:676–683
12. Dandawate CN, Thiruvengadam KV, Kalyanasundaram V, Rajagopal J, Rao TR (1965) Serological survey in Madras city with special reference to chikungunya. Indian J Med Res 53(8):707–714
13. Halstead SB, Scanlon JE, Umpaivit P, Udomsakdi S (1969) Dengue and chikungunya virus infection in man in Thailand, 1962-1964. IV Epidemiologic studies in the Bangkok metropolitan area. Am J Trop Med Hyg 18(6):997–1021
14. Mavalankar D, Shastri P, Bandyopadhyay T, Parmar J, Ramani KV (2008) Increased mortality rate associated with chikungunya epidemic, Ahmedabad, India. Emerg Infect Dis 14(3):412–415. doi:10.3201/eid1403.070720
15. Chretien JP, Anyamba A, Bedno SA, Breiman RF, Sang R, Sergon K, Powers AM, Onyango CO, Small J, Tucker CJ, Linthicum KJ (2007) Drought-associated chikungunya emergence along coastal East Africa. Am J Trop Med Hyg 76(3):405–407
16. Charrel RN, de Lamballerie X, Raoult D (2007) Chikungunya outbreaks—the globalization of vectorborne diseases. New Engl J Med 356(8):769–771. doi:10.1056/NEJMp078013
17. Gerardin P, Guernier V, Perrau J, Fianu A, Le Roux K, Grivard P, Michault A, de Lamballerie X, Flahault A, Favier F (2008) Estimating Chikungunya prevalence in La Reunion Island outbreak by serosurveys: two methods for two critical times of the epidemic. BMC Infect Dis 8:99. doi:10.1186/1471-2334-8-99
18. Outbreak and spread of chikungunya (2007). Weekly epidemiological record/Health Section of the Secretariat of the League of Nations 82 (47), 409–415
19. Mavalankar D, Shastri P, Raman P (2007) Chikungunya epidemic in India: a major public-health disaster. Lancet Infect Dis 7(5):306–307. doi:10.1016/S1473-3099(07)70091-9
20. Parola P, de Lamballerie X, Jourdan J, Rovery C, Vaillant V, Minodier P, Brouqui P, Flahault A, Raoult D, Charrel RN (2006) Novel chikungunya virus variant in travelers returning from Indian Ocean islands. Emerg Infect Dis 12(10):1493–1499. doi:10.3201/eid1210.060610
21. Schuffenecker I, Iteman I, Michault A, Murri S, Frangeul L, Vaney MC, Lavenir R, Pardigon N, Reynes JM, Pettinelli F, Biscornet L, Diancourt L, Michel S, Duquerroy S, Guigon G, Frenkiel MP, Brehin AC, Cubito N, Despres P, Kunst F, Rey FA, Zeller H, Brisse S (2006) Genome microevolution of chikungunya viruses causing the Indian Ocean outbreak. PLoS Med 3(7):e263. doi:10.1371/journal.pmed.0030263
22. Arankalle VA, Shrivastava S, Cherian S, Gunjikar RS, Walimbe AM, Jadhav SM, Sudeep AB, Mishra AC (2007) Genetic divergence of Chikungunya viruses in India (1963-2006) with special reference to the 2005-2006 explosive epidemic. J Gen Virol 88(Pt 7):1967–1976. doi:10.1099/vir.0.82714-0
23. Tsetsarkin KA, Vanlandingham DL, McGee CE, Higgs S (2007) A single mutation in chikungunya virus affects vector specificity and epidemic potential. PLoS Pathog 3(12):e201. doi:10.1371/journal.ppat.0030201
24. Vazeille M, Moutailler S, Coudrier D, Rousseaux C, Khun H, Huerre M, Thiria J, Dehecq JS, Fontenille D, Schuffenecker I, Despres P, Failloux AB (2007) Two Chikungunya isolates from the outbreak of La Reunion (Indian Ocean) exhibit different patterns of infection in the mosquito, *Aedes albopictus*. PLoS One 2(11):e1168. doi:10.1371/journal.pone.0001168
25. Rezza G, Nicoletti L, Angelini R, Romi R, Finarelli AC, Panning M, Cordioli P, Fortuna C, Boros S, Magurano F, Silvi G, Angelini P, Dottori M, Ciufolini MG, Majori GC, Cassone

A, group Cs (2007) Infection with chikungunya virus in Italy: an outbreak in a temperate region. Lancet 370(9602):1840–1846. doi:10.1016/S0140-6736(07)61779-6

26. Angelini R, Finarelli AC, Angelini P, Po C, Petropulacos K, Silvi G, Macini P, Fortuna C, Venturi G, Magurano F, Fiorentini C, Marchi A, Benedetti E, Bucci P, Boros S, Romi R, Majori G, Ciufolini MG, Nicoletti L, Rezza G, Cassone A (2007) Chikungunya in north-eastern Italy: a summing up of the outbreak. Euro Surveill 12(11):E071122.2
27. Grandadam M, Caro V, Plumet S, Thiberge JM, Souares Y, Failloux AB, Tolou HJ, Budelot M, Cosserat D, Leparc-Goffart I, Despres P (2011) Chikungunya virus, southeastern France. Emerg Infect Dis 17(5):910–913. doi:10.3201/eid1705.101873
28. Rianthavorn P, Prianantathavorn K, Wuttirattanakowit N, Theamboonlers A, Poovorawan Y (2010) An outbreak of chikungunya in southern Thailand from 2008 to 2009 caused by African strains with A226V mutation. Intl J Infect Dis 14(Suppl 3):e161–e165. doi:10.1016/j.ijid.2010.01.001
29. AbuBakar S, Sam IC, Wong PF, MatRahim N, Hooi PS, Roslan N (2007) Reemergence of endemic Chikungunya, Malaysia. Emerg Infect Dis 13(1):147–149. doi:10.3201/eid1301.060617
30. Pulmanausahakul R, Roytrakul S, Auewarakul P, Smith DR (2011) Chikungunya in Southeast Asia: understanding the emergence and finding solutions. Intl J Infect Dis 15(10):e671–e676. doi:10.1016/j.ijid.2011.06.002
31. Leparc-Goffart I, Nougairede A, Cassadou S, Prat C, de Lamballerie X (2014) Chikungunya in the Americas. Lancet 383(9916):514. doi:10.1016/S0140-6736(14)60185-9
32. Lanciotti RS, Valadere AM (2014) Transcontinental movement of Asian genotype chikungunya virus. Emerg Infect Dis 20(8):1400–1402. doi:10.3201/eid2008.140268
33. Tsetsarkin KA, Chen R, Leal G, Forrester N, Higgs S, Huang J, Weaver SC (2011) Chikungunya virus emergence is constrained in Asia by lineage-specific adaptive landscapes. Proc Natl Acad Sci U S A 108(19):7872–7877. doi:10.1073/pnas.1018344108
34. Nunes MR, Faria NR, de Vasconcelos JM, Golding N, Kraemer MU, de Oliveira LF, Azevedo Rdo S, da Silva DE, da Silva EV, da Silva SP, Carvalho VL, Coelho GE, Cruz AC, Rodrigues SG, Vianez JL Jr, Nunes BT, Cardoso JF, Tesh RB, Hay SI, Pybus OG, Vasconcelos PF (2015) Emergence and potential for spread of Chikungunya virus in Brazil. BMC Med 13:102. doi:10.1186/s12916-015-0348-x
35. Teixeira MG, Andrade AM, Costa Mda C, Castro JN, Oliveira FL, Goes CS, Maia M, Santana EB, Nunes BT, Vasconcelos PF (2015) East/Central/South African genotype chikungunya virus, Brazil, 2014. Emerg Infect Dis 21(5):906–907. doi:10.3201/eid2105.141727
36. Delisle E, Rousseau C, Broche B, Leparc-Goffart I, L'Ambert G, Cochet A, Prat C, Foulongne V, Ferre JB, Catelinois O, Flusin O, Tchernonog E, Moussion IE, Wiegandt A, Septfons A, Mendy A, Moyano MB, Laporte L, Maurel J, Jourdain F, Reynes J, Paty MC, Golliot F (2015) Chikungunya outbreak in Montpellier, France, September to October 2014. Euro Surveill 20(17)
37. Weaver SC (2014) Arrival of chikungunya virus in the new world: prospects for spread and impact on public health. PLos Negl Trop Dis 8(6):e2921. doi:10.1371/journal.pntd.0002921
38. Weaver SC, Forrester NL (2015) Chikungunya: evolutionary history and recent epidemic spread. Antivir Res 120:32–39. doi:10.1016/j.antiviral.2015.04.016
39. Weaver SC, Lecuit M (2015) Chikungunya virus and the global spread of a mosquito-borne disease. N Engl J Med 372(13):1231–1239. doi:10.1056/NEJMra1406035
40. Leo YS, Chow AL, Tan LK, Lye DC, Lin L, Ng LC (2009) Chikungunya outbreak, Singapore, 2008. Emerg Infect Dis 15(5):836–837. doi:10.3201/eid1505.081390

Chapter 2

Molecular Epidemiology of Chikungunya Virus by Sequencing

Ravindran Thayan, Mohd Apandi Yusof, Zainah Saat, Shamala Devi Sekaran, and Seok Mui Wang

Abstract

Molecular surveillance of Chikungunya virus (CHIKV) is important as it provides data on the circulating CHIKV genotypes in endemic countries and enabling activation of measures to be taken in the event of a pending outbreak. Molecular surveillance is carried out by first detecting CHIKV in susceptible humans or among field-caught mosquitoes. This is followed by sequencing a selected region of the virus which will provide evidence on the source of the virus and possible association of the virus to increased cases of Chikungunya infections.

Key words Molecular surveillance, RT-PCR, Sequencing, Phylogenetic tree

1 Introduction

Molecular epidemiology of Chikungunya viruses is carried out routinely to identify strains of the virus in a geographical location at a given time [1]. This is done by sequencing CHIKV isolated from patients from current infections and comparing these sequences with that of the virus from previous cases or from sequences databases such as Genbank. Molecular epidemiology of CHIKV will be effective if the exercise is done routinely as it will assist in identifying the introduction of new strains. Furthermore, the data may provide some evidence on the association of Chikungunya strains to a sudden increase in Chikungunya cases in places as this may suggest that new strains of the virus have been introduced to the population. This is due to the presence of naïve subpopulation in that area who have not been exposed to the new strain and hence may develop signs and symptoms of Chikungunya infections.

Justin Jang Hann Chu and Swee Kim Ang (eds.), *Chikungunya Virus: Methods and Protocols*, Methods in Molecular Biology, vol. 1426,DOI 10.1007/978-1-4939-3618-2_2, © Springer Science+Business Media New York 2016

Chikungunya virus is found mostly in Africa especially in Eastern, Southern, Central, and Western Africa and Asia including parts of India, Thailand, and Malaysia. Currently there are three genotypes identified, belonging to Asian, East/Central/South African, and West African lineages [2–4]. As humans are the major reservoir of the virus for the mosquito, surveillance for the virus in both humans and mosquitoes is important to prevent outbreaks. This is done by carrying out molecular epidemiology of CHIKV.

There are many publications reporting the importance of molecular epidemiology of CHIKV [1, 5–9]. Their findings illustrate the role of Chikungunya strains causing outbreaks. In the Caribbean island of Saint Martin, two cases of Chikungunya were first reported in December 2013, indicating the start of the first documented outbreak in the Americas [8]. Since then the virus has established itself in the Americas where more than 17,000 suspected and confirmed cases were reported between December 2013 and March 2014. Probably the most significant Chikungunya outbreak of unprecedented magnitude occurred in 2005–2007, affecting countries ranging from the Reunion Island, Mauritius, Seychelles, Comoros, and Southwest India where there were reports of anecdotal deaths, encephalitis, and neonatal infections [4]. The CHIKV strain implicated was genotype East/Central/South Africa (ECSA). In India, CHIKV was implicated during the 2007 outbreak of febrile arthritis, affecting nearly 25,000 people in the Southern State of Kerala [10] while Thailand reported cases of Chikungunya associated with the A226V mutation [11, 12]. Meanwhile, molecular epidemiology of Chikungunya in Malaysia showed that the strains that were circulating in 2008–2009 also belonged to the ECSA genotype (Fig. 1; [13]). Hence, from previous investigations into the molecular epidemiology of CHIKV, the best approach is through sequencing of selected regions of the virus. The most studied region of the CHIKV is the envelope gene 1 (E1). This gene was selected as the variation noted in this region provided information on the mutations associated with environmental pressures. In addition, Sanger sequencing is a method of choice as the target fragment is small. It is more cost-effective to use this approach rather than next generation sequencing (NGS) which is usually used for full genome sequencing.

Fig. 1 Phylogenetic tree of partial glycoprotein E1 sequences of CHIKV inferred using the Neighbor-Joining method from the software MEGA 4. The evolutionary distances were computed using the Maximum Composite Likelihood method. Genotype Asian, Central/East African, and West African are indicated by *square brackets* with O'nyong-nyong virus as an outgroup. 49 CHIKV isolates from Malaysia in 2008 and 2009 are indicated in *red* and *blue* respectively. Representative strains of each genotype obtained from GenBank are labeled using the following format: "Accession number"—"isolate"—"Country of origin"—"Year isolation." Bootstrap values (>75 %) for 1000 pseudoreplicate dataset are indicated at branch nodes [13]

This chapter describes the methods used for the determination of Chikungunya strains. This involves design of primers based on the envelope gene of the virus and subsequent reverse transcriptase polymerase chain reaction of the targeted region. Subsequently, sequencing and data analysis is done to characterize the viruses according to their genotypes. This information is useful to infer molecular epidemiology of CHIKV.

2 Materials

All biological specimens including blood should be considered potentially infectious and should be handled in accordance with the appropriate national/local biosafety practices. In addition collection of blood sample from the study subjects must be approved by the Institutional Human Ethics Board. All biological and clinical waste must be decontaminated before disposal, in accordance with national and local regulations. All solutions must be prepared using ultrapure water and molecular grade reagents.

2.1 Viral RNA Extraction

1. Serum from CHIKV-infected patients or CHIKV-infected cell culture supernatant.
2. Viral RNA extraction kit (QIAGEN) or any commercially available viral RNA extraction kit.
3. RNaseZap Wipes and DNAZap.

2.2 One-Step Reverse Transcriptase Polymerase Chain Reaction (RT-PCR)

1. PCR primer sequences: CHIKV forward primer (Forward 5′ TACCCATTCATGTGGGGC 3′), and CHIKV reverse primer (Reverse 5′ GCCTTTGTACACCACGATT 3′) (*see* **Note 1**; [14]).
2. QIAGEN One-Step RT-PCR Kit or any commercially available RT-PCR reagents.
3. Nuclease-free water.
4. PCR Thermal Cycler.
5. Mini microcentrifuge.

2.3 Agarose Gel Electrophoresis and Purification of PCR Product

1. NuSieve® 3:1 Agarose.
2. Tris EDTA Buffer (1×): 89 mM Tris-borate, 2 mM EDTA.
3. Red Safe™ Nucleic Acid Staining Solution.
4. 100 bp DNA ladder marker.
5. 6× Loading Dye.
6. Agarose gel electrophoresis system.
7. QIAquick® Gel Extraction Kit (Qiagen).
8. QIAquick® PCR Purification Kit (Qiagen).

2.4 Sequencing Reaction

1. PCR product of CHIKV.
2. Forward and reverse primers of the PCR product.
3. BigDye Terminator Ready Reaction Kit V3.1 (Applied Biosystems).
4. Nuclease-free water.
5. Thermal cycler.
6. Mini microcentrifuge.

2.5 Purification of Cycle Sequencing Product

1. Nuclease-free water.
2. 3 M sodium acetate pH 4.6.
3. 95% Ethanol.
4. 70% Ethanol.
5. Hi Di formamide (Applied Biosystems).
6. Centrifuge.
7. Mini microcentrifuge.

3 Methods

Carry out all procedures at room temperature unless otherwise specified. Prepare RT-PCR reaction mixture in a PCR/UV work station. Clean the work bench and pipettes with RNaseZap Wipes and DNAZap to remove all potential RNAase and DNA contamination.

3.1 Viral RNA Extraction

1. All work must be carried out in a biosafety cabinet. Clean all surfaces with RNaseZap before wiping with RNase-free water.
2. Perform viral RNA extraction by using commercially available kit, QIAamp® Viral RNA Kit. Use serum from Chikungunya-infected person or Chikungunya-infected BHK cells.
3. Follow the method as described in the kit insert. Store extracted RNA at −80 °C before use (*see* **Note 2**).

3.2 One-Step Reverse Transcriptase Polymerase Chain Reaction (RT-PCR)

1. Prepare each PCR reaction in a 25 μl reaction volume containing 5 μl of extracted RNA, 1 μl of RT enzyme, 12.5 μl of 2× One-Step RT-PCR Buffer Mix, 0.5 μl of each primers, CHIKV forward (25 μM) and CHIKV reverse primers (25 μM), and nuclease-free water to make up to 25 μl reaction mix. Follow the concentrations of buffer, dNTPs, *Taq* polymerase, and Mg^{2+} as recommended by the manufacturer.
2. Quick spin the PCR tubes in the mini microcentrifuge and load them into a PCR thermal cycler. The thermal cycling conditions consist of a 30 min reverse transcription step at 50 °C, 15 min of initial denaturation at 95 °C, followed by 40 cycles of amplification steps of denaturation at 95 °C for 30 s and

annealing/extension at 54 °C for 60 s, and a final extension at 72 °C for 10 min.

3. Upon completion of the PCR run, store the PCR amplicons at −20 °C for temporary storage or proceed to agarose gel electrophoresis (*see* **Note 3**) to determine the size of PCR amplicons.

3.3 Agarose Gel Electrophoresis and Purification of PCR Product

1. Prepare a 2% gel by carefully transferring 2 g of agarose powder into a clean 250 ml conical flask containing 100 ml of 1× TBE (*see* **Note 4**). Mix gently to ensure all powder comes in contact with TBE buffer.
2. Dissolve the agarose by heating using a microwave at a high-power setting for 2 min (*see* **Note 5**). Mix the flask gently to ensure that all the agarose have completely dissolved. Add 2 μl of Red safe dye into the dissolved gel and mix adequately. Pour the gel gently into suitable gel cast with comb inserted, ensuring no bubbles are formed. Leave the gel cast at room temperature for about 30 min to enable the cast to set.
3. Remove the comb from the gel cast. Mix 10 μl of PCR amplicons of all RT-PCR reactions including positive and negative controls with 1 μl of loading dye (mixing can be done on a parafilm). Load the PCR amplicons and loading dye mixture into individual wells. Also load 5 μg of 100 bp DNA ladder into one of the side wells.
4. Perform agarose gel electrophoresis at 100 V for about 45 min or until the loading dye is nearing the bottom edge of the gel. Stop the electrophoresis run. Transfer the gel into a gel documentation system and capture an image of the gel.
5. Before performing any sequencing reaction, the PCR product must undergo purification to remove excess primers and other PCR reaction mix which can inhibit the chain termination sequencing reaction. If there is only a single band of PCR product present, purify the PCR product directly using QIAquick® PCR Purification Kit. If there is more than 1 band present, then gel purification must be carried out by excising the band of interest before continuing with purification using QIAquick® Gel Extraction Kit.
6. Follow the purification method as described in the kit insert.
7. Use the purified PCR product as a template for subsequent sequencing reaction or keep at −20 °C before use.

3.4 Sequencing Reaction

1. Aliquot 4 μl of Big Dye® Ready Reaction Premix in a 200 μl sterile tube. Then add 2 μl of a single forward or reverse primer (10 pmol/μl), followed by 2 μl of BigDye sequencing buffer.
2. Subsequently, add 1 μl of template (100 ng/μl) and make up the volume to 20 μl with nuclease-free water. Quick spin the reaction tubes in a mini microcentrifuge.

3. Load the reaction tubes into a PCR Thermal cycler. The PCR conditions are as follows: Initial denaturation at 96 °C for 1 min; denaturation at 96 °C for 10 s, hybridization at 50 °C for 5 s, and elongation at 60 °C for 4 min.

3.5 Purification of Cycle Sequencing Product

1. Upon completion of the cycle sequencing, centrifuge the plate at 100 × *g* for 1 min.
2. Add 80 µl of mixture containing 2.55 µl 3 M Sodium Acetate pH 4.6, 54.74 µl 95% ethanol, and 22.71 µl nuclease-free water into each well of the plate.
3. Close the plate with a plate septa to prevent leakage. Mix gently and centrifuge at 100 × *g* for 1 min.
4. Incubate the product at room temperature for 30 min to allow precipitation. Centrifuge the plate again at 3000 × *g* for another 30 min at 4 °C.
5. Discard the supernatant. Add 150 µl of 70% ethanol into each well and rinse the pellet. Vortex the mixture adequately. Centrifuge the plate at 3000 × *g* for 10 min at 4 °C.
6. Discard the supernatant and repeat the ethanol washing step. Briefly centrifuge the plate at 50 × *g* for a minute.
7. Discard the supernatant. Air-dry the pellet using a thermal cycler at 65 °C for 5 min.
8. In preparation for sample injection into the sequencer, add 10 µl of Hi Di formamide into each well and resuspend the pellet. Briefly centrifuge the plate at 100 × *g* for a minute.
9. Denature the samples by heating the plate at 95 °C for 2 min. Then immediately place the plate on ice. Briefly centrifuge the plate for a minute at 200 × *g*.
10. Load all samples into the designated wells in ABI 3130 genetic analyzer DNA sequencer. Proceed to run the reaction using POP 7 polymer 5 cm capillary array.
11. Analyze data using software SeqScape v 5.2.

3.6 Sequence Data Edition

The sequence data derived must be edited by using data from both primer directions of the template to ensure reliability of sequence data.

1. Download suitable sequence data edition software from the internet. For example, CHROMAS is a suitable sequence data editor and can be downloaded from the internet.
2. Open CHROMAS software. Then, open the forward direction of the sequencing reaction raw data followed by the reverse direction of the same sequencing reaction.
3. Using CHROMAS, reverse complement the reverse direction of the sequencing reaction. Now compare both sequences of

the sample, looking for any sequence data that do not match. This is to ensure reliability of the sequence data.

4. Repeat the sequencing experiment if there are mutations that do not match between both sequence data as this suggests error during the sequencing reaction.
5. Once the reliability of the raw sequence data has been established, import these edited raw sequence data into MEGA 6. Use this software to create a phylogenetic tree.

3.7 Construction of Phylogenetic Tree

1. For investigation of molecular epidemiology of viruses, phylogenetic trees must be constructed to reflect the origin of viruses. For that, MEGA 6, a free software, is highly recommended (*see* **Note 6**; [14]).
2. To obtain a reference sequence, perform a nucleotide BLAST analysis from the following website (http://www.ncbi.nlm.nih.gov/BLAST). For analysis, use neighbor-joining method according to the distances between all pairs of the sequences in a multiple alignment.
3. Evaluate the confidence of the sequence clustering by bootstrapping with 1000 replicates.

4 Notes

1. These primers will result in a single PCR product of 294 bp. If other bands appear, further optimization can be carried out by using a gradient PCR. Usually an increase in annealing temperature will reduce the probability of nonspecific bands. If the extra bands are still present, then the best option is to excise the 294 bp sized band and continue with PCR product purification, using gel purification protocol which is available from Qiagen. The primers are designed to be highly specific and produce a single amplicon only. This enables easier PCR amplicon purification.
2. The extracted RNA should be aliquoted into a suitable volume (12 μl suitable for two reactions) to prevent degradation due to several freeze-thawing procedures. The aliquoted RNA must be stored at −80 °C before use.
3. The PCR amplicon should be purified within 1 week after the completion of RT-PCR. This is to reduce the effect of exonuclease activity.
4. Wear heat protecting gloves when handling hot glasswares used for heating agarose. The volume of the buffer used during heating of the agarose should be 10% less than the desired volume to ensure better solubilization of the agarose. Once the agar has been completely dissolved, add adequate volume of 1× TBE to make up to 100 ml of agarose.

5. Ensure that the agarose is completely dissolved and then have the final volume made up to the desired volume.
6. To get started, it is helpful to refer to the tutorial provided in the Mega 6 software [14].

References

1. Singh RK, Tiwari S, Mishra VK, Tiwari R, Dhole TN (2012) Molecular epidemiology of Chikungunya virus: mutation in E1 gene region. J Virol Methods 185:213–220
2. Powers AM, Brault AC, Tesh RB, Weaver SC (2000) Re-emergence of Chikungunya and O'nyong-nyong viruses: evidence for distinct geographical lineages and distinct evolutionary relationships. J Gen Virol 81:471–479
3. Schuffenecker I, Iteman I, Michault A, Murri Frangeul S, Vaney L et al (2006) Genome microevolution of Chikungunya viruses causing the Indian Ocean outbreak. PloS Med 3, e263
4. Powers AM, Logue CH (2007) Changing patterns of Chikungunya virus: re emergence of a zoonotic arbovirus. J Gen Virol 88:2363–2377
5. Laksmipathy DT, Dhanasekaran D (2008) Molecular epidemiology of Chikungunya virus in Vellore District, Tamilnadu, India in 2006. East Afr J Public Health 5(2):122–125
6. Chem YK, Zainah S, Beremdam SJ, Tengku Rogayah TAR, Khairul AH, Chua KB (2010) Molecular epidemiology of Chikungunya virus in Malaysia since its first emergence in 1998. Med J Malaysia 5(1):31–35
7. Noridah O, Paranthaman V, Nayar SK, Masliza M, Ranjit K, Norizah I, Chem YK, Mustafa B, Kumarasamy V, Chua KB (2007) Outbreak of Chikungunya due to virus of Central/East African Genotype 1 in Malaysia. Med J Malaysia 62(4):323–328
8. Bortel WV, Dorleans F, Rosine J, Blateau A, Rousset D, Matheus S, Leparc-Goffart I, Flusin O et al (2014) Chikungunya outbreak in the Caribbean region, December 2013 to March 2014, and the significance for Europe. Euro Surveill 19(13):pii: 20759
9. Laurent P, Roux KL, Grivard P, Bertil G, Naze F, Picard M, Staikowsky F, Barau G, Schuffenecker I, Michault A (2007) Development of a sensitive real time reverse transcriptase PCR assay with an internal control to detect and quantify Chikungunya virus. Clin Chem 53(8):1408–1414
10. Santhosh SR, Dash PL, Parida MM, Khan M, Tiwari M, Lakshmana Rao PV (2008) Comparative full genome analysis revealed E1: A226V shift in 2007 Indian Chikungunya virus isolates. Virus Res 135:36–41
11. Rianthavorn P, Prianantathavorn K, Wuttirattanakowit N, Theamboonlers A, Poovorawan Y (2010) An outbreak of Chikungunya in southern Thailand from 2008-2009 caused by African strains with A226V mutation. Int J Infect Dis 145:161–165
12. Pongsiri P, Praianantathavorn K, Theamboonlers A, Payungporn S, Poorawan Y (2012) Multiplex real time Rt-PCR for detecting Chikungunya virus and dengue virus. Asian Pac J Trop Dis 5(5):342–346
13. Apandi YM, Lau SK, Norfaezah A, Nur Izmawati AR, Liyana AZ, Khairul Izwan H, Zainah S (2011) Epidemiology and molecular characterization of Chikungunya virus involved in the 2008 and 2009 outbreak in Malaysia. J Gen Mol Virol 3(2):35–42
14. Tamura K, Stecher G, Peterson D, Filipski A, Kumar S (2013) MEGA6: molecular evolutionary genetics analysis version 6.0. Mol Biol Evol 30:2725–2729

Chapter 3

Advanced Genetic Methodologies in Tracking Evolution and Spread of Chikungunya Virus

Hapuarachchige Chanditha Hapuarachchi and Kim-Sung Lee

Abstract

Recent advances in genetic methodologies have substantially expanded our ability to track evolution and spatio-temporal distribution of rapidly evolving pathogens. The information gathered from such analyses can be used to decipher host adaptations that shape disease epidemiology. In this chapter, we demonstrate the utilization of freely available resources to track the evolution and spread of Chikungunya virus.

Key words Chikungunya virus, Phylogenetics, Phylogeography, Evolution, Common ancestor, Evolutionary network, Multiple alignment

1 Introduction

The biology of RNA viruses is extremely influenced by their high mutation rates, resulting in a rapid evolutionary process that potentially interacts with disease epidemiology and ecological dynamics on a parallel time scale. Rapid evolution brings about diversity and host adaptation while virus populations are spatially disseminated. The linkage between evolutionary and spatial dynamics of RNA viruses, therefore, allows tracking of their spread based on evolving genetic signatures. The potential of mutation profiles to alter vector specificity, transmissibility, drug sensitivity, and immune responses has an immense impact on virus spread, pathogenicity, and disease control.

Chikungunya virus (CHIKV) is an arthropod-borne RNA virus that has emerged in a new wave of epidemics since 2004 [1, 2]. CHIKV swept across Indian Ocean Islands, Southeast Asia, the Pacific Islands, and the Caribbean, evolving substantially and causing millions of infections. As compared to RNA viruses such as Dengue virus of which the transmission has been consistent and hyperendemic, CHIKV's emergence after a long period of absence has allowed more precise tracking of latter's spread across continents based on mutation profiles [3–5]. Genetic analyses of the

Justin Jang Hann Chu and Swee Kim Ang (eds.), *Chikungunya Virus: Methods and Protocols*, Methods in Molecular Biology, vol. 1426,DOI 10.1007/978-1-4939-3618-2_3,

East, Central, and South African genotype of CHIKV during the last decade with regard to its evolution and geographical spread have demonstrated a remarkable example of convergent molecular evolution that resulted in vector adaptation [6]. A single amino acid substitution in the envelope 1 protein of CHIKV (E1-A226V) that emerged as a result of a convergent evolutionary process has been shown to enhance the virus infectivity of *Aedes albopictus* mosquitoes [7, 8] and thereby, augmenting virus transmission in areas inhabited by *Ae. albopictus*. Evolutionary analyses have shown independent emergence of E1-A226V substitution in at least three geographically distant locations in Indian Ocean Islands, India, and Central Africa. Our analyses have suggested even a fourth occasion of the emergence of E1-A226V substitution among Sri Lankan CHIKV isolates [5]. Interestingly, E1-A226V mutants mainly spread in areas dominated by *Ae. albopictus* in all four locations, exemplifying an adaptation-driven convergent molecular evolutionary process.

The availability of a plethora of analytical methods, abundance of genome sequences defined in time and space as well as increased affordability of computational power have improved our ability to perform evolutionary analyses. Besides conventional phylogenetics, mutation mapping, evolutionary networking [9], and phylogeography [10, 11] have substantially expanded the ability to track evolution and spatio-temporal distribution of rapidly evolving pathogens. Moreover, the advances in in silico structure modeling, molecular characterization through cloning, and understanding of host–virus interactions have tremendously contributed to decipher host adaptations during virus evolution that shape disease epidemiology. In this chapter, we demonstrate the utilization of freely available resources to perform simple phylogenetics, mutation mapping, evolutionary networking, and advanced phylogeography analyses in order to track the evolution and spread of CHIKV. We provide examples of such analyses performed on available CHIKV sequences.

2 Materials

The analyses described in this section are computer-based and all relevant software programs are freely available. Besides the descriptions here, the usage of each software program has previously been described by original developers. They can be accessed through publications and relevant websites cited in respective sections and notes. Therefore, the following descriptions should only be considered as an alternative basic practical guide. It should also be noted that descriptions available in this chapter do not include all functional elements of respective software. It is recommended to refer to tutorials, publications, etc., from original developers of respective software for further details on additional analytical options. As all software programs highlighted in this chapter are freeware, they are not completely devoid of functional defects. One of the recommendations is to try several different versions of the same software if certain error messages are consistently generated by a particular version. Developers constantly update bug fixes and it is strongly encouraged to use the latest versions of respective software.

3 Methods

3.1 Generating Sequences

1. Sequences of CHIKV isolates obtained from clinical specimens can be generated by amplifying overlapping fragments of the genome and subjecting the amplicons to either Sanger dideoxy chain-termination or next-generation sequencing methods (*see* **Note 1**).
2. The pre-sequencing process involves extraction of CHIKV RNA from clinical specimens/isolates, synthesis of complementary DNA (cDNA), amplification of overlapping fragments by polymerase chain reaction (PCR) technology, and purification of amplified PCR products to be sequenced. Commercial kits are available to perform each of these steps. We have previously described a detailed protocol for the whole genome sequencing of CHIKV [5].
3. The PCR amplification and sequencing protocols involve multiple oligonucleotides (primers). The oligonucleotides should flank overlapping regions of CHIKV genome and be designed against conserved segments as far as possible. Even though single pairs of oligonucleotides are used to amplify overlapping fragments, multiple oligonucleotides are required to complete the sequencing of each fragment. We have previously published a complete set of oligonucleotides that can be used to sequence the East, Central, and South African genotype of CHIKV [5].

3.2 Retrieving Sequences from Public Databases

1. Sequences of CHIKV reported from different geographical locations can be downloaded from sequence databases such as GenBank (http://www.ncbi.nlm.nih.gov/genbank/), DNA Data Bank of Japan (http://www.ddbj.nig.ac.jp/), and European Nucleotide Archive (http://www.ebi.ac.uk/ena/). Sequences submitted to any of these databases are shared among each other and uploaded on a daily basis, so that users can download the same dataset by accessing any of these databases. The following steps describe the retrieval of CHIKV sequences from the GenBank database.
2. Using a web browser, go to the National Centre for Biotechnology Information (NCBI) home page at http://www.ncbi.nlm.nih.gov/.

3. Click on "Taxonomy" tab under resources menu. Go to "Taxonomy Browser" under tools.
4. Find CHIKV under "viruses" menu—Alternatively, you may type "Chikungunya virus" in the search tool.
5. Click on "Chikungunya virus" in the appearing window.
6. Find results in the "Entrez records" table in the right-hand side top corner. The total number of nucleotide sequences available in the database is shown under "Nucleotide" row of the table.
7. Click on the number under "nucleotides direct links".
8. This brings you to individual reads of all CHIKV sequences available in the database. Please note that this collection includes partial as well as complete genome sequences.
9. If you need to download only complete genome sequences, type "Chikungunya virus AND complete" in the search window.
10. In order to download all sequences as a batch, click on "Send to" tab.
11. Click on "File" under "Choose Destination" tab and select the "Format" and "Sort by" options. FASTA and GENBANK FULL are commonly used formats.
12. Click on "Create File" tab to download and save sequences. Please note that sequences saved are not aligned.

3.3 Multiple Alignment of Sequences

Multiple sequence alignment is an essential task to be carried out prior to many downstream analyses such as polymorphism analysis, phylogenetic inference, and molecular structure prediction. In most cases, sequence alignment is performed based on homologous genes or sequences, assuming that the set of aligned sequences has diverged and descended from a common ancestor. A batch of sequences derived from different virus strains can be aligned to one another by using a multiple sequence alignment algorithm. The process first performs pairwise alignment between a particular sequence and the remaining sequences before generating the final alignment. Multiple alignment of sequences (*see* **Note 2**) can be performed by using algorithms implemented in Clustal X, Clustal W [12], and MUSCLE [13] which are available either as standalone software or in freeware suites such as BioEdit 7.2.5 software suite [14] and MEGA [15]. The following description demonstrates how to perform multiple sequence alignment of virus sequences using ClustalW algorithm in the BioEdit 7.2.5 software suite.

1. You may download the software suite at http://www.mbio.ncsu.edu/bioedit/bioedit.html.
2. Load the sequences using the "File -> Open" in BioEdit. For example, you may load sequences either downloaded in FASTA

format or saved in the same format as a text editor (.txt) document. Make sure to select the correct file type in the "Open File" panel of BioEdit.

3. BioEdit Sequence Alignment Editor has three main panels: top panel with functional tabs, left lower panel with sequence identities (ID), and right lower panel with respective sequences. Verify the dataset with number of sequences displayed in top panel and sequence IDs in the left lower panel.
4. Click "Accessory Applications ->ClustalW Multiple Alignment" to access the alignment panel.
5. ClustalW multiple alignment panel provides you options to set the alignment parameters. By default, the panel will perform full multiple alignment using a Neighbor-Joining guide tree with 1000 bootstrap iterations.
6. Click "Run ClustalW -> OK" to start the alignment process (*see* **Note 3**).
7. Save the completed alignment and check manually for any errors. If the test sequences represent coding regions, you may toggle from nucleotide to amino acid sequences and check for any stop codons in aligned sequences. The presence of stop codons in coding regions may indicate either alignment or sequence errors. To do this, select all sequences by clicking "Edit -> Select All Sequences" followed by clicking "Translation -> Toggle translation".
8. Change "~" symbols in the alignment to gaps (- symbol) by clicking the "Edit -> Search -> Find/Replace" path (*see* **Note 4**).

3.4 Analysis of Sequence Polymorphisms and Mutations

The rate at which mutations occur in RNA viruses such as CHIKV plays a key role in the generation of diverse virus populations. As CHIKV spreads across the globe, the parallel dynamics of diversity and mutation signatures have enabled researchers to track the spread of CHIKV. More importantly, such information has provided insight into how certain mutations play a role in driving CHIKV epidemics across regions. Essentially, the analysis of sequence polymorphism and mutations, particularly in fast-evolving viruses, forms the basis of understanding virus evolution and its implications on epidemiology. There are many freely available software programs for the analysis of sequence polymorphism. In the following section, we describe the basic method for the mapping of synonymous/non-synonymous mutations in the DnaSP program.

1. The DnaSP program is freely available and can be downloaded from http://www.ub.edu/dnasp/.
2. Load a pre-aligned sequence file into DnaSP through "File -> Open Data File". DnaSP only accepts sequence alignment files in FASTA, MEGA, NBRF/PIR, NEXUS, or PHYLIP formats. All sequences in the alignment must be of equal length.

3. In order to map the synonymous/non-synonymous sites, the coding region needs to be assigned first. To assign coding region, go to “Data -> Assign Coding Region”. As coding region may not necessarily start from the first site, the starting residue can be determined based on a reference sequence. Typically, information about the position of starting codon of a reference sequence can be obtained from the respective GenBank record in which Coding Sequence (CDS) is described under “Features”.
4. In the “Assign Coding Region” window, specify the first and last nucleotides of the coding region and click OK.
5. DnaSP also provides the option to define domain regions of gene sequences. Go to “Data -> Define Domain Sets” to bring up the “Define Domain Regions” panel. Subdomains can be defined by specifying starting and ending sites of a particular domain. Once a domain region is specified, click “Add New Domain” button to label the domain name.
6. To display alignment data, go to “Display -> View Data”. The nucleotide sequence alignment with the corresponding amino acid sequence of the assigned coding region will be displayed in a separate window. Synonymous or non-synonymous changes at specific sites can be highlighted using the “Select Sites/Codons” dropdown menu, located at the bottom right of the alignment data window.
7. To obtain a summary of polymorphisms, go to “Analysis -> Polymorphic Sites”. A window summarizing the number of variable/invariable sites, including singleton variable sites and parsimony informative sites, as well as specific sites of synonymous and non-synonymous changes will be displayed (*see* **Note 5**).

3.5 Construction of a Median Joining Evolutionary Network

Traditionally, the most common way to present evolutionary relationships among homologous sequences is by constructing a phylogenetic tree in the form of a bifurcating tree structure. The phylogenetic tree construction generally assumes that evolution proceeds in a simplistic tree-like fashion. However, the method may not adequately represent the evolutionary history of a population in which events such as recombination, convergent evolution, and microevolution may have taken place. Given these limitations, phylogenetic networks have gained popularity for the visualization of complex evolutionary processes. For instance, use of phylogenetic network in the analysis of CHIKV has shed further light on key evolutionary events that contributed to its re-emergence and spread in South-east Asia [5]. The following methodology describes how a phylogenetic network is constructed using the freely available Network 4.6.1.3 program.

1. Install Network 4.6.1.3 in your computer. The latest version can be downloaded from http://www.fluxus-engineering.com/sharenet.htm.
2. Before constructing the phylogenetic network, select the sequences to be included in the analysis and prepare an alignment. Network 4.6.1.3 accepts alignments in multiple formats. Phylip and nexus formats are preferred as they can be readily processed. Alignments can be prepared in BioEdit software suite (.phy 3.2) and ClustalX (.nxs) software as described in Subheading 3.1 (*see* **Note 6**).
3. Upload the alignment using the "Calculate Network -> Network Calculations ->Median Joining" path.
4. Once the sequence alignment is uploaded, click "Calculate Network".
5. Save the ".out" output file in a preferred file name and destination.
6. Open the output file using the "Draw Network -> File -> Open" path to draw the network. The initial draft may not be adapted to the screen. The software will seek your permission to redraw.
7. Complete the drawing by clicking "Draw ->Finalize"
8. The software panel now displays the complete network with node identifiers and mutational differences. If the network diagram is misaligned in the viewing panel, you may reposition it by using the navigation panel.
9. Display options are available in the display parameters panel. For example, you may unclick "Display Mutated Positions", if you do not intend to display the mutational difference between two nodes. Node size can be adjusted using the "Minimum Node Size" function. A summary of the mutations and haplotypes can be obtained by clicking the "Statistics" tab (*see* **Note 7**).
10. Re-format the branch color, branch thickness, font and nodes by right-clicking on the respective object (*see* **Note 8**).
11. In order to display sequence proportions in a pie chart, obtain the list of isolates in each node by double-clicking on the respective node.
12. Access the pie chart panel by right-clicking on a node (*see* **Note 9**).
13. Save the final version of the phylogenetic network in ".fdi" format using the "File -> Save" path. For publication purposes, diagram files can be saved in PDF format which can subsequently be used to generate image files (Fig. 1).

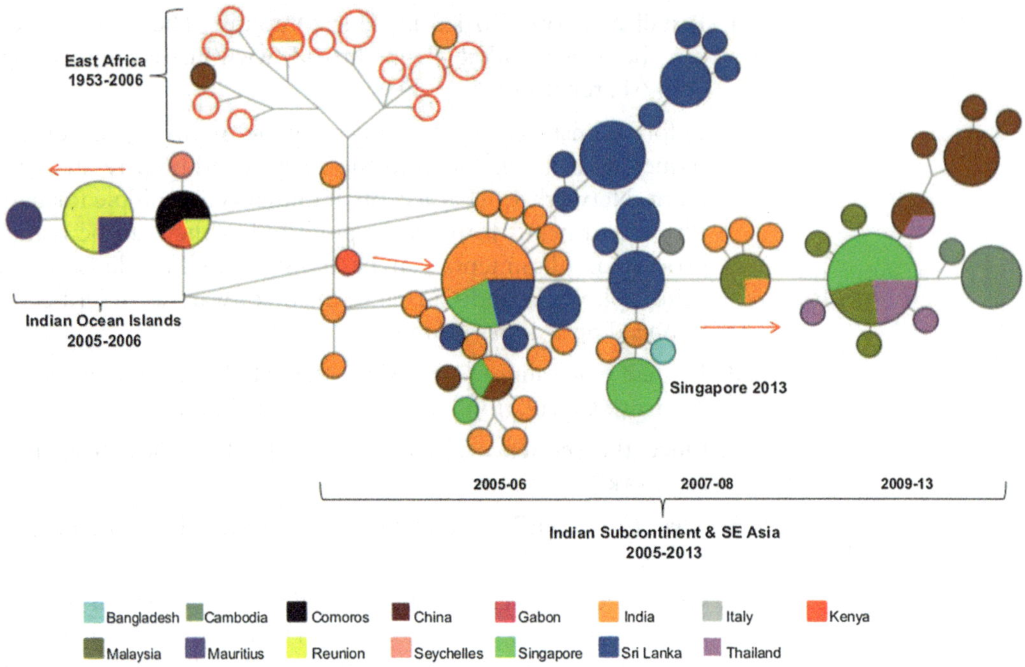

Fig. 1 Median joining network analysis of envelope 1 (*E1*) gene sequences of CHIKV. The network was drawn with Network 4.6.1.3 software using complete *E1* gene sequences ($n = 152$). *Circles* represent either individual isolates or clusters with identical sequences. The diameter of each circle is proportional to the number of isolates within each circle. The length of lines linking circles is not proportional to mutational distance between them. *Red arrows* show the direction of spatio-temporal expansion of CHIKV since 2004. Each *color* represents a country as shown in the figure. *Interpretation notes*: If the time of isolation of each sequence is known, network analysis will be able to demonstrate how a particular pathogen evolved during its spread across regions. For example, the figure illustrates that two CHIKV lineages spread out from Kenya in 2004. One of them established in Indian Ocean Islands and the other lineage went to India. Both lineages caused epidemics in respective locations. However, the Indian Ocean Island lineage did not move out of the neighboring region and its evolution was limited. On the other hand, Indian lineage evolved extensively and swept across Southeast Asia during 2006–2013. As indicated by *red arrows*, CHIKV evolution was progressive during its spread in Southeast Asia. The virus generated unique mutation signatures during this process which can be used to trace the origin of a CHIKV strain upon its emergence in a new location

3.6 Generating a Time-Scaled Bayesian Phylogenetic Tree and Determining the Ancestral Age by Calculating Time to Most Recent Common Ancestor (tMRCA) Using the BEAST Software Suite

Bayesian Evolutionary Analysis by Sampling Trees (BEAST) is a software package that utilizes the Bayesian statistical framework for phylogenetic, phylogeographic, and phylodynamic analyses. BEAST provides a powerful platform that implements Bayesian Markov chain Monte Carlo (MCMC) algorithm for a variety of evolutionary analyses including phylogenetic reconstruction, dating of divergence, phylogeography as well as hypothesis testing. The following descriptions demonstrate the use of BEAST to determine the time of divergence, Bayesian phylogenetic reconstruction, and phylogeography of pathogen populations, demonstrating their outputs by using CHIKV as an example.

1. Download the latest version of BEAST from www.beast2.org [16].
2. Label test sequences with date of sampling (years/months/dates), e.g., SeqID_year.
3. Prepare the alignment of test sequences in NEXUS (.nex) format. This can be done directly in ClustalX. If the alignment has already been prepared in a different format such as FASTA, it can be converted to NEXUS format in ClustalX by clicking "File -> Load Sequences -> Save Sequences As -> Nexus".
4. Open BEAUti by double-clicking the icon in the BEAST folder (*see* **Note 10**).
5. Load the NEXUS alignment by clicking "+" tab in the left bottom corner of BEAUti user interphase. In BEAUti 2, a new panel (Add Partition) appears when "+" tab is clicked. Under the "Select What to Add" drop down menu, select "Add Alignment" and click "OK" (*see* **Note 11**).
6. Click "Tip Dates" tab in the top parameter panel. Click "Use Tip Dates". Specify dates (as in years, months, etc.) and the direction of time calculation in the "Since Sometime in the Past" drop down menu.
7. Click "Guess". Use the "Guess Dates" panel to indicate which section of the sequence ID to be used as sampling dates. For example, to guess dates in the sequence ID format given in step 2, select "Use Everything-> After Last _ symbol" (*see* **Note 12**).
8. Check the panel to make sure that date and height are accurate. Height of the latest date of sampling should be 0 (*see* **Note 13**).
9. Click "Site Model" tab to set the desired substitution model for test sequences (*see* **Note 14**).
10. Select the desired clock model and tree model in "Clock Model" and "Priors" tabs respectively. Either a user defined tree, UPGMA tree or a random tree can be used.
11. Click "+" tab in priors panel. Give a name to the taxon set for which tMRCA is to be calculated and include sequences that need to be included in the tMRCA analysis by selecting sequence IDs from the left panel and moving them to right panel using ">>" tab. Click "OK" (*see* **Note 15**).
12. Set the Markov Chain Monte Carlo (MCMC) parameters using the "MCMC" tab (*see* **Note 16**).
13. Name "Trace Log", "Screen Log", and "Tree Log" log files with desired names.
14. Save parameters by using the "File -> Save As" path. The file will be saved in ".xml" format in the same folder that contains the test alignment.

15. Double-click "BEAST" icon in the BEAST folder. Choose ".xml" file with analytical parameters from the saved destination and click "Run" to start the analysis. If parameter settings are without any error, BEAST will initiate the run (*see* **Note 17**).

3.7 Visualization of BEAST Trace Output

1. Tracer is a MCMC trace analysis tool. The latest version can be downloaded from http://tree.bio.ed.ac.uk/software/. The following description is for version 1.6.0.
2. Click on "Tracer" icon to bring the user interphase.
3. Click "+" icon below the "Trace Files" section in the top left panel. Select the trace log file from the BEAST run.
4. Determine whether the run has achieved adequate independent sample mixing and convergence based on effective sample size (ESS) values (*see* **Note 18**).
5. Click on a parameter in the left panel to visualize estimates in the right panel. Click "Trace" tab in the top right panel to visualize the MCMC trace plot. If a prior was generated for tMRCA as in step 11 in Subheading 3.6, the relevant statistics will also be available in the left panel.

3.8 Construction of the Maximum Clade Credibility Tree Using BEAST Tree Output

1. Double-click "Tree Annotator" icon in the BEAST folder.
2. Adjust the "Burnin" (recommendation: 10 % of number of trees generated in the BEAST analysis) and posterior probability limit (recommendation: use the default value).
3. Select "Maximum Clade Credibility Tree" as the target tree type and "Median Heights" as the node height.
4. Select the tree log file from the BEAST analysis as the input tree file.
5. Define the output maximum clade credibility (MCC) treefile with a different name.
6. Click "Run" to generate the consensus tree.

3.9 Visualization of the Consensus MCC Tree

1. FigTree is a tree visualization software that allows annotating phylogenetic trees with embedded analytical parameters. The latest version of FigTree can be downloaded from http://tree.bio.ed.ac.uk/software/. The following description is for version 1.4.2.
2. Double-click FigTree application icon.
3. Load the MCC tree file using the "File -> Open" path.
4. Annotate the tree that opens in the right panel as required using options available in the left panel (*see* **Note 19**).
5. Add the time scale of the tree by clicking "Scale Axis" (*see* **Note 20**).
6. Save the final tree either in PDF format or as an image file using the "File -> Export" path (Fig. 2).

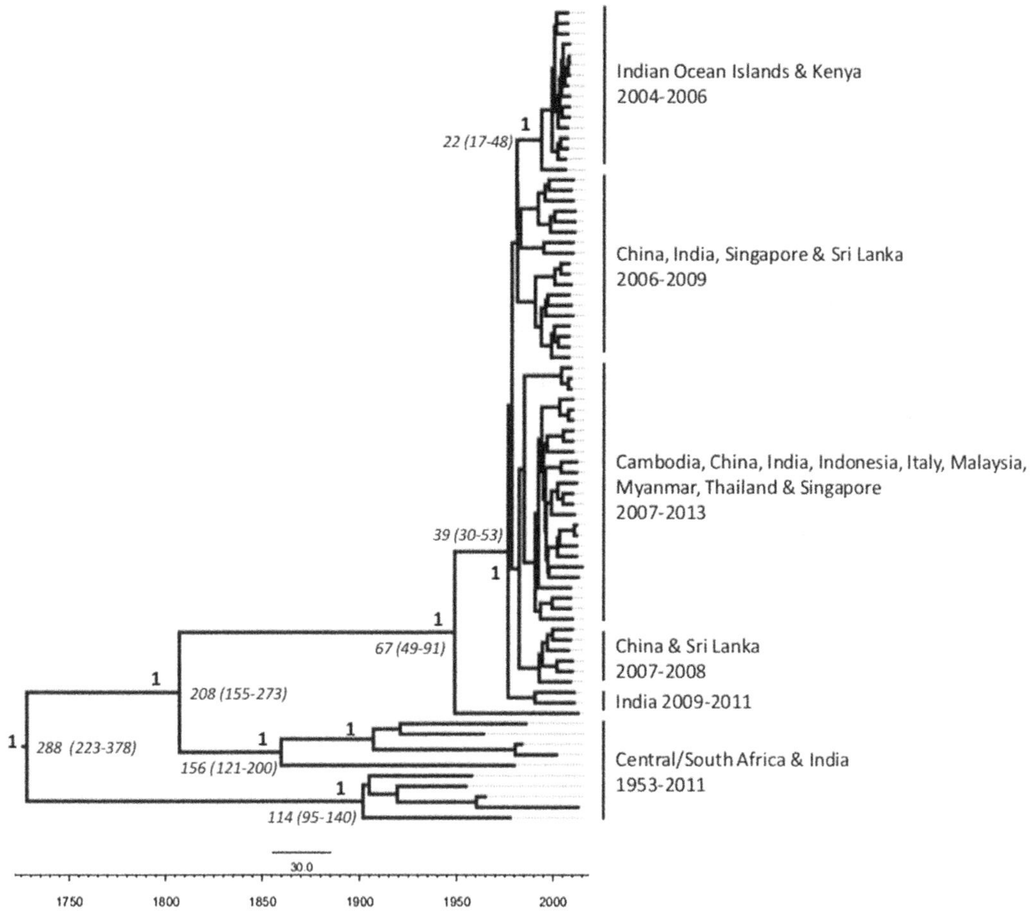

Fig. 2 Time-scaled maximum clade credibility tree of East, Central, and South African genotype of CHIKV. The phylogenetic tree was constructed in the BEAST2 software [16]. The analysis included 78 complete coding sequences of East, Central, and South African genotype of CHIKV reported from 1953 to 2013. The figure was generated in FigTree 1.4.2. Time scale of the tree is shown below. *Numbers in bold* at branch nodes are posterior support values. *Numbers in italics* are ancestral ages (years) at respective branch nodes. The 95 % highest posterior density interval (HPD) values are shown in *brackets*. *Interpretation notes*: Time-scaled trees can be used to determine the likely origin and time of emergence of ancestral strains of virus lineages

3.10 Phylogeography Analysis to Determine Spatio-Temporal Spread of Virus Strains Using the BEAST2 Software

1. Label aligned test sequences with date (years/months/dates) and location of sampling, e.g., SeqID_location_year (*see* **Note 21**).
2. Load the test alignment in NEXUS format into BEAUti as in step 5 of Subheading 3.6.
3. Click "+" sign in left bottom corner of BEAUti user interphase.
4. Add the phylogeography partition by selecting the desired module in the "Select What to Add" drop down menu. The "trait" may be re-named (*see* **Note 22**).
5. Guess location of each sequence from sequence IDs as described in **step** 7 of Subheading 3.6.

6. Define the clock model parameters for the "New Trait" partition.
7. MCMC section includes an additional log for the "Location Tree". Set the logging frequency and name the file as desired.
8. Save ".xml" file and run in BEAST2.
9. Once the BEAST run is complete, check ESS values using the trace log in Tracer to determine sampling and convergence of the analysis.
10. Run "Location Tree" log file in Tree Annotator to generate the consensus MCC tree with spatio-temporal information.

3.11 Visualization of Phylogeography Output Using the SPREAD Software

1. The latest version of SPREAD [17] can be downloaded from http://www.phylogeography.org/SPREAD.html.
2. Double-click SPREAD icon to launch the program.
3. Select the correct phylogeography module (either discrete or continuous) used to generate the ".xml" file in the top panel of SPREAD user interphase. The following description describes how to analyze discrete phylogeography output.
4. Load the MCC location tree file generated in **step 6** of Subheading 3.10 by clicking "Open" tab.
5. Click "Setup" tab to set location coordinates (*see* **Note 23**).
6. Click "Save" to save the location information as a text file and click "Done".
7. Adjust the most recent sampling data according to the test data.
8. Change the state attribute name to that given in **step 4** of Subheading 3.10 (*see* **Note 24**).
9. Set the branch color, thickness, circle color and computational parameters in "branches mapping", "circles mapping" and "computations" tabs respectively (*see* **Note 25**).
10. Define the name of ".kml" output file in "Output" tab.
11. Generate ".kml" file by clicking "Generate" tab in the "Generate KML/plot map" section. The ".kml" output file will be saved in the same folder that contains the location MCC tree file.
12. Click "Plot Map" tab to visualize the phyogeography output on the world map (Fig. 3).
13. If Google Earth software is installed in the computer, double-click on ".kml" output file to launch the phylogeography output in Google Earth.

4 Notes

1. The success of generating whole genomes from clinical specimens depends on viral load in each sample. If the viral load is high enough to generate all overlapping fragments by PCR,

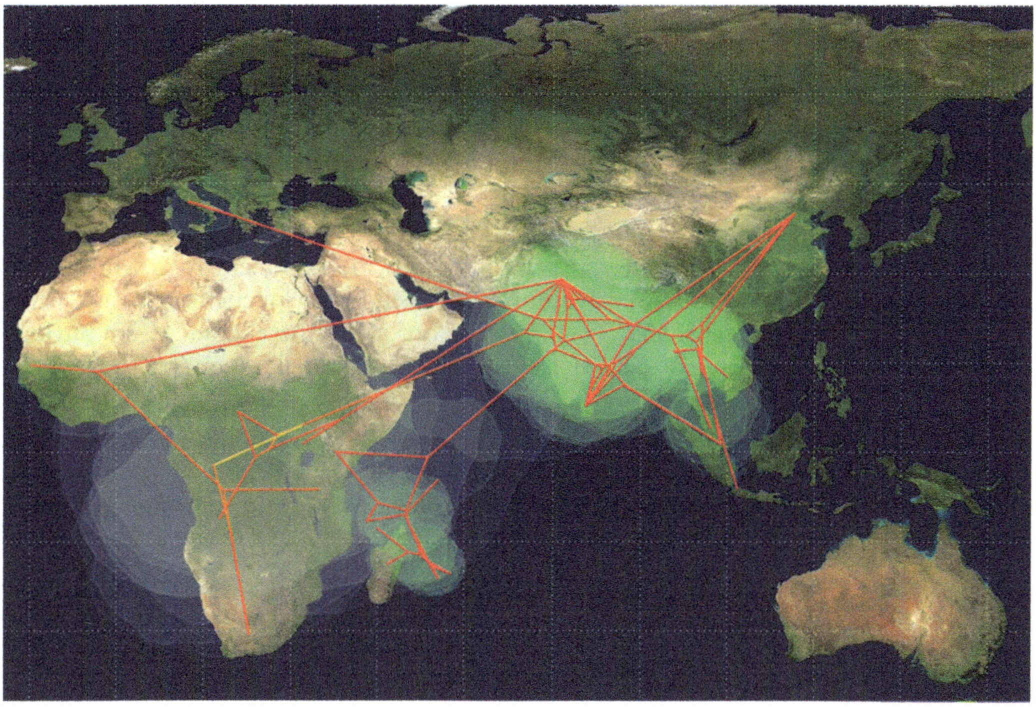

Fig. 3 Phylogeography analysis to demonstrate the spatial expansion of East, Central, and South African genotype of CHIKV. Phylogeography analysis was performed in the BEAST2.2 software using the continuous phylogeography module [16]. The analysis included 78 complete coding sequences of East, Central, and South African genotype of CHIKV reported from 1953 to 2013. The figure was generated in SPREAD 1.0.4 [17]. Branch color was adjusted to node heights (ages), so that the highest and lowest node heights are represented by *yellow* and *red* respectively. *Shaded circular polygons* illustrate number of lineages holding the state (a particular location in this analysis) at a given time. Polygons with the highest diversity are shown in *green shades*. Their radii are proportional to the number of lineages holding the state (i.e., at each location). The temporal scale can be visualized by opening the SPREAD output in Google Earth. *Interpretation notes*: The figure illustrates spatio-temporal distribution of East, Central, and South African genotype of CHIKV from 1953 to 2013. Lines represent the time-scaled maximum clade credibility summary tree superimposed on world map to demonstrate the geographical scale of CHIKV spread. *Yellow lines* represent oldest samples and *red lines* show the latest samples. The analysis demonstrates extensive CHIKV activity during the last decade in Indian Ocean Islands and Southeast Asia. CHIKV diversity in Southeast Asia was relatively higher than that in Indian Ocean Islands as demonstrated by *dark green polygons* over the Southeast Asian region

complete genomes can be generated without isolation of CHIKV from each sample. If not, CHIKV should be isolated from clinical specimens before proceeding with genome sequencing.

2. Accurate alignment of test sequences is an important prerequisite of downstream mutation and phylogenetic analyses. An inaccurate alignment may result in analytical and interpretation errors.

3. The viewing panel displays the progress of the alignment, starting from pairwise alignment. The time needed to complete an

alignment depends on the computational power as well as the number, length, and similarity level of test sequences. The process can be expedited by clicking "Fast algorithm for guide tree" option in the alignment panel. This is useful when test sequences are homogenous and represent the same genotypes/subtypes of a particular virus.

4. Toggling between nucleotide to amino acid sequences in BioEdit converts all gaps in the alignment to ~ symbol. This symbol may not be identified by phylogenetic software such as MEGA. Therefore, before using the alignment for phylogenetic analysis, make sure to convert all ~ symbols to gaps (-).
5. Grouping strains based on their mutation profiles is required to identify genetic signatures of monophyletic virus groups. Genetic signatures fixed in virus strains from different geographical regions can be used to track the movement of viruses from one region to another.
6. Alignments can be saved in BioEdit in two formats of phylip; .phy 3.2 and .phy 4. Only .phy 3.2 is compatible with Network software. Please note that your sequence IDs may be altered according to phylip format.
7. By default, the size of nodes is proportional to the frequency of sequences clustered within each node. Similarly, the distance between two nodes is proportional to the mutational distance between the two. If there are nodes not represented by your sequences, the analytical algorithm generates hypothetical nodes named as median vectors. Median vectors generally represent either hypothetical ancestral strains that are extinct or strains that are extant, but not sampled.
8. In the free version of Network 4.6.1.3, re-formatting of the branch color, branch thickness and font as well as displaying node composition in a pie chart has to be done manually.
9. Displaying nodes in a pie chart format is important when you intend to illustrate the geographical distribution of virus sequences. If the virus spread follows a sequential spatio-temporal pattern, a pie chart node will also be useful to trace the evolutionary path of virus strains.
10. The BEAST folder contains multiple softwares such as BEAUti, BEAST, Tree annotator, Log combiner, and Densi Tree. Of these, BEAUti is used to set parameters for the Bayesian run in BEAST.
11. Do not include symbols such as "&" in alignment file names. Such files will not be uploaded by BEAUti.
12. BEAUti has the ability to guess dates from test sequence IDs pre-formatted to include date information. If the sequence ID has multiple components separated by a character such as "_",

"Split on Character" function can be used to define the section of interest.

13. BEAUti user interphase displays the list of sequences, their sampling data and height. Height refers to the difference between the date of sampling of each sequence and the latest date of sampling among all sequences in the dataset. If there is an error in sequence IDs, BEAUti will not perform date calculation. In such instances, check sequence IDs for format uniformity.
14. If the substitution rate is known, the default value can be altered to the known value. If unknown, let the software calculate based on test sequences by using "Estimate" function.
15. In order to calculate the tMRCA of a monophyletic group of virus strains, individual virus sequences within it need to be defined as one group. This can be done in prior sections in BEAST2. In previous versions of BEAST, there is a separate tab (Taxa) in BEAUti user interphase for this purpose.
16. The chain length can be defined in the MCMC section. The "Trace log", "Screen log", and "Tree log" sections allow to define the frequency of data points in output files. Generally, a minimum of 10, 000 traces and trees are recommended. Therefore, the number that appears in "Log every" section should be adjusted based on the MCMC chain length selected. For example, if the MCMC chain length is 10 million, the program should be asked to log every 1000 chain interval (10,000,000/10,000 = 1000).
17. BEAST will initially check parameter setting for any errors. If there are errors, the process will be terminated with a message indicating the possible section of any error. In such instances, corrections should be done in parameter settings of the ".xml" file using BEAUti. The approximate time taken to complete the analysis can be obtained from the screen log display.
18. Effective sample size (ESS) is an estimate of the posterior distribution of a parameter within the Markov chain. ESS values are directly proportional to the confidence of independent sample mixing and convergence during the MCMC run. In general, an ESS value more than 200 is considered adequate to determine whether a particular parameter is confidently sampled within its posterior distribution. All parameters may not be required to exceed an ESS value of >200, but posterior, likelihood and prior parameters are considered important to meet the ESS cut-off.
19. The font of taxa names can be changed in "Tip Labels" tab. The appearance such as branch colors and thickness can be changed in "Appearance" tab. Taxa names can be annotated by using top panel tabs after selecting "Taxa" tab in the top panel

(Selection Mode). The appropriate time frame of test sequences should be adjusted by inserting the correct time of tree root in "Origin Value".

20. The tMRCA of a particular clade can also be calculated manually using the corrected time scale. The 95 % confidence interval can be obtained by selecting 95 % highest posterior density interval (95 % HPD) in "Node Bars" and "Node Labels" functions.
21. Phylogeography adds time (temporal) and location (spatial) of sampling to conventional Bayesian phylogenetic analysis [10]. As such, the method can be used to understand the spatio-temporal distribution pattern of virus strains within a known time period. Therefore, IDs of test sequences should include the location of sampling information (location name in discrete and latitude and longitude in continuous phylogeography), in addition to time, e.g., SeqID_locationinfo_year.
22. Phylogeography module can automatically be included in the ".xml" file in BEAST2 [16]. There are two phylogeography approaches: discrete and continuous. Their application is described elsewhere [10, 11, 18]. Desired phylogeography module can be incorporated into .xml file as a separate partition once the test sequence alignment is added.
23. Locations of test sequences stated in the alignment will be displayed in the "Location" column of the "Setup Location Coordinates" panel. Insert the latitude and longitude information of each location either manually or by uploading a tab-delimited text file with relevant information.
24. The state attribute name must be the same as that used when generating the ".xml" file. If not, SPREAD will generate an error message. The correct attribute name can be obtained by opening the MCC location tree in any text editor software.
25. A detailed description about these parameter settings is given in http://www.kuleuven.be/aidslab/phylogeography/tutorial/spread_tutorial.html. In our experience, the most stable version of SPREAD is 1.0.4.

Acknowledgement

We acknowledge original developers of software described in this section for their expertise contribution to the advancement of phylogenetic analyses. Their efforts to constantly improve existing software and to make them freely available are greatly appreciated.

References

1. KariukiNjenga M, Nderitu L, Ledermann JP, Ndirangu A, Logue CH, Kelly CH, Sang R, Sergon K, Breiman R, Powers AM (2008) Tracking epidemic Chikungunya virus into the Indian Ocean from East Africa. J Gen Virol 89(Pt 11):2754–2760
2. Schuffenecker I, Iteman I, Michault A, Murri S, Frangeul L, Vaney MC, Lavenir R, Pardigon N, Reynes JM, Pettinelli F, Biscornet L, Diancourt L, Michel S, Duquerroy S, Guigon G, Frenkiel MP, Brehin AC, Cubito N, Despres P, Kunst F, Rey FA, Zeller H, Brisse S (2006) Genome microevolution of chikungunya viruses causing the Indian Ocean outbreak. PLoS Med 3(7):e263
3. Arankalle VA, Shrivastava S, Cherian S, Gunjikar RS, Walimbe AM, Jadhav SM, Sudeep AB, Mishra AC (2007) Genetic divergence of Chikungunya viruses in India (1963-2006) with special reference to the 2005-2006 explosive epidemic. J Gen Virol 88(Pt 7):1967–1976
4. Ng LC, Hapuarachchi HC (2010) Tracing the path of Chikungunya virus—evolution and adaptation. Infect Genet Evol 10(7):876–885
5. Hapuarachchi HC, Bandara KB, Sumanadasa SD, Hapugoda MD, Lai YL, Lee KS, Tan LK, Lin RT, Ng LF, Bucht G, Abeyewickreme W, Ng LC (2010) Re-emergence of Chikungunya virus in South-east Asia: virological evidence from Sri Lanka and Singapore. J Gen Virol 91(Pt 4):1067–1076
6. deLamballerie X, Leroy E, Charrel RN, Ttsetsarkin K, Higgs S, Gould EA (2008) Chikungunya virus adapts to tiger mosquito via evolutionary convergence: a sign of things to come? Virol J 5:33
7. Tsetsarkin KA, Vanlandingham DL, McGee CE, Higgs S (2007) A single mutation in chikungunya virus affects vector specificity and epidemic potential. PLoS Pathog 3(12):e201
8. Vazeille M, Moutailler S, Coudrier D, Rousseaux C, Khun H, Huerre M, Thiria J, Dehecq JS, Fontenille D, Schuffenecker I, Despres P, Failloux AB (2007) Two Chikungunya isolates from the outbreak of La Reunion (Indian Ocean) exhibit different patterns of infection in the mosquito, Aedesalbopictus. PLoS One 2(11), e1168
9. Bandelt HJ, Forster P, Rohl A (1999) Median-joining networks for inferring intraspecific phylogenies. Mol Biol Evol 16(1):37–48
10. Lemey P, Rambaut A, Drummond AJ, Suchard MA (2009) Bayesian phylogeography finds its roots. PLoS Comput Biol 5(9):e1000520
11. Lemey P, Rambaut A, Welch JJ, Suchard MA (2010) Phylogeography takes a relaxed random walk in continuous space and time. Mol Biol Evol 27(8):1877–1885
12. Larkin MA, Blackshields G, Brown NP, Chenna R, McGettigan PA, McWilliam H, Valentin F, Wallace IM, Wilm A, Lopez R, Thompson JD, Gibson TJ, Higgins DG (2007) Clustal W and Clustal X version 2.0. Bioinformatics 23(21):2947–2948
13. Edgar RC (2004) MUSCLE: multiple sequence alignment with high accuracy and high throughput. Nucleic Acids Res 32(5): 1792–1797
14. Hall TA (1999) BioEdit: a user-friendly biological sequence alignment editor and analysis program for Windows 95/98/NT. Nucl Acid Symp Ser 41:95–98
15. Tamura K, Stecher G, Peterson D, Filipski A, Kumar S (2013) MEGA6: molecular evolutionary genetics analysis version 6.0. Mol Biol Evol 30(12):2725–2729
16. Bouckaert R, Heled J, Kuhnert D, Vaughan T, Wu CH, Xie D, Suchard MA, Rambaut A, Drummond AJ (2014) BEAST 2: a software platform for Bayesian evolutionary analysis. PLoS Comput Biol 10(4):e1003537
17. Bielejec F, Rambaut A, Suchard MA, Lemey P (2011) SPREAD: spatial phylogenetic reconstruction of evolutionary dynamics. Bioinformatics 27(20):2910–2912
18. Bloomquist EW, Lemey P, Suchard MA (2010) Three roads diverged? Routes to phylogeographic inference. Trends Ecol Evol 25(11): 626–632

Chapter 4

Synthetic Peptide-Based Antibody Detection for Diagnosis of Chikungunya Infection with and without Neurological Complications

Rajpal S. Kashyap, Shradha S. Bhullar, Nitin H. Chandak, and Girdhar M. Taori

Abstract

Synthetic peptide-based diagnosis of Chikungunya can be an efficient and more accessible approach in immunodiagnostics. Here, we describe the identification of Chikungunya-specific 40 kD protein for development of synthetic peptide-based enzyme-linked immunosorbent assay for the detection of Chikungunya virus-specific antibodies in the patient's sample. The total sodium dodecyl sulfate-polyacrylamide gel electrophoresis protein profile of the patient's sample can be done to identify specific protein bands. The identified proteins can be subjected to liquid chromatography-tandem mass spectrometry (LC-MS/MS) for characterization. After characterization, immunogenic peptides can be designed using softwares and subsequently synthesized chemically. The peptides can be used to develop more specific, sensitive, and simpler diagnostic assay.

Key words One-dimensional electrophoresis, Two-dimensional electrophoresis, LC-MS/MS, Synthetic peptide, Enzyme linked-immunosorbent assay

1 Introduction

The ability to identify potential antigens produced by viral pathogens that are effective diagnostic candidates of disease depends on multiple variables. One of the most important variables is the source of clinical specimen. Clinical specimens from diseased individual can be used for the identification and isolation of a proteome or protein set of interest. The science of protein analysis encompasses a range of analytical technologies including long established approaches such as gel electrophoresis, enzyme-linked immunosorbent assay (ELISA), and liquid chromatography-tandem mass spectrometry (LC-MS/MS; [1, 2]). The analysis of proteins by LC-MS/MS has rapidly evolved in the field of proteomics [3]. Its unique capabilities have significantly enhanced this area of research, which was not possible to be achieved 10 years ago. LC-MS/MS data for

Justin Jang Hann Chu and Swee Kim Ang (eds.), *Chikungunya Virus: Methods and Protocols*, Methods in Molecular Biology, vol. 1426,DOI 10.1007/978-1-4939-3618-2_4, © Springer Science+Business Media New York 2016

a protein are acquired and searched against sequence collections to identify the corresponding peptides and proteins [4]. Once such a protein set is isolated from one-dimensional gel electrophoresis, the individual proteins in the set can be separated using two-dimensional gel electrophoresis or column chromatography. When proteins are separated by gel electrophoresis, individual protein bands or spots can be manually excised from the gel. To identify what protein is in an excised band or spot, gel fragments containing a protein can be subjected to LC-MS/MS. For identification of the protein, the MS data can be used to search against updated protein sequence databases from NCBI and UniProt from amino acid sequences of known proteins. Once the protein sequence is identified, online tools can be used for antigenicity prediction of the identified protein and can be used further in immunodiagnostic assays. Several immunodiagnostic tests for the detection of antibodies have been developed for the detection of viral, bacterial, and parasitic diseases based on synthetic peptides derived from antigenic proteins [5]. Detection of IgM and IgG antibodies in the samples of Chikungunya patients has been reported previously [6]. Studies have been done to detect antibodies using peptides designed with the aid of online softwares [7]. The antigen sequences can be used to obtain peptides with antigenic epitopes on the basis of prediction analysis by online softwares and can be chemically synthesized for validation in ELISA. The protocols outlined in this chapter are illustrated in Fig. 1.

2 Materials

2.1 One-Dimensional Electrophoresis

1. 30 % acrylamide monomer solution: Add 30 g of acrylamide and 0.8 g of bis-acrylamide and make up the volume with distilled water to 100 ml. Store at 2–8 °C in dark (*see* **Note 1**).
2. Separating gel buffer: 1.5 M Tris–HCl, pH 8.8. Add 18.66 g of Tris–HCl in 75 ml of distilled water. Adjust pH to 8.8 with dilute HCl and make up the volume with distilled water to 100 ml.
3. Stacking gel buffer: 0.5 M Tris–HCl, pH 6.8. Add 6.55 g Tris–HCl in 75 ml of distilled water. Adjust pH to 6.8 with dilute HCl and make up the volume with distilled water to 100 ml.
4. Ammonium persulfate (APS): 10 % solution in water. Add 0.1 g of APS to 1 ml distilled water (*see* **Notes 2** and **3**).
5. *N*,*N*,*N*,*N*′-tetramethyl-ethylenediamine (TEMED; *see* **Note 2**).
6. Sample preparation: Take 0.1 g of bromophenol blue (BPB), 1 g of SDS, 2 ml of glycerol and 2.5 ml of stacking gel buffer and make up the volume to 10 ml with distilled water. Determine the protein concentration of the sample and take

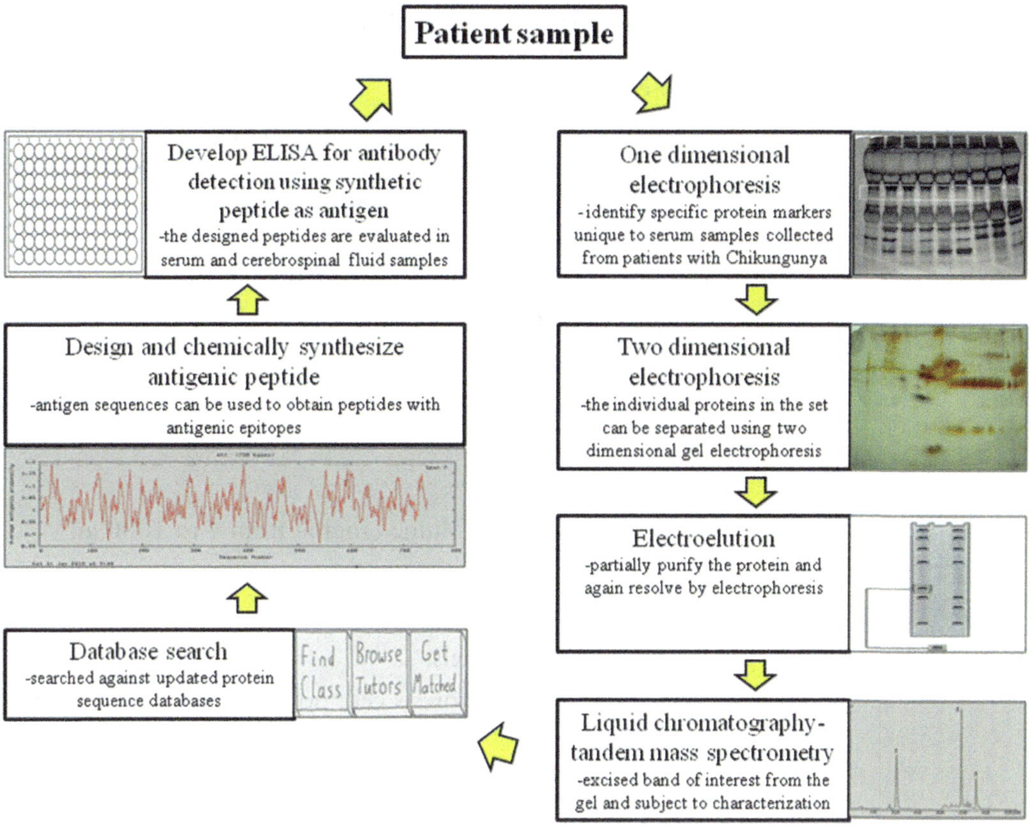

Fig. 1 Schematic overview of identification of specific protein for development of synthetic peptide-based enzyme-linked immunosorbent assay for antibody detection in serum or cerebrospinal fluid samples of Chikungunya patients

the volume of samples equivalent to 40 μg of protein. Now add 10 μl of BPB solution and make the final volume to 30–40 μl (*see* **Notes 4** and **5**).

7. Tank buffer: 25 mM Tris–HCl pH 8.3, 192 mM glycine, 0.1 % SDS. Take 3.03 g of Tris–HCl, 14.4 g of glycine and 1 g of SDS and make up the volume to 1 l with distilled water.
8. Protein stain: 50 % methanol, 10 % acetic acid, 2 g/l coomassie brilliant blue R-250. Weigh 2 g of coomassie brilliant blue R-250. Add 500 ml of methanol and 100 ml of acetic acid. Make up the volume to 1 l with distilled water (*see* **Note 6**).
9. SDS-PAGE electrophoresis cell, power supply appropriate for SDS-PAGE system.

2.2 Two-Dimensional Electrophoresis

1. Rehydration/Sample Buffer: 7 M urea, 4 % 3-[(3-cholamidopropyl) dimethylammonio]-1-propanesulfonate (CHAPS), 60 mM dithiothreitol (DTT), 0.5 % (w/v) Bio-Lyte 3/10 ampholytes, and BPB (0.1 %). Prepare and store at −20 °C (without DTT and BPB). Add 9.3 mg DTT per ml just before use (*see* **Note 7**).

2. Equilibration buffer I: 6 M urea, 2 % SDS, 1.5 M Tris–HCl pH (8.8), 50 % glycerol, and 2 % (w/v) DTT. Add DTT just before use.
3. Equilibration Buffer II: 6 M urea, 2 % SDS, 1.5 M Tris–HCl, pH (8.8), 50 % glycerol, and 2.5 % (w/v) iodoacetamide (IAA). Add IAA just before use.
4. 50 % glycerol solution.
5. PROTEAN isoelectric focusing (IEF) cell Ready Strip immobilized pH gradient (IPG) strips, pH 4–7 (7, 11, 17 cm).
6. IEF Focusing tray with lid (same size as IPG strips).
7. Electrode wicks, precut.
8. Blotting filter papers.
9. Mineral oil.
10. Forceps, pipette for volume ranging from 4 to 1000 μl.
11. Strip plate.
12. Plastic wrap.
13. 8–16 % SDS-PAGE gels.
14. SDS-PAGE electrophoresis cell, power supply appropriate for SDS-PAGE system.
15. Tank buffer.
16. SDS-PAGE protein stain.
17. Destaining solution (40 % methanol, 10 % acetic acid).
18. Disposable rehydration/equilibration trays with lid (same size as IPG strips).
19. High purity water.

2.3 Electroelution

1. Elution buffer: 0.15 M phosphate buffered saline (PBS), pH 7.4.
2. Whole gel eluter system.

2.4 LC-MS/MS Analysis

1. Destain solution: 50 % methanol and 10 % acetic acid.
2. Wash solution: 50 % acetonitrile in 0.1 M Tris–HCl, pH 8.0.
3. Trypsin.
4. Alkylation agent: 100 mM IAA.
5. In-gel digestion agent: 50 mM ammonium bicarbonate buffer, pH 8.5 (*see* **Note 8**).
6. Acetonitrile.
7. 0.5 % acetic acid.
8. A Finnigan LCQ ion trap MS in-line coupled with a high-pressure liquid chromatography (HPLC) system.

9. A 75 μm (ID) × 10 cm length, 3 μm packing C18 capillary column connected to a specially designed nanoSpray device.
10. Mobile phase Solvent A: 2 % acetonitrile, 97.9 % water, 0.1 % formic acid.
11. Mobile phase Solvent B: 90 % acetonitrile, 9.9 % water, 0.1 % formic acid.
12. HPLC grade water.

2.5 Peptide Synthesis

1. 2-(1H-benzotriazol-1-yl)-1,1,3,3-tetramethyluronium hexafluorophosphate (HBTU)/diisopropylethylamine (DIEA) coupling chemistry: for each coupling cycle, use 3 eqiv of N-9-fluorenylmethyloxycarbonyl (Fmoc) amino acid, 6 equiv. of DIEA and 3 equiv. of HBTU.
2. 20 % piperidine in dimethylformamide (DMF) solution.
3. Trifluoroacetic acid (TFA) cleavage solution: TFA/thioanisole/1,2-ethanedithiol/triisopropylsilane/water (70:10:10:1:35).
4. Cold ether.
5. Preparative HPLC by Varian Pro-Star system. Use a linear 0.1 % TFA/acetonitrile/water gradient mixture with 1 % acetonitrile increase per minute with UV detection at 220 nm.

2.6 ELISA

1. 10× PBS: Take 80 g sodium chloride, 2 g potassium chloride, 14.4 g disodium hydrogen phosphate, and 2.4 g sodium dihydrogen phosphate in 800 ml distilled water. Adjust pH to 7.35 and make up the volume with distilled water to 1 l.
2. 1× PBS: Take 100 ml 10× PBS and make up the volume to 1 l with distilled water. Filter through a glass microfiber filter paper and use.
3. Synthetic peptides (25 ng/100 μl): To prepare 10 ml, dissolve 2.5 μg synthetic peptide in 10 ml 1× PBS. Mix well before use.
4. Serum sample dilution (1:200) and CSF sample dilution (1:5): Centrifuge the samples at 268 × *g* for 2 min. Dilute samples in 1× PBS. Mix well and use.
5. 0.5 % bovine serum albumin (BSA): To prepare 10 ml, dissolve 50 mg BSA in 10 ml 1× PBS. Mix thoroughly.
6. Goat-anti-human IgM-horseradish peroxidase (HRP) conjugate: To prepare 1:5000 dilution, add 0.5 μl Goat-anti-human IgM-HRP conjugate in 10 ml 1× PBS. Mix thoroughly.
7. Goat-anti-human IgG-horseradish peroxidase (HRP) conjugate: To prepare 1:10,000 dilution, add 1 μl Goat-anti-human IgG-HRP conjugate in 10 ml 1× PBS. Mix thoroughly.
8. 3,3,5,5′-tetramethylbenzidine (TMB)-substrate. Commercially available in brown bottle stored at 4 °C.

9. 2.5 N sulfuric acid, 98 %. Add 35 ml concentrated sulfuric acid and add 465 ml distilled water to make the total volume to 500 ml.
10. Equipment: rotary shaker, ELISA reader.

3 Methods

3.1 One-Dimensional Electrophoresis

The system of buffers used in the gel system below is that of Laemmli [8].

1. Clean glass plate and assemble the glass plate/gel cassettes in the electrophoresis equipment in a vertical position.
2. Mix the components given in Table 1 in a glass beaker (*see* **Note 9**).
3. Pour the running gel solution and allow it to polymerize.
4. Pour the stacking gel on top of the polymerized gel.
5. After pouring the stacking gel solution, insert comb immediately and allow the gel to polymerize for about 15–30 min.
6. Take out the comb slowly after the stacking gel gets polymerized, and remove the bottom spacer.
7. Wash the wells with tank buffer to remove air bubbles and gel particles.
8. Determine the protein concentration of the sample and load the volume of samples mixed with sample buffer (10 μl) equivalent to 40 μg of protein in each well. Heat the samples to 95–100 °C prior to use.
9. Connect the tank to the power supply and turn the power to 150 V.

Table 1

	Running gel (10 %)	Stacking gel (5 %)
Acrylamide monomer	1.7 ml	0.5 ml
Separating gel buffer	1.3 ml	–
Stacking gel buffer	–	0.4 ml
Distilled water	1.9 ml	2.1 ml
Mix and degas for 5 min before adding the following:		
SDS (10 %)	50 μl	30 μl
APS (10 %)	50 μl	30 μl
TEMED	10 μl	10 μl

10. Continue the run, till the tracking dye reaches the bottom of the gel (*see* **Note 10**).
11. Soak the gel in freshly prepared staining solution for 2 h.
12. Rinse the gel briefly with distilled water and soak in destaining solution and keep on the gel rocker.
13. After 15–20 min change the destaining solution with fresh destaining solution and again keep it on the gel rocker till the protein bands become clearly visible (*see* **Note 11**).
14. Take the photographs of the gel placed on the trans-illuminator.
15. For densitometric analysis, use a gel documentation system.

3.2 Two-Dimensional Electrophoresis

1. Prepare sample volumes according to the strip length as outlined in Table 2.
2. Take out a fresh, clean, dry, and disposable rehydration tray.
3. Add sample diluted in the rehydration buffer (*see* **Note 12**).
4. Take out a readily made IPG strip from −20 °C and by using forceps, peel coversheet off the strip.
5. Gently place the strip gel side down onto the sample, i.e., number should face on upper side and wait for 1 h. Take care that sample do not overflow on the plastic backing (*see* **Note 12**).
6. Overlay each of the strips with 2–3 ml (1.5 ml is sufficient) of mineral oil drop by drop to prevent evaporation during rehydration process.
7. Keep the rehydration tray for about 12–16 h at room temperature.
8. For isoelectric focusing (First dimension), take out dry, clean, oil-free Protean IEF focusing tray of same size as rehydration tray.
9. Using forceps, place the paper wick at both ends of the channels covering the wire electrodes.
10. Pipette 8 μl of nanopure water onto each wick.
11. Using forceps, carefully take out the strips and hold the strip vertically for about 7–8 s on the tissue paper to allow a mineral oil to drain.

Table 2

Strip length	7 cm	11 cm	17 cm
Sample volume (max)	125 μl	185 μl	300 μl
Protein loaded (max)	169 μg	250 μg	405 μg

12. Transfer the IPG strip into the focusing tray while maintaining gel side down.
13. Remember that "+" and "pH range" are always at a left side of the focusing tray.
14. Cover each IPG strip with 1.5–2 ml mineral oil. Then place the lid on the focusing tray.
15. Place the focusing tray into the PROTEAN IEF cell and close the cover.
16. Program the PROTEAN IEF cell using the appropriate 3-step protocol.
17. Press START to initiate electrophoresis run.
18. After completion of electrophoresis run (i.e., 4–5 h), remove the IPG strip from the focusing tray. Hold the strip vertically for about 7–8 s on the tissue paper to allow the mineral oil to drain.
19. To equilibrate IPG strip, transfer the strip into the clean equilibration or rehydration tray with the gel side up.
20. Keep the strips in the equilibration buffer I for 15 min then remove the IPG strip from the equilibration tray. Hold the strip vertically for about 7–8 s on the tissue paper to allow the equilibration buffer I to drain.
21. Repeat **step 20**, firstly with equilibration buffer II and then with SDS buffer.
22. Prepare 12 % gels for SDS-PAGE 15 min before completion of the equilibration process.
23. Slowly place strip on the large glass plate such that strips must face gel side up.
24. Push the strip on the top of the acrylamide gel and overlay with the strip and acrylamide gel with 0.5 % low melting agarose for complete sealing so that all protein from strip should transfer into the acrylamide gel (*see* **Note 13**).
25. Place the whole assembly into 4 °C for 10 min for solidification of agarose.
26. Assemble electrophoretic apparatus for the second dimension, and then carry out second dimension electrophoresis.
27. After electrophoresis is over, proceed with coomassie staining.

3.3 Electroelution

To partially purify the protein, resolve serum by one-dimensional SDS-PAGE and slice the gel portion with protein band and then pre-equilibrate in elution buffer. Electroelute the gel in a whole gel eluter system for 1 h at 30 V. Dialyze the harvested eluate against PBS and measure the protein concentration. Again, resolve this partially purified material by SDS-PAGE, stain the gel with coomassie blue and excise the band from the gel. Use the excised protein band for LC-MS/MS analysis.

3.4 LC-MS/MS Analysis

Destain and wash each gel piece prior to in-gel digestion, then excise protein bands for trypsin digestion after treatment with alkylation agents prior to analysis on LC-MS/MS.

1. Carry out in-gel digestion in 50 mM ammonium carbonate buffer, pH 8.5 at 37 °C for approximately 4 h. Add an equal volume of the digestion buffer, depending on the volume of the gel piece, and usually in a range from 20 to 50 μl. The amount of proteolytic enzyme that is used depends on both the size of the gel piece and the estimated amount of protein within the gel band. Typically, use 200 ng to 1 μg trypsin per gel band.
2. Add acetonitrile, in a volume equal to 3–5 times the volume of the digestion buffer to the digestion mix to extract the peptides.
3. Centrifuge the samples at high speed for 5 min. Transfer the supernatant to a clean microfuge tube with a gel-loading pipette tip and dry in a SpeedVac on medium heat.
4. Add alkylation agents prior to analysis with LC-MS/MS.
5. Dissolve the dried sample in 0.5 % acetic acid for In order to find out a peptide whichLC-MS/MS analysis. Use a Finnigan LCQ ion trap MS in-line coupled with a HPLC system for LC-MS/MS.
6. Connect a 75 μm (ID) × 10 cm length, 3 μm packing C18 capillary column, which is packed in-house, to a specially designed nanoSpray device. The device is capable of delivering a stable electrospray at flow rates of 100–1500 nl/min of the mobile phases that include Solvent A and Solvent B.
7. For this analysis, set the ion trap MS to operate in a data-dependent mode with the Automatic Gain Control turned on.
8. Evaluate the MS/MS data against several internal quality control (QC) standards. After passing the QC standards, load the MS/MS data into the proprietary ProtQuest search engine to search the most recent non-redundant protein database. Then, manually analyze the results from the ProtQuest search.

3.5 Peptide Synthesis

1. Selection and designing of peptide: After the identification of the protein from the database through the data of LC-MS/MS analysis, retrieve the sequence of identified protein from EXPASY proteomic server-UniProtKB/Swiss-Prot. Determine the antigenic peptides on the basis of Kolaskar and Tongaonkar [9] method by using online software titled "Molecular Immunology Foundation-Bioinformatics software".
2. Synthesis of peptides: Prepare the peptides with the following procedure using solid-phase methods employing N-Fmoc/t-Bu protection strategy and chemistry.

(a) Employ the following side chain protection strategy for standard amino acid residues: Asp(OtBu), Glu(OtBu), Arg(Pbf), Lys(Boc), Trp(Boc), Ser(tBu), Thr(tBu), Tyr(tBu), Asn(Trt), Cys(Trt), Gln(Trt), and His(Trt).

(b) Carry out solid phase assembly in a stepwise manner on an AAPPTEC Apex396 multiple peptide synthesizer using HBTU/DIEA coupling chemistry at 0.11–0.15 mmol resin scale (Fmoc Rink resin).

(c) For each coupling cycle, use 3 eqiv of N-Fmoc amino acid, 6 equiv. of DIEA and 3 equiv. of HBTU. The coupling time is 1 h. Carry out Fmoc deprotections with two treatments of a 20 % piperidine in DMF solution, each for 8 min.

3. Peptide cleavage and global deprotections: Cleave/deprotect the target peptides from their respective peptidyl resins by treatment with a TFA cleavage mixture as follows:

(a) Add a solution of TFA/thioanisole/1,2-ethanedithiol/triisopropylsilane/water (70:10:10:1:3:5) (4 ml) to each well in the reaction block, then mix for 3 h.

(b) Collect the TFA solutions from the wells by positive pressure into vials located in a matching block on bottom of the reactor.

(c) Rinse the resins in the wells twice with an additional 0.5 ml of TFA mixture, and combine the rinses with the solutions in the vials.

(d) Then add the cleavage solution to 45 ml of cold ether and precipitate out the peptides.

(e) Wash with ether for 3 times and then dry to yield the crude peptides.

4. Peptide purification: Carry out preparative HPLC on Varian Pro-Star system. Use a linear 0.1 TFA/acetonitrile/water gradient with 1 % acetonitrile increase per minute with UV detection at 220 nm. Collect the desired product eluted in a single 15–20 ml fraction and obtain the final peptides as white powders by lyophilization.

3.6 ELISA

1. In order to find out a peptide which can serve as a diagnostic marker for Chikungunya infection, perform IgM/IgG detection by ELISA protocol. High absorbance value of the test sample as compared to control in the ELISA reader at 450 nm indicates the presence of the antigen-specific antibody in the serum/CSF sample.

2. Coat peptides in micro-wells in the concentration of 25 ng/100 μl in each well and incubate for 3 h at 37 °C.

3. After giving a wash with PBS, block wells with 0.5 % BSA in PBS for 2 h at 37 °C.
4. Wash wells once with PBS and store at 4 °C overnight.
5. Next day add 100 µl of CSF/serum samples to the wells and incubate for 45 min at 37 °C.
6. Wash the wells thrice with PBS and add 100 µl of secondary antibody (Goat-anti-human HRP conjugated antibody, 1:10,000). Incubate for 30 min at 37 °C.
7. In case of IgM antibody detection, add 100 µl of Goat-anti-human IgM HRP-conjugated antibody, 1:10,000 dilution and incubate for 30 min at 37 °C.
8. After washing the wells with PBS, add 100 µl of TMB substrate solution and incubate for 10 min at room temperature.
9. Stop the reaction by adding 100 µl of 2.5 N H_2SO_4 in each well.
10. Read the absorbance of each well at 450 nm.
11. Use receiver operating curve to calculate the cutoff value through comparison between absorbance values in the Chikungunya and non-Chikungunya group. Evaluate the sensitivity and specificity of developed peptide-based antibody (IgG and IgM) detection ELISA test for diagnosis of Chikungunya infection.

4 Notes

1. Acrylamide is a neurotoxin. Therefore, take necessary precautions while using it.
2. Prepare this solution fresh everyday.
3. The APS-TEMED pair is a less efficient polymerization initiator below pH 6 than it is above pH 6. To compensate for this loss of efficiency in buffers below pH 6, increase the concentration of TEMED fivefold.
4. Pure proteins or simple mixtures should be used at 0.5–1 mg/ml concentrations. For more complex samples such as serum, suitable concentrations must be determined by trial and error.
5. SDS-PAGE is a very reproducible procedure. The major variation between different laboratories is the different methods used for the preparation of samples. Improper sample preparation will lead to an improper gel profile.
6. The sensitivity of protein detection by coomassie is near to 0.1 µg of protein. In case the sample of choice contains less amount of target protein, more sensitive methods such as silver staining should be done.

7. Greater concentration of urea of up to 9 M increases the sample density and prevents sample from floating.
8. Sodium phosphate buffer should be used if ammonia in ammonium bicarbonate interferes with the amino acid analysis.
9. The percentage of gel is largely dependent on the size of protein of interest which is to be separated. If protein covers a wide range of molecular weight, then accordingly set the most appropriate percentages of stacking and running gel to be used.
10. If a sample is being analyzed for the first time, stop the run till the dye reaches the bottom of the gel so that the low molecular weight proteins do not migrate out of the gel in the buffer. Determine the length till which the dye has to be run through trial and errors.
11. It takes longer duration to remove excess of coomassie brilliant blue from the gel. Generally, destain solution is replaced at regular intervals to elute the dye from the gel until the bands are clearly visible.
12. Note that there should not be any air-bubble. Take care not to trap air bubbles beneath the gel side down.
13. It is not necessary to overlay with agarose, however, if overlaid, it helps to prevent the gel strip from lifting as the running buffer is added.

References

1. Lequin R (2005) Enzyme Immunoassay (EIA)/Enzyme-Linked Immunosorbent Assay (ELISA). Clin Chem 51:2415–2418
2. Lilley KS, Razzaq A, Dupree P (2002) Two-dimensional gel electrophoresis: recent advances in sample preparation, detection and quantitation. Curr Opin Chem Biol 6:46–50
3. Shi Y, Xiang R, Horváth C, Wilkins JA (2004) The role of liquid chromatography in proteomics. J Chromatogr A 1053:27–36
4. Fenyo D (2000) Identifying the proteome: software tools. Curr Opin Biotechnol 11: 391–395
5. Gómara MJ, Haro I (2007) Synthetic peptides for the immunodiagnosis of human diseases. Curr Med Chem 14:531–546
6. Kashyap RS, Morey SH, Chandak NH et al (2010) Detection of viral antigen, IgM and IgG antibodies in cerebrospinal fluid of Chikungunya patients with neurological complications. Cerebrospinal Fluid Res 7:12
7. Morey SH, Kashyap RS, Purohit HJ et al (2010) An approach towards peptide-based antibody detection for diagnosis of Chikungunya infection. Biomarkers 15:546–552
8. Laemmli UK (1970) Cleavage of structural proteins during the assembly of the head of bacteriophage T4. Nature 227:680–685
9. Kolaskar AS, Tongaonkar PC (1990) A semi-empirical method for prediction of antigenic determinants on protein antigens. FEBS Lett 276:172–174

Chapter 5

Expression and Purification of E2 Glycoprotein from Insect Cells (Sf9) for Use in Serology

Chong Long Chua, I-Ching Sam, and Yoke Fun Chan

Abstract

Chikungunya virus (CHIKV) is a mosquito-borne arbovirus which poses a major threat to global public health. Definitive CHIKV diagnosis is crucial, especially in distinguishing the disease from dengue virus, which co-circulates in endemic areas and shares the same mosquito vectors. Laboratory diagnosis is mainly based on serological or molecular approaches. The E2 glycoprotein is a good candidate for serological diagnosis since it is the immunodominant antigen during the course of infection, and reacts with seropositive CHIKV sera. In this chapter, we describe the generation of stable clone Sf9 (*Spodoptera frugiperda*) cells expressing secreted, soluble, and native recombinant CHIKV E2 glycoprotein. We use direct plasmid expression in insect cells, rather than the traditional technique of generating recombinant baculovirus. This recombinant protein is useful for serological diagnosis of CHIKV infection.

Key words Chikungunya, Recombinant protein, E2 glycoprotein, Immunodominant antigen, Serology, Enzyme-linked immunosorbent assay, Sf9

1 Introduction

Chikungunya virus (CHIKV) is a positive-strand RNA alphavirus, with three circulating genotypes—Asian, East Central/South Africa (ECSA), and West African. CHIKV has become a major public health concern worldwide following its emergence as a cause of large outbreaks capable of spreading to new areas [1]. In response to the need to diagnose this previously neglected disease, several enzyme-linked immunosorbent assays (ELISA) have been described. The use of recombinant proteins as antigens rather than inactivated virus would be safer, easier to standardize, and may improve specificity. The virus particle is enclosed by the capsid and the envelope glycoproteins E1 and E2. These structural proteins are highly immunogenic during the course of infection and are the target sites of antibodies production [2]. The immunodominant properties of E2 glycoprotein are well known, as it is recognized by CHIKV-seropositive sera in acute, early, and late convalescent

Justin Jang Hann Chu and Swee Kim Ang (eds.), *Chikungunya Virus: Methods and Protocols*, Methods in Molecular Biology, vol. 1426,DOI 10.1007/978-1-4939-3618-2_5, © Springer Science+Business Media New York 2016

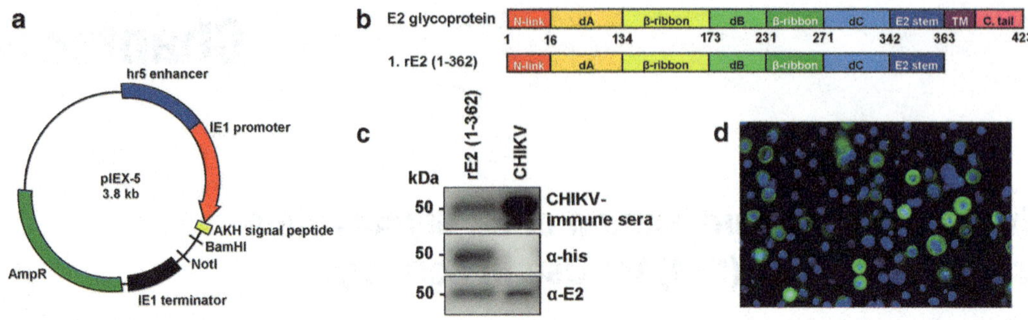

Fig. 1 Cloning strategy of recombinant CHIKV E2 glycoprotein in pIEX-5. (**a**) Vector map of pIEX-5 with *Bam*H1 and *Not*1 restriction sites. (**b**) Schematic representation of the E2 glycoprotein with the numbers indicating the amino acid positions. Transmembrane region (TM) and cytoplasmic tail (C. tail) are not expressed. *dA* domain A, *dB* domain B, *dC* domain C, *C. arch* central arch. (**c**) Seroreactivity of CHIKV immune sera against recombinant CHIKV E2 glycoprotein [rE2 (1–362)] and purified CHIKV virions (CHIKV) at 1:1000 dilution. Mouse anti-His (α-His, at 1:2000 dilution) and mouse monoclonal anti-E2 (α-E2, clone B-D2(C4) at 1 μg/ml) are used as controls. (**d**) Immunostaining of stable cells with E2 glycoprotein expression using mouse monoclonal anti-E2 antibody. DAPI was used to counterstain nuclei. Magnification: 20×

phases [3–5]. CHIKV E2 glycoprotein mediates the interaction of virus binding to the cell surface receptor. Numerous neutralizing epitopes are distributed on domain A and domain B, which lead to robust production of neutralizing antibodies [6, 7]. Thus, recombinant E2 glycoprotein would be useful for development of immunodiagnostics for acute disease and for seroprevalence surveys of past infection.

Functional studies of recombinant E2 glycoprotein have been carried out in bacterial, insect, and mammalian expression systems [8–10]. Among these, recombinant insect expression systems are often used in research, diagnostics, and vaccine development. The recombinant proteins from insect cells are more similar to native virus particles in terms of accurate protein folding, including formation of disulphide bonds and post-translational processing such as glycosylation [11]. One of the major benefits of using an insect cell expression system is the preservation of antigenicity and reactivity of the recombinant proteins which is similar to those of native virus particles. Therefore, this platform has been successfully used for the development of a virus-like particle (VLP) vaccine which is safe and able to effectively induce protective immunity [12].

This chapter describes direct plasmid expression in insect cells, rather than the traditional technique of baculovirus-mediated insect cell expression. Direct expression in insect cells allows rapid and high-yield protein production, while the generation of recombinant baculovirus is time-consuming and laborious. The clone amplicons inserted into pIEX-5 (Fig. 1a), a vector encoding the signal sequence of adipokinetic hormone (AKH), allows secretion of the recombinant protein into the media along with a His Tag coding sequence. Generation of stable clones is achieved with co-transfection of a pIEX-5 expression cassette with the pIE1-neo plasmid under selection of G418 sulfate (Geneticin) resistance. Protein purification can be performed directly from the media and the purified recombinant protein may be used as the antigen in serological diagnosis of CHIKV infection.

2 Materials

The E2 glycoprotein sequence is derived from virus strain MY/08/065, of the ECSA genotype, which was isolated in Malaysia in 2008 (GenBank accession number FN295485) [13]. The transmembrane region and cytoplasmic tail of E2 glycoprotein are not included in the expression cassette to ensure soluble protein expression into the media (Fig. 1b).

2.1 Construction of Insect Expression Cassette

1. CHIKV-containing cell supernatant.
2. QIAamp Viral RNA extraction kit.
3. SuperScript III Reverse Transcriptase kit.
4. Random primers.
5. Q5 high-fidelity DNA polymerase.
6. Primers (restriction sites are underlined).
   (a) E2-1 F: GCGGATCCTAGCACCAAGGACAACTTCAAT
   (b) E2-362R: GCGCGGCCGCCAGCTCATAATAATACAG AAT
7. 10 mM dNTPs.
8. *Bam*HI and *Not*I.
9. PCR clean-up kit.
10. T4 DNA ligase.
11. pIEX-5 insect expression vector.
12. TOP10F’ competent *Escherichia coli*.
13. LB broth and LB agar with 100 μg/ml ampicillin.
14. Endotoxin-free plasmid extraction kit.
15. Thermocycler.
16. Agarose gel electrophoresis equipment and consumables.

2.2 Maintenance of Sf9 Cells and Transient Expression of E2 Glycoprotein

1. 15 ml tube, T-75 cm^2 flask.
2. 6-well plate.
3. Sf9 insect cells.
4. BacVector Insect Cell medium.
5. Cellfectin II reagent.

2.3 Stable Expression of E2 Glycoprotein in Insect Cells

1. T-25 cm^2 flask.
2. pIE1-neo vector.
3. BacVector Insect Cell medium supplemented with 0.5 mg/ml G418 sulfate.
4. BacVector Insect Cell medium supplemented with 5 % heat-inactivated fetal bovine serum (FBS) and 0.5 mg/ml G418 sulfate.
5. BacVector Insect Cell medium supplemented with 5 % heat-inactivated FBS and 1 mg/ml G418 sulfate.

2.4 Stable Cell Cryopreservation

1. Cell freezing container.
2. Sterile 1.8 ml cyrovials.
3. Freezing medium: BacVector Insect cell medium with 10 % heat-inactivated FBS, 20 % DMSO, 1 mg/ml G418 sulfate. To prepare 10 ml freezing medium, use 7 ml media, 2 ml DMSO, 1 ml FBS, and G418 sulfate to a final concentration of 1 mg/ml.

2.5 Purification of Native Secreted CHIKV-E2

1. Filter Minisart plus 0.45 μm syringe filter.
2. His-Tag Purification Resin.
3. 1× His-Tag binding buffer: 50 mM NaH_2PO_4, 300 mM NaCl, pH 8.0.
4. 1× His-Tag elution buffer: 50 mM NaH_2PO_4, 300 mM NaCl, 250 mM imidazole, pH 8.0.
5. Polypropylene column.
6. Amicon Ultra-15 centrifugal filter units, for buffer exchange.
7. 99 % glycerol.

2.6 Serological Diagnosis of Chikungunya Virus

1. 96-well Maxisorp ELISA plate.
2. Coating buffer: 0.05 M carbonate–bicarbonate buffer, pH 9.6.
3. Washing buffer: 1× Phosphate buffered-saline with 0.05 % Tween 20 (1× PBST).
4. Blocking buffer: 3 % BSA in 1× PBST.
5. Antibody diluents: 1 % BSA in 1× PBST.
6. Secondary antibodies, rabbit anti-human IgG-horseradish peroxidase.
7. TMB microwell peroxidase substrate.
8. Stop solution: 1 M phosphoric acid, H_3PO_4.

3 Methods

3.1 Construction of Insect Expression Cassette

1. Extract the CHIKV viral RNA according to the manufacturer's instructions and reverse-transcribe into cDNA using random primers.

2. Add and mix 8.5 μl ddH_2O, 2 μl 10 mM dNTP, 0.5 μl random primers, and 2 μl viral RNA.
3. Heat the tube at 65 °C for 5 min and place immediately on ice for at least 3 min.
4. Add and mix 4 μl of 5× reaction buffer, 1 μl DTT, 1 μl RNase out, and 1 μl Superscript III reverse transcriptase.
5. Incubate in a thermocycler with the following parameters: 25 °C for 15 min, 50 °C for 60 min, and 70 °C for 15 min.
6. Amplify the E2 genes using high-fidelity polymerase and clean up the amplicons (*see* **Notes 1** and **2**). The amplicons are integrated with restriction enzyme sites *Bam*H1 and *Not*1 to allow directional cloning into pIEX-5 vector (Fig. 1a).
7. Put together a 50 μl reaction mixture in a PCR tube. The mixture comprises 10 μl 5× Q5 reaction buffer, 1 μl 10 mM dNTP, 1 μl 30 μM forward primer, 1 μl 30 μM reverse primer, 1 μl cDNA template, 35.5 μl ddH_2O, and 0.5 μl Q5 high-fidelity DNA polymerase.
8. Run the reaction in a thermocycler with the following parameters: initial denaturation of 98 °C for 30 s; 30 cycles of 98 °C for 10 s, 56 °C for 30 s and 72 °C for 35 s; final extension of 72 °C for 2 min.
9. Check the amplified DNA with 1 % agarose electrophoresis and perform DNA clean-up.
10. Perform the digestion on the amplicons and vector, clean up and ligate the amplicons and vector in a single reaction tube.
11. Put together a 50 μl reaction mixture in a PCR tube. The mixture comprises 1 μg of DNA or vector, 5 μl of 10× buffer, 1 μl *BamH*1, 1 μl *Not*1, and water to a final volume of 50 μl.
12. Perform the digestion for 1 h at 37 °C.
13. Check the digested products with 1 % agarose electrophoresis and perform DNA clean-up.
14. Put together a 10 μl ligation mixture in a PCR tube. The mixture comprises 1 μl 10× ligase buffer, 0.5 μl digested vector, 4.5 μl digested DNA fragment, 3 μl ddH_2O, and 1 μl ligase.
15. Incubate the mixture at 4 °C overnight.
16. Transform the ligation mixture into competent cells (*see* **Note 3**).
17. Thaw the competent cells on ice. Incubate 4 μl of ligation mixture with the cells for 15 min.
18. Heat-shock the cells at exactly 42 °C for 1 min and immediately place on ice for 3 min.
19. Recover the cells by adding LB broth and incubate at 37 °C for 1 h.

20. Spin the cells at 1000 × *g* for 5 min and plate the cell pellet on an agar plate with ampicillin as selection marker. Incubate at 37 °C overnight.
21. Select the positive clones and extract the plasmid for restriction enzyme analysis and sequencing.

3.2 Maintenance of Sf9 Cells and Transient Expression of E2 Glycoprotein

1. Thaw the cyropreserved Sf9 cells at 37 °C for 2 min. Disinfect the tube with 70 % ethanol before transfer into a biosafety cabinet.
2. Transfer the whole Sf9 cell suspension from the vial in drops into 10 ml of pre-warmed (28 °C) BacVector medium in a 15 ml tube.
3. Centrifuge the tube at 200 × *g* for 5 min. Resuspend the cell pellet gently with 10 ml media and transfer into a 75 cm^2 flask.
4. Maintain the flask at 28 °C in the absence of CO_2. Cells should reach 90–100 % confluency by the fourth to sixth day. Passage the cells at least 2 times before proceeding with transfection. Seed the cells at 3×10^6 cells in 10 ml media in each passage.
5. Dislodge the healthy cells in the flask and transfer cells into a 50 ml centrifugal tube (*see* **Note 4**).
6. Count the cell number and seed cells at 8×10^5 cells in 2 ml/well in a 6-well plate.
7. Incubate the cells at room temperature for 15 min to allow cell attachment to the plate.
8. Prepare the transfection complexes.
9. Place 200 μl of Sf9 growth medium in a 1.5 ml sterile tube.
10. Add 1 μg plasmid DNA. Mix gently by pipetting.
11. Add 8 μl Cellfectin II reagent. Mix gently by pipetting.
12. Incubate at room temperature for 30 min.
13. Add the transfection complex in drops to different areas of the well.
14. Gently rock the plate to distribute the complexes evenly.
15. Incubate for 48–72 h at 28 °C.
16. Collect the cell supernatant directly (50 μl) for western blotting. The protein expression can be verified by mouse anti-His or anti-CHIKV E2 monoclonal antibodies (Fig. 1c).

3.3 Stable Expression of E2 Glycoprotein in Insect Cells

1. When the cells have reached 90 % confluency after 4–5 days, dislodge the cells and collect them in a 50 ml centrifugal tube.
2. Count the cell number and seed cells at 2×10^6 cells in 5 ml in a 25 cm^2 tissue culture flask. Prepare a flask without transfection which serves as a negative control.

3. Incubate the cells at room temperature for 15 min to allow cell attachment to the plate.
4. Prepare the transfection complexes.
5. Place 500 μl of Sf9 growth medium in a 1.5 ml sterile tube.
6. Add 3.2 μg plasmid DNA + 0.8 μg pIE1-neo vector. Mix gently by pipetting.
7. Add 32 μl Cellfectin II reagent. Mix gently by pipetting.
8. Incubate at room temperature for 30 min.
9. Add the transfection complex into the flask.
10. Gently rock the plate to distribute the complexes evenly.
11. After 24 h, add G418 to a final concentration of 500 μg/ml in the media and continue the incubation for another 2–3 days. There should be no surviving cells in the non-transfected flask (*see* **Note 5**).
12. Replace the media every 5 days in the presence of 5 % heat-inactivated FBS and 0.5 mg/ml G418. Continue this step for 2–3 weeks (*see* **Note 6**).
13. When there are many colonies of stable clones (reaching 30–40 % confluency), dislodge the cells and transfer the cell suspension into a new flask in the presence of FBS and 1 mg/ml G418 and allow the cells to reach confluency.
14. Propagate the cells in the presence of 1 mg/ml G418 only (*see* **Note** 7).
15. Examine the protein expression before cryopreservation by western blotting from the cell culture supernatant (Fig. 1c) or by performing immunofluoresence assay on stable cells (Fig. 1d) using anti-His or anti-CHIKV E2 monoclonal antibodies (*see* **Note 8**).

3.4 Cryopreservation of Stable Cells

1. Pre-cool the cell freezing container with 100 % isopropyl alcohol at 4 °C.
2. After the cells have reached 70–80 % confluency, dislodge the cells and collect them in a 50 ml centrifugal tube.
3. Discard the media.
4. Count the cell number and adjust the cell number to $6–7 \times 10^6$ cells/ml.
5. Prepare the freezing medium. Add 500 μl of cell suspension to sterile cryogenic vials.
6. Add another 500 μl of freezing medium and mix by pipetting. Immediately put the vials on ice.
7. Place all the vials in the pre-cooled cell freezing container and immediately store at −80 °C overnight. Transfer all the vials into liquid nitrogen for long-term storage.

3.5 Purification of Native Secreted CHIKV-E2

1. Expand the stable cells by adding 3×10^6 cells in 10 ml/75 cm^2 flask in the presence of 1 mg/ml G418.
2. Harvest the cell media after the cells reach confluency of 90–100 %. Collect all the media in a 50 ml centrifugal tube.
3. Spin at 4000 × *g* for 10 min to remove cell debris and filter the media through a 0.45 μm pore size membrane.
4. Mix the media and 1× binding buffer at 1:2 ratio (50 ml media: 100 ml binding buffer) without adjusting the pH (*see* **Note 9**).
5. Incubate the whole mixture with 1 ml activated resin at 4 °C with constant stirring for 1 h.
6. Transfer the resin into a polypropylene column.
7. Allow the lysate to drain through the column by gravity.
8. Wash the column five times with 10 ml of binding buffer from the top of the resin and drain the unbound material through the column.
9. Elute the recombinant protein by adding 5 ml of elution buffer to the top of the column and collect the eluted material from the bottom of the column.
10. Check the eluted protein by western blotting.
11. Perform buffer exchange if necessary, using 1× binding buffer and concentrate the recombinant protein to around 0.5–1 ml (*see* **Note 10**).
12. For long-term storage, mix glycerol with the purified protein to a final concentration of 50 % and keep at −20 °C.
13. Quantitate the protein using a bicinchoninic acid (BCA) protein assay or equivalent.

3.6 Serological Diagnosis of Chikungunya Virus

1. Dilute the recombinant E2 glycoprotein with coating buffer to 1 μg/ml and coat the plate with 100 μl per well (*see* **Note 11**).
2. Incubate the plate at 37 °C for 1 h or 4 °C overnight.
3. Wash the plate four times and dry the plate by tapping onto thick paper towels.
4. Block the plate with blocking buffer and incubate at 37 °C for 1 h.
5. Wash the plate four times and tap dry.
6. Dilute the test sera (1:500–1:1000 dilution) in antibody diluent. Add 100 μl of diluted serum to each well. Each test serum sample should be performed in duplicate/triplicate. Include the controls (two positive controls—high and low, at least four negative controls and a blank without primary antibodies). Incubate the plate at 37 °C for 1 h (*see* **Note 12**).
7. Wash the plate four times and tap dry.

8. Add 100 μl of diluted secondary antibodies into each well (rabbit anti-human IgG-HRP at 1:5000 dilution). Incubate the plate at 37 °C for 1 h.
9. Wash the plate four times and tap dry.
10. Develop the plate with TMB substrate with 100 μl in each well. Incubate at room temperature with constant rocking for 5 min.
11. Terminate the reaction by adding 100 μl of stop solution into each well.
12. Determine the optical density (OD) at 450 nm using a microplate reader with 630 nm as the reference wavelength.
13. Correct the final optical density reading by adjusting for the background absorbance ($OD = OD_{450\ nm} - OD_{630\ nm}$).
14. Establish the cut-off value from the mean OD obtained from negative controls plus three standard deviations.
15. Determine the index value (IV), which is the ratio of mean sample OD to cut-off value. An $IV > 1$ is considered seropositive, while an $IV \leq 1$ is considered seronegative (*see* **Note 13**).

4 Notes

1. The PCR product of CHIKV E2 (1–362) should be 1103 bp.
2. If the vector does not encode a signal peptide, the cloned amplicons can integrate with the E3 signal sequence region. A start codon should be added upstream of the E3 sequence [9].
3. Keep the bacterial clones with the correct expression construct in glycerol stock. Grow the bacteria in LB broth with ampicillin supplement until the OD_{600} reaches 0.6 (log-phase, OD to be measured with a spectrophotometer), then transfer 1 ml of culture into a 2 ml screw-capped tube with 225 μl sterile glycerol. Store at −80 °C.
4. The TriExSf9 cell line (Merck Millipore) can be used as an alternative for transient insect protein expression, and must be used together with TriEx cell medium. Cellfectin II can be used for this cell line. However, high protein expression levels can be achieved with TransIT-Insect Transfection reagent (Mirus). For transfection, add 1.6×10^6 cells in 2 ml of TriEx Insect cell medium per well in a 6-well plate; then mix 2 μg of plasmid and 5 μl TransIT-Insect Transfection reagent into another 600 μl of TriEx Insect cell medium. Note that TransIT-Insect Transfection reagent is not compatible with BacVector Insect cell medium, and is therefore not suitable for use with Sf9 cells.

5. A G418 kill curve should be performed if the Sf9 cells and media are obtained from different sources.
6. Heat-inactivated FBS is added upon selection of G418 to boost the growth of stable cells.
7. The generated stable cell line is a heterogeneous population of cells with varying levels of protein expression. To obtain a stable cell line with higher expression, clone the cells by a limiting dilution method, by adding approximately 1–2 cells/well in a 96-well plate in the presence of FBS to boost the cell colony growth.
8. The stable cell line is suitable for use to passage number 20–25, after which the protein expression drops. Make more vials for cryopreservation.
9. The media has an acidic pH and contains electron-donating groups that potentially inhibit the binding of His-tagged proteins to resin. In some circumstances, adjusting the pH of media in the presence of binding buffer can result in a white, cloudy salt precipitate; therefore, dilution of media with 2× binding buffer is strongly discouraged. The optimal binding conditions would be dilution of media with 1× binding buffer at pH 8.0. This may still result in some precipitation, but should enable successful purification of the recombinant protein directly from the media. Other methods of purification such as size-exclusion chromatography (gel-filtration) can be considered.
10. It is strongly recommended to perform the buffer exchange with 1× binding buffer to keep the protein stable.
11. The recombinant His-tagged proteins expressed from the insect cell can be used for ELISA with plates coated with nickel or cobalt.
12. Positive and negative control serum should be first confirmed by a specific seroneutralization assay, particularly to rule out the possibility of cross-reactivity with other alphaviruses.
13. The calculation of the index values is as described by O'Shaughnessy et al. [14].

Acknowledgement

This work was supported by the European Union Seventh Framework Program (Integrated Chikungunya Research, grant agreement no. 261202), University Malaya (UMRG grant RG526-13HTM and PPP grant PG114-2012B) and the Ministry of Higher Education, Malaysia (FRGS grant FP035-2015A).

References

1. Rougeron V, Sam IC, Caron M et al (2015) Chikungunya, a paradigm of neglected tropical disease that emerged to be a new health global risk. J Clin Virol 64:144–52.
2. Sourisseau M, Schilte C, Casartelli N et al (2007) Characterization of reemerging chikungunya virus. PLoS Pathog 3:e89
3. Kowalzik S, Xuan NV, Weissbrich B et al (2008) Characterisation of a chikungunya virus from a German patient returning from Mauritius and development of a serological test. Med Microbiol Immunol 197: 381–386
4. Kam YW, Lum FM, Teo TH et al (2012) Early neutralizing IgG response to Chikungunya virus in infected patients targets a dominant linear epitope on the E2 glycoprotein. EMBO Mol Med 4:330–343
5. Chua CL, Chan YF, Sam IC (2014) Characterisation of mouse monoclonal antibodies targeting linear epitopes on Chikungunya virus E2 glycoprotein. J Virol Methods 195: 126–133
6. Strauss JH, Strauss EG (1994) The alphaviruses: gene expression, replication, and evolution. Microbiol Rev 58:491–562
7. Voss JE, Vaney MC, Duquerroy S et al (2010) Glycoprotein organization of Chikungunya virus particles revealed by X-ray crystallography. Nature 468:709–712
8. Tripathi NK, Priya R, Shrivastava A (2014) Production of recombinant Chikungunya virus envelope 2 protein in Escherichia coli. Appl Microbiol Biotechnol 98:246–271
9. Metz SW, Geertsema C, Martina BE et al (2011) Functional processing and secretion of Chikungunya virus E1 and E2 glycoproteins in insect cells. Virol J 8:353
10. Cho B, Jeon BY, Kim J et al (2008) Expression and evaluation of Chikungunya virus E1 and E2 envelope proteins for serodiagnosis of Chikungunya virus infection. Yonsei Med J 49:828–835
11. Metz SW, Pijlman GP (2011) Arbovirus vaccines; opportunities for the baculovirus-insect cell expression system. J Invertebr Pathol 107:S16–S30
12. Metz SW, Gardner J, Geertsema C et al (2013) Effective Chikungunya virus-like particle vaccine produced in insect cells. PLoS Negl Trop Dis 7:e2124
13. Sam IC, Loong SK, Michael JC et al (2012) Genotypic and phenotypic characterization of Chikungunya virus of different genotypes from Malaysia. PLoS One 7:e50476
14. O'Shaughnessy L, Carr M, Crowley B et al (2011) Recombinant expression and immunological characterisation of proteins derived from human metapneumovirus. J Clin Virol 52:236–243

Chapter 6

Diagnostic Methods for CHIKV Based on Serological Tools

Paolo Gaibani, Maria Poala Landini, and Vittorio Sambri

Abstract

This chapter presents the most commonly used serological methods for the diagnosis of Chikungunya virus (CHIKV) infection in humans. CHIKV is a mosquito-borne *Alphavirus* widely distributed in the tropical and subtropical regions of Africa, Asia, and America. CHIKV infection in human causes acute febrile illness frequently accompanied by severe joint pain. Most of the infected patients may develop chronic arthralgia that may persist for several months or years. Laboratory diagnosis of CHIKV infection is mainly based on molecular and serological tests. The serological tests represent a valuable tool for diagnosis and epidemiological studies. Enzyme-linked immunosorbent assay (ELISA) and immunofluorescence assay (IFA) are simple, rapid, and sensitive techniques widely used for the diagnosis of CHIKV infection. However, these methods represent a screening tool and often require confirmation by a second-line assays. Serum virus neutralization assay is more specific than ELISA and IFA tests and is considered a confirmatory test. Neutralization assay is employed to determine the titer of virus neutralizing antibodies against CHIKV in patients' sera. The basis of microneutralization assay (MNA), results interpretation, and procedures will be illustrated in this chapter.

Key words Chikungunya virus, Serological methods, ELISA, IFA, Microneutralization

1 Introduction

Chikungunya virus (CHIKV) is a mosquito-borne togavirus belonging to the *Alphavirus* genus. CHIKV is an enveloped, spherical virus of approximately 50–70 nm in diameter that contains a single-stranded, positive-sense RNA genome of ~12 kb [1]. CHIKV was isolated for the first time from blood of a febrile patient in Eastern Africa (Tanzania and Mozambique) during an outbreak in 1952–1953. The name "Chikungunya" derives from the Makonde language and means "that which bends up," thus referring to the bent posture assumed by CHIKV-infected persons suffering of severe arthralgia [2]. CHIKV is transmitted to human by the bite of infected mosquitoes mainly belonging to the *Aedes* genus [3]. While *A. aegypti* is considered the principal vector responsible for urban transmission of CHIKV, *A. albopictus* has

Justin Jang Hann Chu and Swee Kim Ang (eds.), *Chikungunya Virus: Methods and Protocols*, Methods in Molecular Biology, vol. 1426,DOI 10.1007/978-1-4939-3618-2_6, © Springer Science+Business Media New York 2016

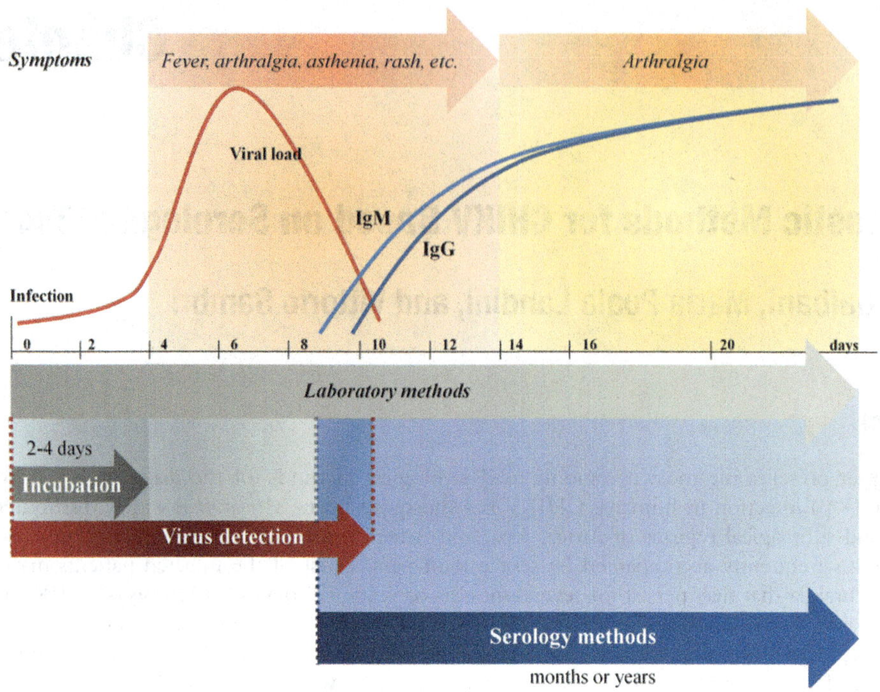

Fig. 1 CHIKV infection timeline, symptoms, and laboratory methods

been shown to efficiently transmit the virus. *Aedes* spp. mosquitoes show a wide distribution across tropical, subtropical, and temperate regions.

CHIKV infection is characterized by a rapid onset of febrile disease accompanied by a wide range of symptoms [4]. Clinical manifestations occur commonly 2 to 4 days after transmission by the bite of infected mosquitoes (Fig. 1). In contrast to other arbovirus infections such as dengue fever (DF), CHIKV infections are often symptomatic (75–95 %) [1]. The symptoms in the early stage of infection are high fever (>38.5 °C), asthenia, and arthralgia. The joint pain is usually symmetric and localized to arms (fingers, wrists, and elbows) and legs (ankles and toes). Additional symptoms include myalgia, headaches, and rash. Symptoms can resolve within 1–2 weeks since onset of the disease (Fig. 1), whereas joint pain can persist for several months or years in most (50–70 %) of CHIKV-infected patients [5].

CHIKV infection should be suspected on the basis of clinical and epidemiological criteria [3]. However, diagnosis of CHIKV infection can be difficult due to the clinical manifestations that overlap many other infectious diseases [6]. In particular, CHIKV infection often reproduces the symptoms of DF in the early stage of the infection. Moreover, CHIKV and dengue viruses show an intersecting distribution in Africa, Asia, and America.

Based on these considerations, an efficient and accurate laboratory diagnosis of CHIKV infection should be mandatory for the appropriate clinical management. At the same time, an efficient diagnosis of CHIKV infection could be considered of primary importance to prevent and control the transmission of CHIKV in nonendemic areas.

The laboratory diagnosis of CHIKV infection can be achieved by following two different strategies: the detection of virus RNA and the identification of the specific immune response by serological methods. The appropriate time of specimen collection and of the use of the most suitable diagnostic methodology result is crucial for accurate diagnosis of CHIKV infection. Consequently, a full knowledge of the kinetic and pathogenesis of CHIKV infection, including the duration of viremia and the raise of the host immune response, should be considered when selecting the appropriate diagnostic tests [7]. The virus RNA is found in plasma or sera specimens between 2 and 4 days since the onset of fever; this phase corresponds to the period of incubation [3]. The period of incubation is followed by an abrupt onset of fever accompanied by several symptoms (Fig. 1). In early stage of infection, viremia can reach high viral load (up to 10^9 copies per ml) and typically lasts from 4 to 6 days after onset of symptoms [8]. During the viremic phase, viruses can be isolated from sera or plasma with different cell cultures (Vero or C6/36 cells). Virus isolation is rarely used in the diagnosis of CHIKV infection, while it is mainly applied for epidemiological studies [9]. The diagnosis of CHIKV infection in the acute phase of infection is typically performed by detection of viral RNA in plasma or sera by RT-PCR. The viral RNA can be detected by different molecular methods, such as nested and real-time PCRs [8]. Serological tests are the most common methods used for the laboratory diagnosis of CHIKV infection and hold a reliable use both in acute and in convalescent sera. Antibodies against CHIKV appear few days later since onset of the clinical signs and may persist for several months or years. The first antibodies detected in CHIKV-infected patients during acute phase are immunoglobulin M (IgM), which commonly appears 4–6 days after the onset of symptoms (Fig. 1). After 6–7 days since the appearance of symptoms, the IgG antibodies become detectable in sera and viral RNA tends to disappear rapidly [2].

The serological testing algorithm for the diagnosis of CHIKV infection is usually based on the detection of IgM and IgG antibodies by reliable, fast, and easy-to-use screening tests followed by a second-line serological test to confirm the results [8]. Several serological assays have been developed and the large majority show high reliability and specificity. The common first-line serological techniques frequently used for CHIKV diagnosis are the enzyme-linked immunosorbent (ELISA) and indirect immunofluorescence (IFA) assays.

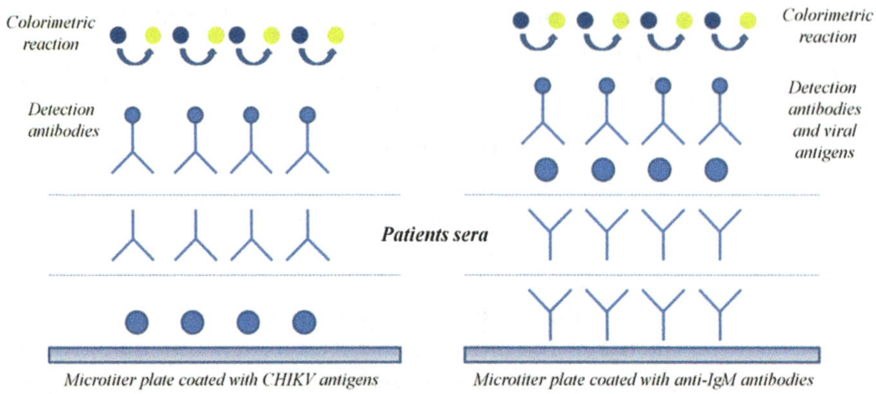

ELISA	MAC-ELISA
Detection of IgM or IgG antibodies	Detection of specific IgM antibodies
Good sensitivity and specificity	High sensitivity and specificity
Acute or convalescent sera	Acute or convalescent sera
Fast results (< 2 h)	Fast results (< 2-3 h)

Fig. 2 Schematic representation of antibodies detection by i-ELISA and MAC-ELISA

ELISA is a rapid and sensitive method largely used for the detection of anti-CHIKV antibodies. The most common tests used for the diagnosis of CHIKV infection are IgM antibody-capture ELISA (MAC-ELISA) and indirect ELISA (i-ELISA) for the detection of type M (IgM) and type G (IgG) immunoglobulin, respectively [8]. The major differences between MAC-ELISA and i-ELISA are shown in Fig. 2. On the MAC-ELISA, human IgM is captured by monoclonal antibody on a microtiter plate. Then, the addition of CHIKV antigens and detector antibody consents the enzymatic reaction. On the i-ELISA, CHIKV-specific antigens are attached onto the plate. The presence of IgM or IgG antibodies are revealed by addition of type-specific secondary antibody conjugated with an enzyme, usually horseradish peroxidase. Following the addition of substrate, the color change is measured by an ELISA microplate reader. The optical density (OD) corresponds to the measurement of primary antibodies and is evaluated as positive or negative following the interpretation criteria. Some commercial kits are now available and most of them show acceptable specificity and sensitivity values. A list of different commercially available ELISA tests is shown in Table 1. Serum represents the sample of choice and a dilution of 1/100 is generally used for diagnosis.

The detection of IgM antibodies is considered a marker of a recent CHIKV infection [3]. In detail, acute viral infection could be considered if the serum sample is collected within the 15 days before the onset of symptoms (Fig. 3). At the same time, detection

Table 1
Commercial diagnostic tests available for serological diagnosis of CHIKV infection

Company	IgM	IgG	Test
Euroimmun	+	+	ELISA
Novatec	+	/	IgM-capture ELISA
Novatec	/	+	ELISA
IBL	+	/	IgM-capture ELISA
IBL	/	+	IgG-capture ELISA
Abcam	+	+	ELISA
DRG[a]	+	/	IgM-capture ELISA
DRG*	/	+	ELISA
GenWay	+	/	IgM-capture ELISA
GenWay	/	+	ELISA
Standard diagnostic	+	+	ELISA
Euroimmun	+	+	IFA

[a]This kit is intended for research use only

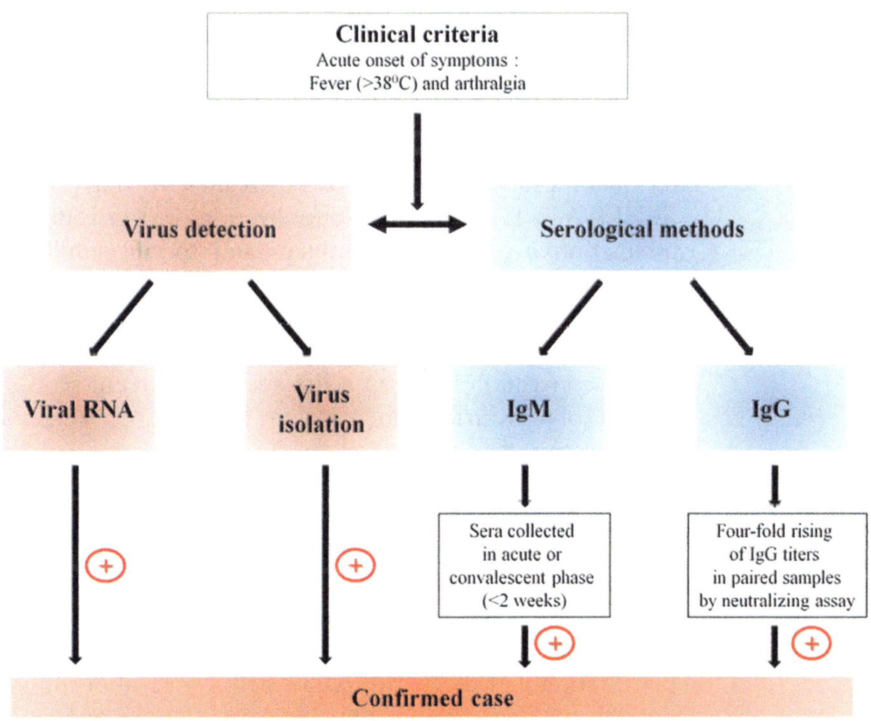

Fig. 3 Algorithm for the diagnosis of acute Chikungunya virus infection

of specific IgG antibodies indicates a recent or past CHIKV infection. Indeed, anti-CHIKV IgG antibodies persist for several months to years in infected patients [2]. A fourfold or greater increase of IgG antibodies titer between acute and convalescent paired sera may be used to confirm a recent CHIKV infection (Fig. 3). Previous studies demonstrated that the commercially available ELISA showed high sensitivity (ranging from 82 to 88 %) and specificity (ranging from 82 to 97 %) in specimens collected during convalescent phase of CHIKV infection [10, 11].

The main disadvantages of ELISA are: false-positive results due to the cross-reactivity with other alphavirus such as Ross river virus (RRV), Barmah Forest virus (BFV), and Sindbis virus [12]; and low sensitivities (4–20 %) in sera collected during the acute phase of the infection, thus corresponding to 5–7 days after illness onset [11, 13].

IFA is an accurate and reliable technique widely used for the detection of specific anti-CHIK antibodies. IFA reveals the presence of type-specific antibodies against CHIKV by detecting the presence of virus antigens in infected cells. The ligation of the secondary antibodies conjugated with fluorescent dye is observed under UV microscope. IFA commonly has highest sensitivity and specificity in respect to ELISA. Previously study demonstrated that commercial IFA showed a specificity ranging from 75 to 100 % in sera collected since 5–6 days of the beginning of the CHIKV disease [11]. IFA suffers of few disadvantages such as: the method is laborious and requires a well-trained laboratory personnel; the microscopic examination is cumbersome and interpretations of the results are quite subjective; and the lack of interlaboratory standardization.

However, the neutralization assay represents the confirmatory test that is usually performed downstream following positive results by ELISA or IFA (Fig. 3). Neutralization of virus is defined as the interaction between viral antigens and specific antibodies, thus resulting in a block of the infection. The identification of neutralizing antibodies may be used for the diagnosis of CHIKV infection. Indeed, confirmation of a recent infection is identified either by seroconversion between acute and convalescent sera or by revelation of a fourfold increase in neutralizing antibodies titer between paired sera.

In neutralization test, virus and serum are mixed under suitable conditions and then inoculated into cell culture. After several days, inhibition of the biological effect observed in infected cells is investigated with different methods [14]. The microneutralization assay (MNA) evaluates the neutralizing antibody titers in sera against CHIKV. These techniques show high specificity and sensitivity. The main disadvantages of MNA are: labor-intensive; low number of samples that is possible to process per run; and requirement of a Biosafety level 3 laboratory (BSL-3) for the handling of live virus. In this chapter, we provide a detailed MNA protocol.

2 Materials

1. Incubator, 37 °C, 5 % CO_2.
2. Inverted microscope for cells observation.
3. Water bath, 56 °C.
4. Cell culture flasks.
5. 96-well microtiter plates (U-bottom).
6. 96-well microtiter plates (flat-bottom).
7. Cell counter.
8. Multichannel pipette.
9. Sterile micropipette tips.
10. Pipette aid.
11. D-MEM with 5 % fetal bovine serum (FBS; *see* **Note 1**), 1 % Penicillin/Streptomycin, and 1 %l-Glutamine.
12. Freezer, –80 °C for storage of virus.
13. Freezer –20 °C for storage of serum.
14. Trypsin-EDTA solution: 0.05 % Trypsin, 0.02 % EDTA.
15. PBS, pH 7.2.
16. Vero cells in monolayers.
17. Trypan blue stain (0.4 %).
18. Positive control (PC) serum sample (*see* **Note 2**).
19. Negative control (NC) serum sample (*see* **Note 2**).
20. Patient sera samples.
21. Chikungunya virus (CHIKV) cultured in Vero cells (*see* **Note 3**), harvested at a high titer of >4 Log10 pfu/ml (*see* **Note 4**).

3 Methods

A summary of MNA protocol is shown in Fig. 4.

3.1 Preparation of Sera Samples

In each assay, include a positive (PC) and negative (NC) serum controls and patient sera to be tested.

1. Heat inactivate sera in water bath (*see* **Note 5**). Add 50 μl of serum diluents (D-MEM with FBS, Glutamine and antibiotics) to all wells of microtiter plate (U-bottom).
2. Add 40 μl of serum diluent to the first well of each column (A1, A2, A3, etc.).
3. Add 10 μl of heat-inactivated serum to the first wells considering one sera per column (A1, sera 1; A2, sera 2; A3, sera 3; …, A11, positive control; A12, negative control).

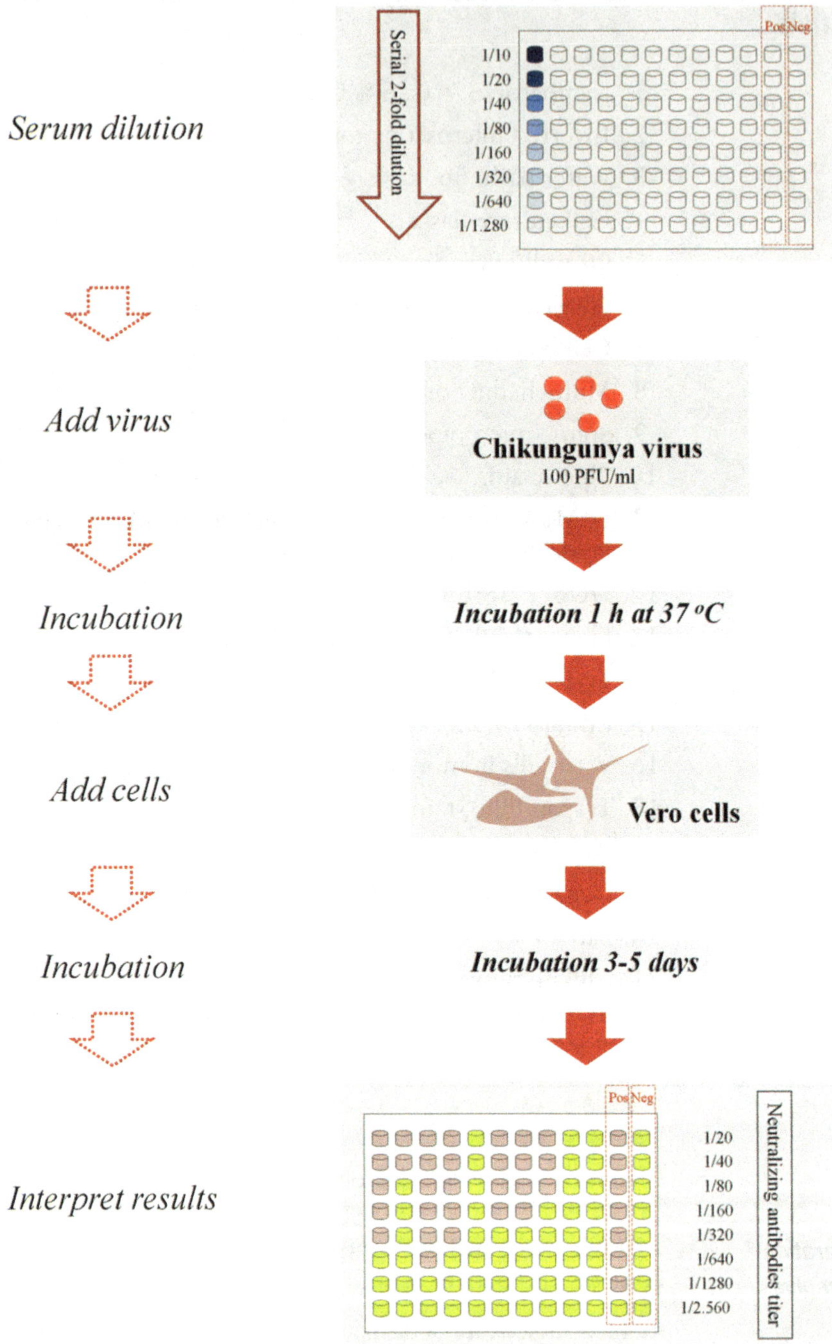

Fig. 4 Schematic representation of microneutralization assay (MNA)

4. Perform twofold serial dilutions of each serum by transferring 50 µl of diluted sera to the next wells, raw to raw (example A1, B1, C1....A2, B2, C2....A3, B3, C3...). At each transfer, pipette five to ten times. The serum dilution in first well is 1/10 and in last well is 1/1280.

3.2 Preparation of Virus–Sera Mixture

All the following procedures must be performed inside a laminar flow biosafety cabinet located in a Biosaftey level 3 (BSL3) environment.

1. Prepare a dilution from virus stock with serum diluents, as previously described in Subheading 3.1, to a final concentration of 100 pfu/ml (*see* **Note 6**).
2. Add 50 µl of virus to all wells including positive and negative controls and tested sera. The serum dilution in first well is now 1/20 and in last well is 1/2560.
3. Mix the plates by gentle shaking.
4. Incubate the microtiter plate at 37 °C for 1 h.

3.3 Preparation of Cell Culture

1. Visualize cells under microscope to evaluate viability and the confluence of monolayer. Cells need to be 95–100 % confluent prior to infection. Growth medium for cells culture consists of D-MEM supplemented with antibiotics, l-Glutamine, and FBS.
2. Incubate cells monolayer into the flask with trypsin–EDTA.
3. Wash cells two times with PBS and resuspend in growth medium.
4. Determine the cell count and viability (*see* **Note 7**).
5. Dilute the cells in growth medium to a final concentration of 1×10^5 per ml.
6. Plate 100 µl of cell suspension to each well of the 96-well microtiter plate (flat-bottom) corresponding to 1×10^4 Vero E6 cells per well.

3.4 Infection of Cells with Neutralized Virus

1. Add 100 µl of virus-serum mix into each well containing cells (*see* **Note 8**).
2. Incubate the infected cells at 37 °C in 5 % CO_2 for five days until cytopathic effect (CPE) is visible (*see* **Note 9**). Observe the presence or absence of CPE of each well and register the data. A valid assay must show CPE in all wells containing negative control sera and no CPE for positive control corresponding to the neutralizing antibodies titer.

3.5 Calculations of the Results

The antibodies neutralizing titer is defined as the reciprocal of the highest dilution of serum at which the CPE is neutralized. (Example. Serum 1 shown CPE in F1, G1, H1 whereas no CPE is shown in A1, B1, C1, D1, F1. The resulting neutralizing antibodies titer is 1/320)

4 Notes

1. FBS should be inactivated in water bath at 56 °C for 30 min before use.
2. Positive control (PC) consists of sera with neutralizing antibodies against CHIKV. Optimal positive control could be serum sample collected from CHIKV-infected patient during convalescent phase [5]. Negative control (NC) is serum collected from person not exposed to CHIKV.
3. Chikungunya stock virus should be at low passage level (≤5 passages).
4. Endpoint dilution assay could be used as alternative to the plaques counting. The 50 % tissue culture infectious dose ($TCID_{50}$) is simpler and less laborious than that of using plaque count [15].
5. The serum sample should be inactivated in a water bath at 56 °C for 30 min. Inactivation of sera is necessary to degrade all complement proteins present in sera samples that may alter MNA results.
6. Virus stock should not be freeze–thawed for more than 4–5 times. Repeated freezing and thawing could decrease the viral titer. It is recommended to freeze several aliquots of virus.
7. Cells viability could be determined in a cell chamber with Trypan blue staining [16].
8. Optional: Transfer the virus-serum mixture into a 96 μ-well plate using a multichannel pipette.
9. Although plaque reduction neutralization assay is the gold standard for CPE determination [17], a simpler and less laborious method used for identification of CPE could be achieved using crystal violet staining followed by microscopic examination with an inverted microscope and observation by color changing of the medium [18, 19].

References

1. Morrison TE (2014) Reemergence of chikungunya virus. J Virol 88:11644–11647
2. Weaver SC, Lecuit M (2015) Chikungunya virus and the global spread of a mosquito-borne disease. N Engl J Med 372:1231–1239
3. Burt FJ, Rolph MS, Rulli NE, Mahalingam S, Heise MT (2012) Chikungunya: a re-emerging virus. Lancet 379:662–671
4. Schwartz O, Albert ML (2010) Biology and pathogenesis of chikungunya virus. Nat Rev Microbiol 8:491–500
5. Moro ML, Grilli E, Corvetta A, Silvi G, Angelini R, Mascella F, Miserocchi F, Sambo P, Finarelli AC, Sambri V, Gagliotti C, Massimiliani E, Mattivi A, Pierro AM, Macini P, Study Group "Infezioni da Chikungunya in Emilia-Romagna" (2012) Long-term Chikungunya infection clinical manifestations after an outbreak in Italy: a prognostic cohort study. J Infect 65:165–172
6. Fact Sheet 327: Chikungunya." World Health Organization (WHO). Accessed from

http://www.who.int/mediacentre/factsheets/fs327/en/

7. Chusri S, Siripaitoon P, Silpapojakul K, Hortiwakul T, Charernmak B, Chinnawirotpisan P, Nisalak A, Thaisomboonsuk B, Klungthong C, Gibbons RV, Jarman RG (2014) Kinetics of chikungunya infections during an outbreak in Southern Thailand, 2008-2009. Am J Trop Med Hyg 90:410–417
8. Cavrini F, Gaibani P, Pierro AM, Rossini G, Landini MP, Sambri V (2009) Chikungunya: an emerging and spreading arthropod-borne viral disease. J Infect Dev Ctries 3:744–752
9. Kucharz EJ, Cebula-Byrska I (2014) Chikungunya fever. Eur J Intern Med 23:325–329
10. Prat CM, Flusin O, Panella A, Tenebray B, Lanciotti R, Leparc-Goffart I (2014) Evaluation of commercially available serologic diagnostic tests for Chikungunya virus. Emerg Infect Dis 20:2129–2132
11. Yap G, Pok KY, Lai YL, Hapuarachchi HC, Chow A, Leo YS, Tan LK, Ng LC (2010) Evaluation of Chikungunya diagnostic assays: differences in sensitivity of serology assays in two independent outbreaks. PLoS Negl Trop Dis 4, e753
12. Pialoux G, Gaüzère BA, Jauréguiberry S, Strobel M (2007) Chikungunya, an epidemic arbovirosis. Lancet Infect Dis 7:319–327
13. Blacksell SD, Tanganuchitcharnchai A, Jarman RG, Gibbons RV, Paris DH, Bailey MS, Day NP, Premaratna R, Lalloo DG, de Silva HJ (2011) Poor diagnostic accuracy of commercial antibody-based assays for the diagnosis of acute Chikungunya infection. Clin Vaccine Immunol 18:1773–1775
14. Sambri V, Capobianchi MR, Cavrini F, Charrel R, Donoso-Mantke O, Escadafal C, Franco L, Gaibani P, Gould EA, Niedrig M, Papa A, Pierro A, Rossini G, Sanchini A, Tenorio A, Varani S, Vázquez A, Vocale C, Zeller H (2013) Diagnosis of west nile virus human infections: overview and proposal of diagnostic protocols considering the results of external quality assessment studies. Viruses 5:2329–2348
15. Reed LJ, Muench H (1983) A simple method of estimating fifty percent endpoints. Am J Hygiene 27:493–497
16. Doyle A, Griffiths JB, Newell DG (1995) Cell and tissue culture: laboratory procedures. John Wiley, Chichester
17. WHO (2007) Guidelines for plaque reduction neutralization testing of human antibodies to dengue virus. Accessed from http://whqlibdoc.who.int/hq/2007/who_ivb_07.07_eng.pdf
18. Hierholzer JC, Bingham PG (1978) Vero microcultures for adenovirus neutralization tests. J Clin Microbiol 7:499–506
19. Greig AS (1969) A serum neutralization test for infectious bovine rhinotracheitis based on colour reaction and cytopathic effects in cell culture. Can J Comp Med 33:85–88

Chapter 7

Utilization and Assessment of Throat Swab and Urine Specimens for Diagnosis of Chikungunya Virus Infection

Chandrashekhar G. Raut, H. Hanumaiah, and Wrunda C. Raut

Abstract

Chikungunya is a mosquito-borne infection with clinical presentation of fever, arthralgia, and rash. The etiological agent Chikungunya virus (CHIKV) is generally transmitted from primates to humans through the bites of infected *Aedes aegypti* and *Aedes albopictus* mosquitoes. Outbreaks of Chikungunya occur commonly with varied morbidity, mortality, and sequele according to the epidemiological, ecological, seasonal, and geographical impact. Investigations are required to be conducted as a part of the public health service to understand and report the suspected cases as confirmed by laboratory diagnosis. Holistic sampling at a time of different types would be useful for laboratory testing, result conclusion, and reporting in a valid way. The use of serum samples for virus detection, virus isolation, and serology is routinely practiced, but sometimes serum samples from pediatric and other cases may not be easily available. In such a situation, easily available throat swabs and urine samples could be useful. It is already well reported for measles, rubella, and mumps diseases to have the virus diagnosis from throat swabs and urine. Here, we present the protocols for diagnosis of CHIKV using throat swab and urine specimens.

Key words Chikungunya virus, Throat swab, Urine, Arthralgia

1 Introduction

Chikungunya is a mosquito-borne common infection which causes fever, arthralgia, and rash. Causative agent Chikungunya virus, an *Alphavirus* of the *Togaviridae* family, is generally transmitted from primates to humans through *Aedes aegypti* and *Aedes albopictus* mosquitoes [1, 2]. Chikungunya virus is a single-stranded, positive-sense RNA, enveloped virus. It has a genome consisting of a linear, approximately 11.8 kb RNA molecule which is surrounded by a 60–70 nm diameter capsid and a phospholipid envelope [3–5]. During past years, life-threatening complications due to Chikungunya infections were very rarely reported. However, from the recent worldwide outbreaks, there have been several reports of unusually severe complications and deaths [6–8]. Also recovered patient may have varied degree of joints pain for months together.

Justin Jang Hann Chu and Swee Kim Ang (eds.), *Chikungunya Virus: Methods and Protocols*, Methods in Molecular Biology, vol. 1426,DOI 10.1007/978-1-4939-3618-2_7, © Springer Science+Business Media New York 2016

Chikungunya virus infection may cause typical, severe or fatal presentations, and it affects all age groups. The detection of Chikungunya virus is based on isolation of virus from blood specimens obtained from viremic patients or in infected tissue specimens obtained from blood-feeding arthropods, which are time-consuming. Sometimes, it becomes difficult to collect blood samples especially from infants, children, and individuals to whom access is limited. In such a scenario, an alternative and easily accessible specimen instead of blood would be necessary. Throat swab and urine samples are easier to collect and they are body fluids with low concentrations of immunoglobulins. It has been postulated that in urine, large macromolecules such as immunoglobulin M (IgM) antibodies cannot pass through the glomerular filter under normal conditions. However, monomeric IgM proteins (67,000 kDa) have been detected in post-renal sources and not in the glomerular filter [9, 10]. Throat swab as such was not reported for diagnosis of Chikungunya but utility of saliva and urine for diagnostic testing has been reported for many viral infectious diseases [11–15].

In developing countries like India where facilities for virological diagnosis are often limited, the ability to rapidly diagnose the infections in an outbreak situation is even troublesome. In this chapter, we provide protocols for the rapid confirmation of Chikungunya virus infection in an outbreak situation by using easily accessible specimens (throat swab and urine) by employing IgM Capture ELISA, conventional reverse transcriptase polymerase chain reaction (RT-PCR), and virus isolation in appropriate cell lines.

2 Materials

2.1 Virus Detection

1. Samples—throat swab and urine (*see* **Note 1**).
2. Chikungunya virus diagnostic primers.
   - Forward primer: 5′-CAACTTGCCCAGCTGATCTC-3′
   - Reverse primer: 5′-GGATGGCAAGACTCCACTCT-3′
3. Gel Electrophoresis apparatus and consumables.
4. Gel documentation system.
5. Qiagen OneStep RT-PCR Kit: OneStep RT-PCR Enzyme Mix, 5× RT-PCR buffer, 10 mM dNTP mix, RNase-free water.
6. Positive control RNA (*see* **Note 2**).
7. Protector RNase inhibitor (40 U/μl).
8. Aerosol-resistant Micropipette tips (2, 10, and 100 μl, *see* **Note 3**).
9. Micropipette sets—P10, P20, P100 (separate sets for Pre-PCR and PCR use).
10. QIAmp viral RNA extraction kit.

2.2 IgM Antibody Detection by ELISA

1. Suspected Chikungunya infection specimens (*see* **Note 4**).
2. ELISA plate/strip holder.
3. Microplate washer.
4. Bread box (Plastic box with cover).
5. Incubator (37 °C), humidified without CO_2.
6. Microplate reader with 450 nm filter.
7. High grade distilled water (2.5 l).
8. Positive control samples (*see* **Note 5**).
9. Negative control samples (*see* **Note 5**).
10. Human Chikungunya IgM Capture ELISA kit: positive control, negative control, wash buffer concentrate 20×, anti-human IgM coated strips, Anti-Chikungunya monoclonal antibodies-biotinylated, Avidin HRP, 3,3′,5,5′-Tetramethylbenzidine (TMB) substrate, stop solution.

2.3 In Vitro Virus Isolation

1. Monolayer of cell culture—C6/36.
2. Media: Mitsuhashi-Maramorosch (MM) medium, 10 % fetal bovine serum, 1 % penicillin–streptomycin.
3. Throat swab suspension in MEM (*see* **Note 1**). Centrifuged and supernatant filtered through a 0.22 μm filter.
4. Urine sample centrifuged and supernatant filtered through a 0.22 μm filter.

2.4 In-Vivo Virus Isolation

1. Infant mice of 2–3 days old.
2. Disposable needle 26 G (3/4 in.).
3. Insulin syringe or quarter cc syringe.
4. Bovine albumin phosphate saline (BAPS) 0.75 % with 1 % penicillin-streptomycin.
5. Throat swab suspension and urine samples (*see* **Note 6**).

3 Methods

3.1 Virus Genome Detection by Molecular Technique and Sequence Analysis [16–18]

1. Extract total RNA from 140 μl of throat swab suspension and urine samples using the QIAamp viral RNA kit according to the manufacturer's protocol.
2. Elute RNA in 50 μl of nuclease-free water.
3. Prepare PCR protocol for required number of samples to be tested including positive and two negative controls (one each for cell control and RNase-free water). Label required 0.2 ml thin-walled PCR tubes and arrange them according to the protocol. Thaw the required working primers and enzymes, mix well, spin down, and keep in 4 °C cold pack.

5. Preparation of master-mix: the quantity of Master-Mix preparation depends on the number of samples and controls to be tested. Total reaction volume is 25 μl which includes 23 μl of Master-Mix and 2 μl of extracted RNA sample. Each master mix reaction contains 13.755 μl RNase-free water, 5.0 μl RT-PCR Buffer (5×), 1.0 μl dNTP Mix (10 mM), 1.0 μl each of forward and reverse primer (20 μM).
6. Vortex and wait for 2 min before adding enzyme Qiagen OneStep enzyme mix—1.0 μl, RNase Inhibitor (40 U/μl) —0.25 μl.
7. Vortex and dispense 23 μl of master mix into 0.2 ml thin-walled PCR tubes and keep reaction tubes immediately in ice. Carry out template (extracted RNA) addition in designated PCR room.
8. Add 2 μl of extracted RNA from the given samples, a negative control (RNase-free water) and positive control (positive RNA) to the appropriate PCR tube as per protocol. Total reaction volume is 25 μl. After addition of template RNAs, close the caps, vortex and spin down in and keep reaction tubes in ice bucket.
9. Follow the thermal cycling condition below:
   - Holding stage: RT reaction at 50 °C for 30 min.
   - Holding stage: RT inactivation/DNA polymerase activation at 95 °C for 15 min.
   - Cycling stage: 40 cycles
   - Cycling stages: Step 1—Denaturation at 94 °C for 30 s, Step 2—Annealing at 57 °C for 30 s, Step 3—Extension at 72 °C for 30 s.
   - Final extension at 72 °C for 10 min.
   - Hold at 4 °C.
10. Analyze 10 μl of each product along with both positive and negative controls by agarose gel electrophoresis utilizing 1.5 % agarose gel, and stain gels with ethidium bromide at 5 μg/ml. Use 100-bp DNA ladder. Visualize gel by documentation system using the software Quantity One version 4.6.3 to determine the band size.
11. Use RT-PCR positive product for nucleotide sequencing and phylogenetic analysis. Do the sequencing of few representative cases of RT-PCR positive specimens. Compare the nucleotide sequences with other CHIKV strains reported from different geographic regions and other reference sequences. Do the nucleotide homology percentages using the sequence similarity search tool BLAST (Basic Local Alignment Search Tool). Align the DNA sequences using Clustal X 1.83 software and

make a phylogenetic tree using MEGA 6 [19]. Analyze the strength of a phylogenetic tree by bootstrap using 1000 random samplings.

3.2 IgM ELISA Serological Test [8]

1. Aliquot the specimens, throat swab and urine for detection of Chikungunya IgM antibodies. Follow the standard procedure as per the kit protocol. Use throat swab and urine undiluted (*see* **Note 5**) for the test procedure having appropriate positive and negative controls (*see* **Note 6**).
2. Dilute "Wash Buffer Concentrate" from 20× to 1× with high grade distilled water. For ten samples (eight clinical samples and two controls), 70 ml of wash buffer is required. However, make 100 ml of wash buffer and attach the bottle to an automated ELISA washer.
3. Label and arrange required number of micro-centrifuge tubes as per the specimen numbers. Carry out the work in the Biosafety cabinet.
4. Remove required number of "Anti-Human IgM coated strips" (*see* **Note 7**). Number the test strips as 1, 2, 3 ... and positive and negative controls as PC and NC respectively.
5. Wash the strips three times with 1× Wash buffer (300 μl per well). Do not allow the wells to dry.
6. Transfer 50 μl of samples from the deep well plate to respective wells as per the protocol on ELISA sheet using multichannel pipette.
7. Add 50 μl of "CHIK IgM Positive control" and "Negative Control" to respective wells (DO NOT DILUTE THE CONTROLS). (*see* **Note 8**).
8. Cover a plate with aluminum foil. Keep the plate in a closed humidified box inside the incubator and incubate a plate at 37 °C for 1 h.
9. At the end of the incubation, wash the plate five times with 1× Wash buffer. Tap the plate after the last wash on a tissue paper to remove the traces of wash buffer.
10. Add 50 μl "CHIK antigen" to each well of the plate (CAUTION! Take out the CHIK antigen vial from refrigerator, add 50 μl to each well and immediately put the vial back in the refrigerator. DO NOT EQUILIBRATE CHIK ANTIGEN TO ROOM TEMPERATURE; *see* **Note 9**).
11. Cover the plate with aluminum foil. Keep the plate in a closed humidified box inside the incubator and incubate a plate at 37 °C for 1 h.
12. At the end of the incubation, wash the plate five times with 1× Wash buffer. Tap the plate after the last wash on a tissue paper to remove the traces of wash buffer.

13. Add 50 μl of "Anti-CHIK monoclonal antibody (Biotin labeled)" to each well.
14. Cover the plate with aluminum foil. Keep the plate in a closed humidified box inside the incubator and incubate the plate at 37 °C for 1 h.
15. At the end of the incubation, wash the plate five times with 1× Wash buffer. Tap the plate after the last wash on a tissue paper to remove the traces of wash buffer.
16. Add 50 μl of "Avidin–HRP" to each well.
17. Cover the plate with aluminum foil. Keep the plate in a closed humidified box inside the incubator and incubate a plate at 37 °C for 30 min.
18. At the end of the incubation, wash the plate five times with 1× Wash buffer. Tap the plate after the last wash on a tissue paper to remove the traces of wash buffer.
19. Add 100 μl of liquid "TMB Substrate (TMB/H_2O_2)".
20. Incubate at room temperature in dark for 10 min.
21. Stop the reaction exactly after 10 min by adding 100 μl of "Stop Solution".
22. Measure the absorbance at 450 nm as early as possible (maximum 10 min; *see* **Note 10**).

3.3 In Vitro Virus Isolation

1. Inoculate the samples onto *Aedes albopictus* C6/36 cell lines [20].
2. Incubate the inoculated cell lines at 30 °C and examine daily up to 7 days for the appearance of cytopathic effect (CPE).
3. If cultures show no CPE, pass to the next passage. If there is no CPE after three passages, then record it as CPE negative.
4. If virus-induced CPE observed, consider it as positive for virus growth.
5. Harvest the virus from CPE positive cells culture by freezing and thawing.
6. Take the cell culture supernatant for RNA extraction and further viral genome confirmation by RT-PCR method.

3.4 In Vivo Virus Isolation [21]

1. Take normal healthy mouse group of age 2–3 days.
2. Take sample inoculum in a quarter cc or one cc syringe.
3. Anesthetize the mice with vaporized ether.
4. Inoculate 20 μl sample intracerebrally to infant mice with 26 G (3/4 in.) needle and 1 cc syringe.
5. Observe the mice for sickness daily.
6. Remove the brains (*see* **Note 11**) aseptically of sick mice as the source of virus.

7. Make 10 % suspension of the brains in suitable diluents. For example, for an infant brain of 0.2 g, add 1.8 ml of the diluents (0.75 % BAPS) and grind in a mortar with the aid of a pestle. Homogenize the suspension.
8. Transfer the suspension in a sterile tube and centrifuge at 10,173 × *g* for 30–40 minutes at a temperature of 4 °C.
9. Remove the supernatant fluid with the aid of micropipette or needle syringe and transfer to a sterile tube placed in an ice bath.
10. Label the tubes properly and store at −80 °C as a wet pool of virus stock.
11. Confirm the isolated virus by given RT-PCR method.

4 Notes

1. Collect throat swabs in virus transport medium (Minimum Essential Medium). Store urine and throat swab samples in −20 °C till further utilization.
2. Culture supernatant of CHIKV-infected cells is a good source of positive control. Store RNA at −80 °C.
3. Aerosol-resistant tips have better pipetting accuracy than ordinary tips.
4. Collect samples during 4–10 post onset days of infection.
5. Take throat swab and urine from healthy and Chikungunya-infected individuals as the known negative and positive controls respectively.
6. Earlier serological studies have shown that the concentration of immunoglobulins in saliva and urine was low compared to serum. Therefore, use undiluted throat swab and urine specimens. For virus isolation, centrifuge throat swab suspension and urine before passing the respective supernatants through 0.22 μm filters.
7. Anti-human IgM-coated strips target to capture IgM from human species only, no other species of mammalian.
8. As we are targeting immunoglobulin M (IgM) molecules against CHIKV from the given samples, hence a positive control CHIKV IgM is required.
9. CHIKV antigen is used to target Immunoglobulin M (IgM) molecules against CHIKV from the given samples.
10. In ELISA reaction final output is color based. The color intensity fades over time, hence document the results as soon as possible.

11. Sample was inoculated intracerebrally hence virus growth in the brain of sick mice is expected. So harvesting of brains as the source of virus is advisable.

Acknowledgement

This work supported by the National Institute of Virology, Pune, the National Vector Borne Disease Control Programme, New Delhi, and the State Health Department, Government of Karnataka. Thanks are due to Mr. D. P. Sinha and M. J. Manjunath for systematic technical works.

References

1. Jupp PG, McIntosh BM (1988) Chikungunya virus disease. In: Monath TP (ed) The arboviruses: epidemiology and ecology. CRC, Boca Raton, FL, pp 137–157
2. Khan AH, Morita K, Mdel MD et al (2002) Complete nucleotide sequence of Chikungunya virus and evidence for an internal polyadenylation site. J Gen Virol 83:3075–3084
3. Simon F, Savini H, Parola P (2008) Chikungunya: a paradigm of emergence and globalization of vector- borne diseases. Med Clin North Am 92:1323–1343
4. Pialoux G, Gaüzère BA, Jauréguiberry S, St. robel M (2007) Chikungunya, an epidemic arbovirosis. Lancet Infect Dis 7:319–327
5. Powers AM, Logue CH (2007) Changing patterns of Chikungunya virus: re-emergence of a zoonotic arbovirus. J Gen Virol 88:2363–2377
6. Raut CG, Rao NM, Sinha DP, Hanumaiah H, Manjunath MJ (2015) Chikungunya, dengue, and malaria co-infection after travel to Nigeria, India. Emerg Infect Dis 21:908–909
7. Robin S, Ramful D, Le Seach F, Jaffar-Bandjee MC, Rigou G, Alessandri JL (2008) Neurologic manifestations of pediatric Chikungunya infection. J Child Neurol 23:1028–1035
8. Shaikh NJ, Raut CG, Sinha DP, Manjunath MJ (2015) Detection of Chikungunya virus from a case of encephalitis, Bangalore, Karnataka State. Ind J Med Micro 33:454–455
9. Guyton A (1992) Tratado de fisiología me´dica, 8th edn. Raven, New York, NY
10. Tencer J, Frick IM, Oquist BM, Alm P, Rippe B (1998) Size selectivity of the glomerular barrier to high molecular weight proteins: upper size limitations of shunt pathways. Kidney Int 53:709–715
11. Perry KR, Parry US, Vandervelde ME, Mortimer PP (1992) The detection in urine specimens of IgG and IgM antibodies to hepatitis A and hepatitis B core antigens. J Med Virol 38:265–270
12. Nitsan C, Fuchs E, Margalith M (1994) Antibodies to HIV-1 and to CMV, in serum and urine of HIV-1 and CMV infected individuals. AIDS Res Hum Retroviruses 10(S):98
13. Martínez P, Ortiz de Lejarazu R, Eiros JM, De Benito J, Rodríguez-Torres A (1996) Urine samples as a possible alternative to serum for human immunodeficiency virus antibody screening. Eur J Clin Microbiol Infect Dis 15:810–813
14. Elsana SE, Sikuler E, Yaari A et al (1998) HCV antibodies in saliva and urine. J Med Virol 55:24–27
15. Takahashi S, Machikawa F, Noda A, Oda T, Tachikawa T (1998) Detection of immunoglobulin G and A antibodies to rubella virus in urine and antibody responses to vaccine-induced infection. Clin Diagn Lab Immunol 5:24–27
16. Yergolkar PN, Tandale BV, Arankalle VA, Sathe PS, Sudeep AB, Gandhe SS et al (2006) Chikungunya outbreaks caused by African genotype, India. Emerg Infect Dis 12:1580–1583
17. Parida MM, Santhosh SR, Dash PK, Tripathi NK, Lakshmi V, Mamidi N, Shrivastva A, Gupta N, Saxena P, Pradeep Babu J, Lakshmana Rao PV, Kouichi M (2007) Rapid and real-time detection of chikungunya virus by reverse transcription loop-mediated isothermal amplification assay. J Clin Microbiol 45:351

18. Lakshmi V, Neeraja M, Subbalaxmi MV, Parida MM, Dash PK, Santhosh SR et al (2008) Clinical features and molecular diagnosis of Chikungunya fever from South India. Clin Infect Dis 46:1436–1442
19. Tamura K, Stecher G, Peterson D, Filipski A, Kumar S (2013) MEGA6: Molecular Evolutionary Genetics Analysis version 6.0. Mol Biol Evol 30:2725–2729
20. Singh KRP (1967) Cell cultures derived from larvae of *Aedes albopictus* (Skuse) and *Aedes aegypti* (L.). Curr Sci 36: 506–508
21. Raut CG, Deolankar RP, Kolhapure RM, Goverdhan MK (1996) Susceptibility of laboratory-bred rodents to the experimental infection with Dengue-2 virus. Acta Virol 40(3):143–146

Part II

Cell Culture, Virus Replication, and Cellular Responses

Chapter 8

Propagation of Chikungunya Virus Using Mosquito Cells

Swee Kim Ang, Shirley Lam, and Justin Jang Hann Chu

Abstract

Chikungunya virus (CHIKV) is a mosquito-borne alphavirus that transmits in between a mosquito host vector to a primate host and then back to the mosquito host vector to complete its life cycle. Hence, CHIKV must be able to replicate in both host cellular systems that are genetically and biochemically distinct. The ability to grow and propagate the virus in high titers in the laboratory is fundamentally crucial in order to understand virus replication in different host cellular systems and many other CHIKV research areas. Here, we describe a method on CHIKV propagation using C6/36, a mosquito cell line derived from *Aedes albopictus* in both serum-containing and serum-free media.

Key words CHIKV, Virus propagation, Mosquito cells, C6/36, Serum-free

1 Introduction

Chikungunya virus (CHIKV) is an alphavirus of the *Togaviridae* family. Its genome comprises a single positive-stranded RNA of around 11.8 kb containing a 5′-methylguanylate cap, two open reading frames, and a 3′-polyadenylate tail [1, 2]. It encodes for four nonstructural proteins (nsP1–4), three structural proteins (C, E1, and E2), and two small peptides (E3 and 6K). CHIKV is maintained in two distinct transmission cycles: a sylvatic cycle that is mainly confined within African countries involving forest-dwelling *Aedes* mosquitoes and nonhuman primates on a smaller epidemic scale [3]; and a human–mosquito–human cycle involving urban mosquitoes *Aedes aegypti* and *Aedes albopictus* being the transmission vectors during large epidemics [4].

Chikungunya has re-emerged as a disease of global significance following major outbreaks with millions of infections reported since 2005 [5]. Despite causing public health threat globally, there is currently no vaccine or effective therapeutics toward CHIKV infections. Therefore, extensive research to understand CHIKV pathogenesis as well as CHIKV diagnosis and drug development is needed to better control CHIV epidemics. Generation of virus

Justin Jang Hann Chu and Swee Kim Ang (eds.), *Chikungunya Virus: Methods and Protocols*, Methods in Molecular Biology, vol. 1426,DOI 10.1007/978-1-4939-3618-2_8, © Springer Science+Business Media New York 2016

infectious particles in the laboratory is fundamentally essential in many areas of CHIKV research. Since CHIKV is able to replicate both in vertebrate and invertebrate cellular systems that are biochemically and genetically divergent in nature, CHIKV infectivity in different established cell lines derived from mammalian and mosquito origins have been tested [2, 6]. Cells in general possess innate defense mechanisms, e.g. type-I interferon response in mammalian cells [7] and RNA interference (RNAi) response in insect cells [8, 9] to suppress virus replication. However, Vero cells (African green monkey kidney epithelia) lacks type-I interferon, making them highly permissive for CHIKV infection and therefore commonly used for CHIKV propagation [10]. Similarly, C6/36 (a genetically homogenous isolate of Singh's *Aedes albopictus* cells) lacks functional RNAi response [11] where Dcr2 fails to cleave double-stranded viral RNA molecules into small RNA effector molecules which in turn silence complimentary RNA molecules [12]. CHIKV infection in C6/36 only causes mild cytopathic effect (CPE) and apoptosis compared to that of Vero cells [13], making it a better cell line of choice for persistent CHIKV infection to produce higher titer of virus progeny.

In this chapter, we will describe protocols for culturing C6/36 cells, followed by CHIKV infection on the cells for virus propagation in a detailed step-by-step manner. To ensure high virus titer production, tips will be given on the steps that require special attention on the methods described.

2 Materials

2.1 Culture of C6/36 Cells

1. *Aedes albopictus* C6/36 cells.
2. Leibovitz (L15) media containing 10 % heat-inactivated fetal calf serum (FCS).
3. 1× Trypsin-EDTA solution: 0.05 % trypsin, 0.02 % EDTA, 0.1 % glucose, 0.8 % NaCl, 0.04 % KCl, 0.058 % $NaHCO_3$.
4. 1× Phosphate Buffered Saline (PBS).
5. T75 cm^2 flasks.

2.2 Virus Propagation

1. Chikungunya virus.
2. C6/36 cells at 80 % confluency.
3. L15 media containing 2 % FCS.
4. L15 media containing 0.2 % BSA.
5. 1× PBS.

3 Methods

3.1 Culture of C6/36 Cells

1. Thaw 2 vials (1 ml each) of cryo-preserved C6/36 stock culture (*see* **Note 1**).
2. Aliquot 10 ml of L15 media supplemented with 10 % heat-inactivated FCS into a sterile T75 cm^2 flask and add thawed cells into the flask (*see* **Note 2**).
3. Mix well by gently pipetting up and down several times to dissociate clumped cells while preventing the formation of excessive air bubbles in the media. Lay the flask flat and swirl gently to equally distribute the cells for adherence.
4. Alternatively, upon thawing, suspend cells in 10 ml of L15 media and centrifuge at 300 × g for 10 min. Decant the media carefully and gently resuspend cell pellet in 10 ml of L15 media before adding the cell suspension into a T75 cm^2 flask (*see* **Note 3**).
5. Incubate the cells at 28 °C overnight in the absence of CO_2.
6. On the next day, check the cells under light microscope. Most of the cells will have attached and there may be some dead cells. Replace the media with fresh L15 supplemented with 10 % heat-inactivated FCS.
7. Incubate at 28 °C for 5–6 days until ~80 to 100 % confluency.
8. To expand cells, aspirate and decant the culture media from the flask, wash the cells once with 1× PBS followed by the addition of 2 ml trypsin-EDTA solution sufficient to cover the cell monolayer.
9. Swirl gently to allow trypsin to be in contact with cells for ~2 min (*see* **Note 4**).
10. Add 8 ml of L15 media supplemented with 10 % heat-inactivated FCS. If the cells have yet to dislodge after 5 min, replace with fresh trypsin and/or gently hit the flask against your palms.
11. Resuspend cells by pipetting up and down several times to avoid cell clumping (*see* **Note 5**).
12. Add 3 ml of cells to 6 ml of L15 media (1:3 dilution) per flask and incubate at 28 °C for 3–5 days until ~80 % confluency (*see* **Note 6**).

3.2 Virus Propagation

1. Thaw 3 vials (1 ml each) of cryo-preserved CHIKV from −80 °C to room temperature or 37 °C (*see* **Note 7**).
2. Aspirate and decant culture media from T75 cm^2 flasks and wash C6/36 cell monolayer with 1× PBS.

3. Infect the cells with 1 ml of virus/flask at a multiplicity of infection (MOI) of at least 0.1 (*see* **Note 8**). Swirl gently to ensure equal distribution of virus to the cells.
4. Incubate the cells at 37 °C for 1.5 h, with gentle swirling at 15 min interval (*see* **Note 9**).
5. After 1.5 h, add 9 ml of L15 media supplemented with 2 % heat-inactivated FCS to the flask (*see* **Note 10**). Incubate CHIKV-infected cells at 28 °C for 3–5 days until the cells are observed to be unhealthy (*see* **Note 11**).
6. Alternatively, to have the virus cultured serum-free, incubate CHIKV-infected cells at 28 °C for 1 day in L15 media supplemented with 2 % heat-inactivated FCS. Then, remove media, wash twice with 1× PBS and then top up with 10 ml L15 media supplemented with 0.2 % BSA. Incubate for 3–4 days until the cells are observed to be unhealthy.
7. Collect viral supernatant from the flasks into a sterile centrifuge tube and centrifuge at 1000 × *g* for 10 min at 4 °C (*see* **Note 12**).
8. Clarified supernatant containing virus can now be used directly for quantification by viral plaque assays. Alternatively, virus supernatant should be stored as 1 ml aliquots in cryo-preservation tubes at −80 °C (*see* **Note 13**).

4 Notes

1. C6/36 cells must be thawed rapidly and completely in a water bath set at the cell's normal growth temperature, i.e. 28 °C. This is to ensure maximal recovery of cell viability upon thawing in the presence of dimethyl sulfoxide (DMSO), which is a cryo-preservant [14]. During thawing, ensure cryovials are half-submerged in the water bath to prevent water from leaking in or contaminating screw caps of any loosen vials.
2. For adherent cell line like C6/36, it is optimal to have at least 3×10^4 cells/cm^2 per flask [14]. The routine sub-culturing ratio of C6/36 cells is 1:5. However for thawing purpose, this ratio can be increased to 1:10 where more media can dilute the toxic effect of DMSO. A higher dilution ratio can also initiate rapid cell growth as some cells might be lost due to damage from thawing.
3. This additional centrifuge step removes DMSO in the thawed media which helps in cell recovery.
4. Trypsin enables cells to detach from one another and from the substratum by digesting extracellular matrix protein and cell junction proteins, while EDTA disaggregates cells by chelating calcium ions. Excessive trypsinization is detrimental to cell

recovery and may undesirably lead to cell death. Therefore, it is important to use the smallest possible volume of trypsin solution at 0.025–0.05 w/v and the shortest exposure trypsinization time to detach the cells from the flask. Once the cells round up and begin to dislodge from the monolayer, trypsin activity must be inhibited immediately by adding fresh media containing 5–10 % FCS to the cell-trypsin solution [15].

5. After detachment from a monolayer culture, C6/36 cells have the tendency to form clusters if they are not well-dissociated by pipetting forces. Therefore, it is important to dissociate the clusters into single cells in order for the cells to have better growth and adaptation to the fresh culture environment [16].
6. Sufficient cells of at least 80 % confluency is optimal to achieve a high-titer CHIKV progeny since C6/36 cells are able to support persistent CHIKV infection [13, 17].
7. Thawing of frozen virus should be performed in water bath set at 37 °C so that the viral supernatant is equilibrated for optimal cell absorption upon adding to the cell monolayer.
8. With a MOI of 0.1, the volume of virus (volume from virus stock: L15 media dilution) to add per flask can be calculated by taking the expected cell density of C6/36 cells at the time of confluency divided by the virus titer. If possible, use a higher MOI to allow multiple rounds of CHIKV infection and achieve a culture of higher virus titer.
9. Regular swirling of the culture flask allows CHIKV to be well-distributed which helps in virus adsorption to the cell surface.
10. Heat-inactivated 2 % FCS is sufficient to maintain cell growth and metabolism so that CHIKV propagation can occur for the next few days.
11. Due to the presence of certain host factors in C6/36 cells, these mosquito cells do not exhibit apparent CPE and apoptosis after CHIKV infection [11, 13].
12. Low-speed centrifugation causes dead cells to pellet down and thus helps to purify the viral supernatant.
13. Repeated freeze–thawing should be minimized as this may decrease virus titer.

References

1. Khan AH, Morita K, Parquet Md Mdel C, Hasebe F, Mathenge EG, Igarashi A (2002) Complete nucleotide sequence of chikungunya virus and evidence for an internal polyadenylation site. J Gen Virol 83:3075–3084
2. Solignat M, Gay B, Higgs S, Briant L, Devaux C (2009) Replication cycle of chikungunya: a re-emerging arbovirus. Virology 393: 183–197
3. McIntosh BM, Jupp PG, dos Santos I (1977) Rural epidemic of chikungunya in South Africa with involvement of Aedes (Diceromyia) furcifer (Edwards) and baboons. S Afr J Sci 73:267–269

4. Pulmanausahakul R, Roytrakul S, Auewarakul P, Smith DR (2011) Chikungu-nya in Southeast Asia: understanding the emergence and finding solutions. Int J Infect Dis 15:671–676
5. Weaver SC, Forrester NL (2015) Chikungunya: evolutionary history and recent epidemic spread. Antiviral Res 120:32–39
6. Wikan N, Sakoonwatanyoo P, Ubol S, Yoksan S, Smith DR (2012) Chikungunya virus infection of cell lines: analysis of the East, Central and South African Lineage. PLos One 7(1): e31102
7. Her Z, Malleret B, Chan M, Ong EK, Wong SC, Kwek DJ, Tolou H, Lin RT, Tambyah PA, Renia L, Ng LF (2010) Active infection of human blood monocytes by chikungunya virus triggers an innate immune response. J Immunol 184:5903–5913
8. Myles KM, Wiley MR, Morazzani EM, Adelman ZN (2008) Alphavirus-derived small RNAs modulate pathogenesis in disease vector mosquitoes. Proc Nat Acad Sci 105:19938–19943
9. Myles KM, Morazzani EM, Adelman ZN (2009) Origins of alphavirus-derived small RNAs in mosquitoes. RNA Biol 6:387–391
10. Desmyter J, Melnick JL, Rawls WE (1968) Defectiveness of interferon production and of rubella virus interference in a line of African Green Monkey kidney cells (Vero). J Virol 2:955–961
11. Brackney DE, Scott JC, Sagawa F, Woodward JE, Miller NA, Schilkey FD, Mudge J, Wilusz J, Olson KE, Blair CD, Ebel GD (2010) C6/36 Aedes albopictus cells have a dysfunctional antiviral RNA interference response. PLoS Neglect Trop Dis 4(10):e856
12. Scott JC, Brackney DE, Campbell CL, Bondu-Hawkins V, Hjelle B, Ebel GD, Olson KE, Blair CD (2010) Comparison of dengue virus type-2-specific smalls from RNA interference-competent and –incompetent mosquito cells. PLoS Negl Trop Dis 4(10):e848
13. Li YG, Siripanyaphinyo U, Tumkosit U, Noranate N, A-nuegoonpipat A, Tao R, Kurosu T, Ikuta K, Takeda N, Anantapreecha S (2013) Chikungunya virus induces a more moderate cytopathic effect in mosquito cells than in mammalian cells. Intervirology 56(1): 6–12
14. Morris CB (2007) Cryopreservation of animal and human cell lines. In: Day JG, Stacey GN (eds) Cryopreservation and freeze-drying protocols, methods in molecular biology, vol 368. Humana Press, Totowa, NJ, pp 227–236
15. Richardson A, Fedoroff S (2009) Tissue culture procedures and tips. In: Doering LC (ed) Protocols for neural cell culture, Springer protocols handbooks. Humana Press, Totowa, NJ, pp 375–390
16. Morita K, Igarashi A (1989) Suspension culture of *Aedes albopictus* cells for Flavivirus mass production. J Tissue Cult Meth 12(3):35–36
17. Tripathi NK, Shrivastava A, Dash PK, Jana AM (2011) Detection of dengue virus. In: Stephenson JR, Warnes A (eds) Diagnostic virology protocols, methods in molecular biology, vol 665. Humana Press, Totowa, NJ, pp 51–64

Chapter 9

Infectious Viral Quantification of Chikungunya Virus—Virus Plaque Assay

Parveen Kaur, Regina Ching Hua Lee, and Justin Jang Hann Chu

Abstract

The plaque assay is an essential method for quantification of infectious virus titer. Cells infected with virus particles are overlaid with a viscous substrate. A suitable incubation period results in the formation of plaques, which can be fixed and stained for visualization. Here, we describe a method for measuring Chikungunya virus (CHIKV) titers via virus plaque assays.

Key words Plaque assays, Chikungunya virus, Virus titration, Carboxymethyl cellulose, Aquacide, Crystal violet

1 Introduction

Detection of viral load or infection level is essential in any virology-based study. There are numerous methods that are currently employed for this, including direct quantification of infectivity through plaque assays, tissue culture infective dose ($TCID_{50}$) assays, and immunofluorescence focus assays. In addition, a host of indirect methods measuring viral particles, genome, or proteins are also available. These include flow cytometry, transmission electron microscopy, quantitative RT-PCR (qRT-PCR), hemagglutination assays, western blot, and ELISA. Despite being more laborious and time-consuming than some modern technologies, plaque assays remain the gold standard for the detection of viruses [1]. Plaque assays were first developed in 1952 by Renato Dulbecco as a method to quantify animal viruses, modeling after the bacteriophage plaque assay [2, 3]. Plaque assays rely on the ability of viruses to cause cytopathic effects (CPE) in host cells as a result of infection. For alphaviruses including CHIKV, CPE usually manifest as apoptosis in mammalian cells, although infection of neurons with Semliki Forest virus (SFV) is known to result in necrosis [4].

Unlike the indirect methods of detecting viral loads, plaque assays allow the quantification of infectious virus particles. This is

Justin Jang Hann Chu and Swee Kim Ang (eds.), *Chikungunya Virus: Methods and Protocols*, Methods in Molecular Biology, vol. 1426,DOI 10.1007/978-1-4939-3618-2_9, © Springer Science+Business Media New York 2016

important in applications where differentiation between viable and nonviable virions is necessary, for instance, during investigation of potential viral inhibitors, titration of virus pools, and generation of clonal virus populations. Plaque assays are also especially important in studies of viral evolution. In samples of purified SFV particles, the virus particle to plaque-forming unit (pfu) ratio is 2.5, suggesting a high rate of infectivity [5]. However, Bruton et al. reported that repeated passaging of SFV in baby hamster kidney (BHK) cells resulted in a 4 $\log_{10}$ decrease in infectious titer due to the accumulation of defective viral particles that interfered with replication, corresponding to only a 1 $\log_{10}$ decrease in total number of particles [6]. Defective particles did not display any differences in structural protein compared to infectious particles [6]. This suggests that utilizing indirect methods (e.g. ELISA, hemagglutination) to detect viral load may result in an incomplete picture of replication kinetics. Therefore, the importance of plaque assays as a tool in virology cannot be undermined.

While plaque assays work on the assumption that one infectious particle results in one plaque, the titer obtained is more likely to be only a proportion of the total number of infectious particles [7]. One reason for this is that virus particles may exist in aggregates and may not be normally distributed in the sample [7]. The detection threshold of plaque assays is also higher than molecular methods like qRT-PCR [8]. As such, absence of plaques does not necessarily indicate absence of infection and investigators are urged to select detection assays that are suited to their experimental needs [7]. Plaque assays can be performed only for viruses that exhibit CPE and form plaques. For species or mutant strains that display CPE but do not form plaques, $TCID_{50}$ assays can be employed to quantify infectivity [9]. $TCID_{50}$ assays, however, take a longer time due to the requirement for complete cell death in infected monolayers [10]. Immunofluorescence focus assays, on the other hand, are useful for viruses that produce neither CPE nor plaques, as well as for quicker generation of results [11]. The requirement for specific antibodies and fluorescence microscopes in immunofluorescence focus assays may however significantly increase the costs of the study [11]. The sensitivities and detection limits for all three methods (plaque assays, $TCID_{50}$ assays, and immunofluorescence focus assays) has been found to vary according to virus species [11–13].

In this chapter, we describe a method for performing plaque assays to quantify infectious CHIKV titers. We will outline the protocols for infection of cell monolayers with tenfold serial dilutions of virus-containing samples, followed by overlaying of infected monolayers with media containing a high molecular weight substrate like agarose or carboxymethyl cellulose. The solid or semi-solid overlay media function to limit viral spread, ensuring that viruses egressing from one cell will go on to infect neighboring cells instead of being randomly dispersed in the media. Over time, a localized patch of cell death (apoptosis) develops as a result of infection, forming a plaque. Each plaque is therefore assumed to originate from one infectious virus particle. Plaques can be visualized by staining cell monolayers with crystal violet or neutral red.

2 Materials

Prepare all solutions using ultrapure water (distilled water which undergoes reverse osmosis to attain a sensitivity of 18 MΩ cm at 25 °C). Prepare all cell culture media in a biosafety cabinet (BSC) to ensure sterility.

1. RPMI-1640 medium supplemented with 10 % FCS. Weigh 2 g $NAHCO_3$ and dissolve in 1 l of autoclaved ultrapure water. Dissolve 1 bottle of RPMI-1640 powder into the solution and mix well. To prepare 500 ml of RPMI with 10 % fetal calf serum (FCS), filter 450 ml of RPMI solution through a 0.22 μM Corning filter and add 50 ml of FCS. Store at 4 °C.
2. RPMI-1640 medium supplemented with 2 % FCS. As above but filter 490 ml of RPMI solution and add 10 ml of FCS.
3. Overlay media: 1 % Aquacide II in RPMI with 2 % FCS. Weigh 5 g of Aquacide II (*see* **Note 1**) and dissolve in 250 ml of ultrapure water (*see* **Note 2**). Prepare two bottles of this solution and autoclave (*see* **Note 3**). Weigh 2 g $NAHCO_3$ and dissolve in 500 ml of autoclaved ultrapure water. Dissolve 1 bottle of RPMI-1640 powder into the solution and mix well. Filter 240 ml of RPMI solution into 1 bottle containing 250 ml autoclaved Aquacide II solution using a 0.22 μM Corning filter and add 10 ml of FCS. Repeat with the second bottle containing 250 ml Aquacide II. Shake well to mix. Store at 4 °C.
4. 10× Trypsin-EDTA: 0.5 % trypsin, 0.2 % EDTA, 1 % glucose, 8 % NaCl, 0.4 % KCl, 0.58 % $NaHCO_3$. Add 80 g NaCl and 4 g KCl to 1 l ultrapure water and autoclave. Cool solution to room temperature and add 10 g d-glucose, 5.8 g $NaHCO_3$, 5 g trypsin and 2 g EDTA. Stir at room temperature until dissolved. Store aliquots of 10 ml at −20 °C (*see* **Note 4**).
5. 1× Trypsin-EDTA. Add 10 ml of 10× trypsin-EDTA to 90 ml autoclaved ultrapure water. Store at 4 °C (*see* **Note 5**).
6. 10× phosphate buffered saline (PBS): 1.37 M NaCl, 27 mM KCl, 43 mM Na_2HPO_4, 14.7 mM KH_2PO_4. Weigh 400 g NaCl, 72 g Na_2HPO_4, 12 g KH_2PO_4 and 10 g KCl and add to 4.5 l of ultrapure water. Adjust pH to 7.2 with HCl and NaOH. Top up solution to 5 l using ultrapure water. Autoclave and store at room temperature.

7. 1× PBS. Add 50 ml 10× PBS to 450 ml ultrapure water and autoclave. Store at 4 °C.
8. 4 % Paraformaldehyde in PBS. Weigh 40 g of paraformaldehyde (*see* **Note 6**) and add to 1 l of 1× PBS. Heat to 60 °C and stir until dissolved (*see* **Note 7**). Store at 4 °C.
9. Crystal violet solution: 0.2 % crystal violet, 3.2 % paraformaldehyde. Mix 2 ml absolute ethanol with 8 ml 1× PBS. Weigh 0.1 g crystal violet powder (*see* **Note 8**) and add to ethanol-PBS solution. Add 40 ml of 4 % paraformaldehyde and mix. Store at room temperature.
10. Baby hamster kidney (BHK21) cells.
11. CHIKV-containing cell supernatants.
12. Consumables: 10 ml Serological pipettes, hemocytometer, counter, 24-well plates, micropipettes (P20, P200, and P1000), centrifuge tubes, autoclaved glass bottles, pipette tips, microcentrifuge tubes, microcentrifuge tube racks, waste bucket, and disinfectant.
13. Equipment: BSC, light microscope, water bath, vortex, orbital shaker, and CO_2 incubator.

3 Methods

3.1 Harvesting of Samples for Plaque Assay

1. Cell culture supernatants can be harvested from culture plates and stored in microcentrifuge tubes at −80 °C if plaque assays are to be done on a separate day. Alternatively, the entire culture plates (with cell monolayers) can be stored at −80 °C (*see* **Note 9**).

3.2 Seeding of BHK21 Cells on 24-Well Plates

1. Warm all cell culture media, PBS, and trypsin-EDTA in a water bath to 37 °C prior to cell culture.
2. Remove cell culture media from a confluent T75 flask of BHK21 cells (*see* **Note 10**).
3. Wash cell monolayer once with 5 ml 1× PBS (*see* **Note 11**).
4. Add 2 ml 1× trypsin-EDTA and incubate at room temperature for 1–3 min (*see* **Note 12**).
5. Gently knock the side of the flask to dislodge cells.
6. Add 8 ml of RPMI with 10 % FCS to neutralize trypsin-EDTA. Mix well by pipetting to ensure complete neutralization of trypsin-EDTA and even dispersion of cells (*see* **Note 13**).
7. Pipette 20 μl of BHK21 cell suspension onto a hemocytometer and count cells under a light microscope using a counter. Calculate the appropriate dilution factor for a seeding density

Table 1
Arrangement of sample dilutions on 24-well plates

Sample 1 10^{-1}	Sample 1 10^{-2}	Sample 1 10^{-3}	Sample 1 10^{-4}	Sample 1 10^{-5}	Sample 1 10^{-6}
Sample 1 10^{-1}	Sample 1 10^{-2}	Sample 1 10^{-3}	Sample 1 10^{-4}	Sample 1 10^{-5}	Sample 1 10^{-6}
Sample 2 10^{-1}	Sample 2 10^{-2}	Sample 2 10^{-3}	Sample 2 10^{-4}	Sample 2 10^{-5}	Sample 2 10^{-6}
Sample 2 10^{-1}	Sample 2 10^{-2}	Sample 2 10^{-3}	Sample 2 10^{-4}	Sample 2 10^{-5}	Sample 2 10^{-6}

that would result in a confluency of >80 % after overnight incubation (*see* **Note 14**).

8. Dilute cells in RPMI with 10 % FCS in a centrifuge tube or an autoclaved glass bottle, depending on number of plates to be seeded. Pipette to ensure even dispersion of cells (*see* **Note 15**).
9. Pipette 1 ml of diluted cell suspension into each well in 24-well plates (*see* **Note 16**).
10. Leave plaque assay plates in the BSC or the bench top for at least 15 min to allow cells to settle (*see* **Note 17**).
11. Incubate plates overnight at 37 °C with 5 % CO_2 to allow attachment.

3.3 Plaque Assays

1. Perform tenfold serial dilutions of harvested samples in RPMI supplemented with 2 % FCS (*see* **Note 18**). A good starting point would be to prepare dilutions of up to 10^{-6} (*see* **Note 19**).
2. Decant media from each 24-well plate and add in 100 μl of diluted samples into each well (*see* **Note 20**). For each sample, inoculate 2 wells per dilution (duplicates) as shown in Table 1. Include at least three wells for negative controls (inoculated with RPMI with 2 % FCS) during each set of plaque assays.
3. Incubate plaque assay plates for 90 min at 37 °C with 5 % CO_2 to allow virus adsorption.
4. After the incubation period, wash cell monolayers twice with 1 ml of PBS per well (*see* **Note 21**).
5. Add 1 ml of overlay media per well and incubate for approximately 3 days at 37 °C with 5 % CO_2 (*see* **Note 22**).
6. Observe plaques under a light microscope to ensure that plates are ready to be stained (Fig. 1).
7. Remove overlay media and stain plates with crystal violet solution (500 μl/well) for visualization and counting of plaques (*see* **Note 23**). Shake plates on an orbital shaker at room temperature overnight (*see* **Note 24**).

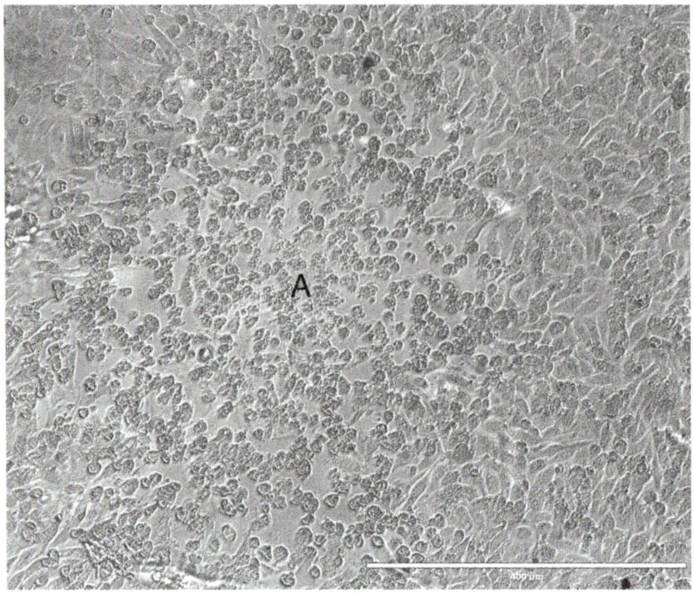

Fig. 1 A CHIKV plaque (A) viewed at 10× magnified corresponds to a localized region of cell death. Plaques at this size are ready to be stained. The scale bar corresponds to 400 μm

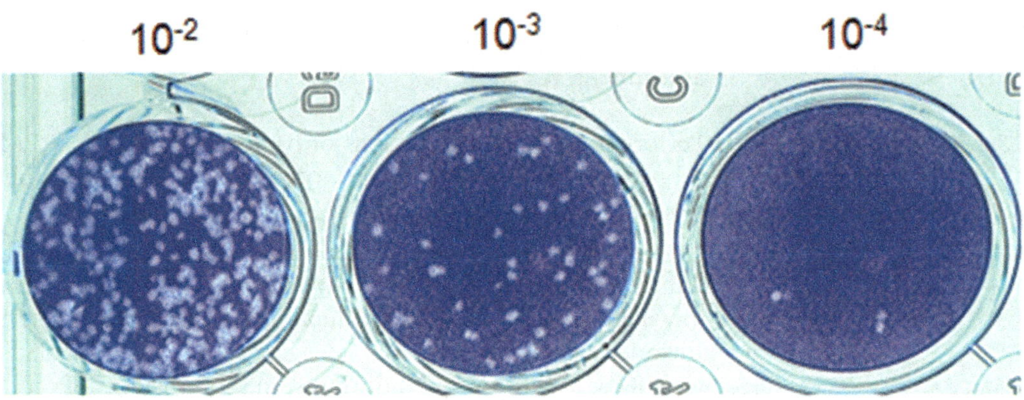

Fig. 2 CHIKV plaques after fixing and staining with crystal violet. Wells with plaques that are well spaced out (10^{-3} below) should be counted

8. Destain plates and wash them under running water to remove residual stains (*see* **Note 25**). Dry plates in a drying incubator or on the benchtop.
9. To determine virus titer, count wells with 10–100 plaques (*see* **Note 26** and Fig. 2). Virus titers can be expressed as plaque forming unit per milliliter (pfu/ml) after adjusting for appropriate dilution factors. For example, if 42 plaques are observed in the well with 10^{-3} dilution, virus titer $= 42 \times 10^3 \times 10 = 4.2 \times 10^5$ pfu/ml (*see* **Note 27**).

4 Notes

1. Aquacide II is the sodium salt of carboxymethyl cellulose (CMC). It has a molecular weight of 500,000 and is highly viscous. If Aquacide II is unavailable, alternative overlay media (e.g. CMC or agarose) with similar viscosity may be used, but further optimization is necessary to determine the appropriate percentage of these alternatives in the final overlay media.
2. Do not dissolve all the powder at once. Aquacide II will not easily dissolve and the formation of large clumps of powder on the sides of the bottle may lead to burnt residues during autoclaving. Instead, add about 1/3 of the powder to water and shake vigorously to mix before adding the next 1/3 and so on. The solution will not be homogenous, and there will be small clumps of powder suspended within the water. These clumps will dissolve during autoclaving to give a viscous solution with a yellow tint.
3. Use 500 ml bottles for this step since an equal volume of media needs to be added to the autoclaved Aquacide.
4. Trypsin undergoes autolysis in solution and slowly loses its enzymatic activity over time, especially upon repeated freeze–thaw cycles and storage at temperatures above −20 °C. Storage at −20 °C allows trypsin-EDTA to be stored for up to 1 year. In addition, aliquoting in small volumes eliminates the need for repeated freeze–thawing.
5. In order to ensure that the enzymatic activity remains at sufficient levels to dislodge cells, prepare small volumes (100 ml) of 1× trypsin-EDTA and store at 4 °C for use in cell culture. Larger volumes should only be prepared if trypsin-EDTA is to be rapidly used up (within 4 weeks).
6. Paraformaldehyde powder is a poison and should be weighed out in the fume hood. If the weighing scale is unable to function well in the fume hood due to the high internal ventilation, an empty capped falcon tube can be tared on the weighing scale first. Paraformaldehyde powder can be added into the falcon tube in the fume hood and the capped falcon tube can be brought to the bench top to be weighed.
7. Paraformaldehyde fumes are toxic and heating of the solution should be done in the fume hood. Ensure that the bottle cap is not tightly capped. Bottle cap should be placed loosely on the mouth of the bottle to allow fumes to escape without resulting in over-evaporation.
8. Crystal violet powder is an irritant, corrosive, carcinogenic, and an environmental toxin. It should be weighed out in the fume hood, as described in **Note 6**.

9. Since CHIKV is an enveloped virus, virus samples can only be frozen and thawed once. Virus titer decreases with every cycle of freeze–thawing. If a virus pool needs to be titered, supernatants should be collected, aliquoted, and frozen in vials. Three vials can be removed, thawed for plaque assays, and the results averaged so as to reflect the titer for that pool.
10. For plaque assays, any cell line that is able to support CHIKV infection and result in plaque formation can be utilized. BHK21 cells are easily cultured with a rapid growth rate and production of well-defined plaques upon CHIKV infection. If BHK21 cells are unavailable, green monkey kidney (Vero) cells can be used instead. If alternative cell lines are used, the cell densities and incubation periods stated in this protocol would have to be modified and specifically optimized to suit the growth rates of these substitute cell lines.
11. The washing step is essential to remove all traces of FCS, which contains inhibitors of trypsin. Add PBS to the bottom corner of the flask and not directly onto the cell monolayer. Directly dispensing PBS onto cells may cause cells to detach from the monolayer. Gently rock the flask back and forth a few times to wash the cells before discarding the PBS.
12. Repeated warming of trypsin-EDTA at 37 °C results in a loss of enzymatic activity. As such, the incubation times required to dislodge cells become progressively longer, even taking longer than 5 min in some cases. During incubation with trypsin-EDTA, the side of the flask should be gently knocked against the palm of the hand to check whether cells are sufficiently trypsinized. If trypsinization is sufficient, cells should detach in clumps upon gentle knocking and the cell monolayer will be visibly disrupted. In addition, cells will be rounded when observed under a light microscope. When this occurs, the flask should be gently knocked until the entire monolayer is dislodged. If trypsinization is insufficient (i.e. gentle knocking does not disturb the monolayer), the incubation time should be lengthened before knocking the flask further. Vigorous banging of the flask should be avoided as this may increase cell death due to mechanical stress. Flasks can also be placed in a 37 °C incubator to speed up trypsinization.
13. BHK21 cells may clump if the cell suspension is not mixed properly. In addition, trypsinizing cells for too long or using freshly prepared trypsin may result in visible clumps upon neutralization. In such cases, pipetting numerous times until the clumps become smaller or nonexistent should suffice to obtain a cell suspension that can be used for seeding.

14. A seeding density of 50,000–70,000 cells per well would be likely to produce a confluency of at least 80 % after overnight incubation.
15. Thorough mixing is necessary to ensure even seeding. If cells are seeded in clumps, cell growth will be observed in patches, with some regions of the well being over-confluent and some regions being too sparse. This may lead to inconsistencies in plaque sizes and inaccuracies in results. As such, it is important to mix the cell suspension well both before and during seeding.
16. The pipetting method has to be optimized to ensure even seeding of cells. Some researchers set the mouth of the serological pipette onto the bottom each well and dispense 1 ml of cell suspension at a relatively fast rate. Others prefer to angle the pipette to the side of the well, employing a steadier rate of pipetting. Yet another method involves slowly dripping the cell suspension into different regions of the well until 1 ml has been dispensed. Pipetting methods differ among researchers and individual researchers should explore varying methods to find one that is easy for them. It is important that cells are seeded evenly, without clumping in the middle or one side of the well to ensure reproducibility and accuracy of results.
17. Moving 24-well plates immediately after seeding may also result in clumping of cells. Leaving them in the BSC to allow cells to settle is ideal. However, the internal airflow of some BSCs produces significant vibrations on the BSC floor, which may result in unnecessary agitation of the cell suspension and cell clumping. In such cases, plates should be taken out and placed on a stable bench top for cells to settle. After cells are settled to the bottom of each well, plates can be viewed under a light microscope to observe seeding density and whether cells are uniformly distributed prior to incubation in the CO_2 incubator.
18. It is essential to mix well between dilutions, either by pipetting or by vortexing. We have found that CHIKV particles tend to be "sticky" and that mixing well allows a more even infection of cells.
19. The dilution range for subsequent experiments may be modified and reduced appropriately (to only 3 or 4 dilutions instead of 6) based on the results obtained from earlier experiments.
20. Again, ensure that samples are mixed well before being added into wells. Also, be sure not to decant too many plates at a time to prevent drying out of cell monolayers. A single pipette tip may be used for inoculating all the dilutions of one sample by starting with the highest dilution and working down to the lowest.

21. When removing PBS during the washing steps, decant PBS into a waste bucket instead of pipetting. This will conserve pipettes and shorten the time taken for the assay. If serological pipettes are used, ensure that PBS is removed from wells with highest dilutions first. Also, avoid using the same serological pipette for more than one sample. This is to prevent carryover of virus particles across different samples, which may make titer results less reliable. Likewise, when dispensing PBS or overlay media, ensure that pipette tip does not touch any of the wells to prevent cross-contamination of samples.
22. The period of incubation is closely tied to the size of the plaques for each CHIKV strain. The longer the incubation time, the larger the plaque size. The ideal incubation period will allow plaques to form large enough to be seen and counted individually upon staining, but not so large as to merge and become difficult to distinguish from one to another. We found a 3-day incubation period to be suitable for all the strains of CHIKV (ECSA genotype) in our lab. If incubation period is to be drastically increased (e.g. 6 days instead of 3 days), the seeding density for 24-well plates needs to be reduced so as to prevent BHK21 cells from dying due to overgrowth. Plaques can be observed under a light microscope to determine the appropriate incubation period (Fig. 1).
23. The amount of crystal violet added does not have to be strictly measured out. As long as there is enough solution to cover the cell layer without drying out, cells will stain well.
24. The crystal violet solution is stored at room temperature and can be recycled. Once the stain starts to look faint on plates after a 1-day incubation, replace with freshly prepared solution. If fresh solution is used, incubation times can be shortened to 2 h.
25. Crystal violet solution can be decanted through a glass funnel into a bottle for reuse. Due to the presence of paraformaldehyde, which serves as a fixative in the crystal violet solution, this step should be done in a fume hood especially if there are many plates to be destained. Any spills of crystal violet solution may be removed with 70 % denatured ethanol.
26. Different CHIKV strains have different plaque sizes. The countable range would therefore need to be modified accordingly. Ideally, wells that are used for counting should have plaques that are well spaced out, and not too few (e.g. <10) in number.
27. Since only 100 μl is inoculated per well, the dilution factor is multiplied by a factor of 10 to reflect virus titer per ml.

References

1. Leland DS, Ginocchio CC (2007) Role of cell culture for virus detection in the age of technology. Clin Microbiol Rev 20:49–78
2. Dulbecco R (1952) Production of plaques in monolayer tissue cultures by single particles of an animal virus. Proc Natl Acad Sci U S A 38:747–752
3. Harrison GE (2001) Making DNA from RNA: the strange life of the retrovirus. In: Carol AG (ed) Operators and promoters: the story of molecular biology and its creators. University of California Press, California, p 297
4. Glasgow GM, McGee MM, Sheahan BJ et al (1997) Death mechanisms in cultured cells infected by Semliki Forest virus. J Gen Virol 78:1559–1563
5. Marsh M, Helenius A (1980) Adsorptive endocytosis of Semliki Forest virus. J Mol Biol 142:439–454
6. Bruton CJ, Kennedy SI (1976) Defective-interfering particles of Semliki Forest virus: structural differences between standard virus and defective-interfering particles. J Gen Virol 31:383–395
7. Percival SL, Wyn-Jones P (2014) Methods for the detection of waterborne viruses. In: Percival SL, Yates MV, Williams D et al (eds) Microbiology of waterborne diseases: microbiological aspects and risks, 2nd edn. Academic, California, pp 458–460
8. Zhang WD, Evans DH (1991) Detection and identification of human influenza viruses by the polymerase chain reaction. J Virol Methods 33:165–189
9. Nadgir SV, Hensler HR, Knowlton ER et al (2013) Fifty percent tissue culture infective dose assay for determining the titer of infectious human herpesvirus 8. J Clin Microbiol 51:1931–1934
10. Delogu I, Pastorino B, Baronti C et al (2011) In vitro antiviral activity of arbidol against Chikungunya virus and characteristics of a selected resistant mutant. Antiviral Res 90: 99–107
11. Gonzalez-Hernandez MB, Bragazzi Cunha J, Wobus CE (2012) Plaque assay for murine norovirus. J Vis Exp 66:e4297
12. Smither SJ, Lear-Rooney C, Biggins J et al (2013) Comparison of the plaque assay and 50% tissue culture infectious dose assay as methods for measuring filovirus infectivity. J Virol Methods 193:565–571
13. Butchaiah G (1988) Infectivity assay of bovine rotavirus: evaluation of plaque and end-point methods in comparison with immunofluorescent cell assay. Acta Virol 32:60–64

Chapter 10

Detection and Quantification of Chikungunya Virus by Real-Time RT-PCR Assay

Seok Mui Wang, Ummul Haninah Ali, Shamala Devi Sekaran, and Ravindran Thayan

Abstract

Real-time PCR assay has many advantages over conventional PCR methods, including rapidity, quantitative measurement, low risk of contamination, high sensitivity, high specificity, and ease of standardization (Mackay et al., Nucleic Acids Res 30:1292–1305, 2002). The real-time PCR system relies upon the measurement of a fluorescent reporter during PCR, in which the amount of emitted fluorescence is directly proportional to the amount of the PCR product in a reaction (Gibsons et al., Genome Res 6:995–1001, 1996). Here, we describe the use of SYBR Green I-based and TaqMan® real-time reverse transcription polymerase chain reaction (RT-PCR) for the detection and quantification of Chikungunya virus (CHIKV).

Key words SYBR Green I, TaqMan®, Real-time RT-PCR, Chikungunya virus, Detection, Quantification

1 Introduction

The diagnosis of CHIKV infection is mostly based on either virus isolation, genomic detection using RT-PCR, or detection of virus-specific antibodies [3, 4]. The presence of viral RNA can be detected by RT-PCR during the early viremic phase of illness (days 0 to 7) [5, 6]. However the classic serological methods such as hemagglutination inhibition, complement binding, immunofluorescence, and ELISA can be done from day 2 or day 3 onwards up to day 15 after the onset of fever [7]. IgM usually can persist for several weeks to 3 months. Currently there is no specific treatment or commercially available vaccine for CHIKV infection; hence an early diagnosis of the disease will allow better patient management, control of outbreaks, and epidemiological studies [1]. Virus quantification will also enable monitoring of disease progression and measuring the efficacy of vaccine candidates.

Real-time PCR allows collection of data as the reaction is proceeding, hence allowing a more accurate diagnostic tool for rapid

Justin Jang Hann Chu and Swee Kim Ang (eds.), *Chikungunya Virus: Methods and Protocols*, Methods in Molecular Biology, vol. 1426,DOI 10.1007/978-1-4939-3618-2_10,

detection and quantification of CHIKV. The principle of real-time PCR is based on the measurement of fluorescence during DNA amplification. The amount of emitted fluorescence is proportional to the amount of PCR product [2]. There are three main fluorescence-monitoring systems for DNA amplification: (1) hydrolysis probes, (2) hybridizing probes, and (3) DNA-binding agents [8]. Hydrolysis probes include TaqMan® probe which utilizes the 5′ exonuclease activity of *Taq* polymerase [9], while hybridizing probes include molecular beacons [10] and scorpions [11, 12]. The simplest and most economical format is SYBR Green I dye, which binds specifically to double-stranded DNA.

This chapter describes the protocols of SYBR Green I-based and TaqMan® real-time RT-PCR for the detection and quantification of CHIKV using primers targeting the E1 structural gene region. The major difference between the TaqMan® and SYBR Green I dye detection chemistries is that the TaqMan® chemistry uses a fluorogenic probe to enable the detection of a specific PCR product as it accumulates during PCR cycles. It requires tailor-made sequences which are expensive. Conversely, SYBR Green I dye chemistry will detect all double-stranded DNA, including the nonspecific reaction products and primer dimers. It is cheaper and simpler to use and can be easily applied to already establish PCR assays. However, as SYBR Green I binds to any double-stranded DNA, the assays require careful optimization of the PCR conditions and a clear differentiation between specific and nonspecific PCR products using analysis of melting curves. The choice of detection chemistry used in real-time PCR depends on the overall assay design, cost, and feasibility.

Both real-time PCR methods described in this chapter have been tested for CHIKV isolated from human, monkey, and mosquito samples [13]. The assay could serve as an excellent epidemiological tool for laboratory diagnosis and confirmation of CHIKV infection. In addition, it could also be used to study the association of viral load with disease severity.

2 Materials

All biological specimens should be considered potentially infectious and should be handled in accordance with the appropriate national and local biosafety practices. All solutions must be prepared using ultrapure water and molecular grade reagents.

2.1 Oligonucleotide Sequences

1. PCR primer sequences: CHIKV forward primer (5′-CTCATACCGCATCCGCATCAG-3′), and CHIKV reverse primer (5′-ACATTGGCCCCACAATGAATTTG-3′) (*see* **Note 1**; [13]).
2. T7 promoter sequence (5′-TAATACGACTCACTATAGGGC TCATACCGCATCCGCATCAG-3′) (*see* **Note 1**; [13]).

3. TaqMan® probe sequence (5′-HEX-TCCTTAACTGTGA CGGCATGGTCGCC-BHQ-3′) (*see* **Note 1**; [13]).

2.2 Viral RNA Extraction

1. CHIKV-infected cell culture supernatant.
2. AccuPrep® Viral RNA Extraction Kit (Bioneer) or any commercially available viral RNA extraction kit.
3. RNaseZap Wipes and DNAZap.

2.3 Conventional RT-PCR

1. AccessQuick™ RT-PCR System (Promega) or any commercially available RT-PCR reagents.
2. PCR thermal cycler.

2.4 Agarose Gel Electrophoresis and Purification of PCR Product

1. NuSieve® 3:1 Agarose.
2. Tris-borate-EDTA buffer (1×): 89 mM Tris-borate, 2 mM EDTA.
3. Red Safe™ Nucleic Acid Staining Solution.
4. 100 bp DNA ladder.
5. 6× Loading dye.
6. Agarose gel electrophoresis system.
7. QIAquick PCR Purification Kit (Qiagen) or any commercially available nucleic acid purification kit.

2.5 Synthesis of RNA by In Vitro Transcription

1. PCR product of CHIKV.
2. MAXIscript® In Vitro Transcription Kit (Ambion) or any commercially available in vitro transcription (IVT) kit.
3. 5 M Ammonium acetate.
4. 100% Ethanol.
5. 70% Ethanol.
6. Nuclease-free water.

2.6 One-Step SYBR Green I Real-Time RT-PCR

1. iScript™ One-Step RT-PCR Kit with SYBR® Green premix (Bio-Rad) or any commercially available one-step RT-PCR kit.
2. Real-time PCR thermal cycler.

2.7 One-Step TaqMan® Real-Time RT-PCR

1. QuantiTect® Probe RT-PCR Kit (Qiagen) or any commercially available one-step RT-PCR kit.
2. Real-time PCR thermal cycler.

3 Methods

Carry out all procedures at room temperature unless otherwise specified. Prepare RT-PCR reaction mixture in a PCR/UV work station. Clean the work bench and pipettes with RNaseZap Wipes and DNAZap to remove all potential RNase and DNA contamination (*see* **Note 2**; [14]).

3.1 Viral RNA Extraction

1. Extract RNA from CHIKV-infected culture supernatant by using AccuPrep® Viral RNA Extraction Kit (Bioneer).
2. Follow the method as described in the kit insert (*see* **Note 3**). Store extracted RNA at −70 °C.

3.2 Preparation of PCR Product for In Vitro Transcription

1. Prepare PCR product from extracted CHIKV RNA by using AccessQuick™ RT-PCR System. Use CHIKV reverse primer for reverse transcription, while T7 promoter sequence and CHIKV reverse primer set for PCR amplification.
2. Prepare each PCR reaction in a 25 μl reaction volume containing 5 μl of RNA template, 0.25 μl of RNA transcriptase, 12.5 μl of AccessQuick™ Master Mix 2×, 0.2 μM of each primer, and nuclease-free water to bring the volume to 25 μl.
3. Quick spin the PCR tubes in the mini microcentrifuge and load them into a PCR thermal cycler. The thermal profile consists of a 30-min reverse transcription step at 50 °C, 15 min of polymerase activation at 95 °C, followed by 35 cycles of PCR at 95 °C of denaturation for 30 s, 55.8 °C of annealing for 30 s, 72 °C of extension for 30 s, and a final extension at 72 °C for 10 min.
4. Upon completion of the PCR run, store the PCR products at −20 °C or proceed to analysis by agarose gel electrophoresis.

3.3 Agarose Gel Electrophoresis and Purification of PCR Product

1. Prepare a 3% (W/V) agarose gel by mixing 3 g of agarose in 100 ml of 1× Tris-borate-EDTA buffer (TBE). Dissolve the agarose by heating using a microwave oven or a hot plate. Once all agarose has completely dissolved, add 2 μl of Red safe dye into the dissolved gel and mix gently. Pour the gel into the mold and immediately place the sample well-forming comb in position.
2. When the gel has set, remove the comb and place the gel in the electrophoresis apparatus. Add sufficient buffer to fill the electrode chambers and cover the gel.
3. Mix 5 μl of PCR products with 1 μl of 6× loading dye on a parafilm, and load the PCR mixture into individual wells. Include 100 bp DNA ladder as molecular weight marker. Run the gel at 80 V for 1 h or until the dye front has reached the bottom of the gel.
4. Stop the electrophoresis run, transfer the gel into a gel documentation system, and capture image of the gel. The RT-PCR reaction will result in a single PCR product of 149 bp (*see* **Note 4**).
5. Following RT-PCR, purify the PCR product by using QIAquick® PCR Purification Kit (Qiagen). Follow the method

as described in the kit insert (*see* **Note 3**). Store purified PCR product at −20 °C. The purified PCR product can then be used as DNA template for subsequent IVT (*see* **Note 5**).

3.4 In Vitro Transcription

1. Perform IVT step by using MAXIscript® In Vitro Transcription Kit (Ambion). Follow the method as described in the kit insert.
2. Add 1 μg of purified CHIKV DNA template into a transcription reaction mixture containing 2 μl of 10× T7 transcription buffer, 4 μl of rNTPs (10 mM ATP, CTP, GTP, UTP), 2 μl of T7 enzyme mix, and nuclease-free water to bring the volume to 20 μl (*see* **Note 6**). Gently mix the reaction and incubate at 37 °C for 1 h (*see* **Note 7**).
3. Add 1 μl of DNase I (2 U/μl) to the IVT reaction and incubate for 15 min to remove the remaining DNA template.
4. Add 1 μl of 0.5 M EDTA. Incubate at 65 °C for 10 min to heat inactivate the DNase I. Then place the IVT reaction on ice for 1 min.

3.5 Purification of RNA Transcripts

1. RNA transcripts can be purified by either ammonium acetate/ethanol precipitation or by using QIAquick® PCR Purification Kit to remove potentially inhibitory or interfering components.
2. For precipitation with ammonium acetate/ethanol, add 30 μl of nuclease-free water to IVT reaction to bring the volume to 50 μl. Add 5 μl of 5 M ammonium acetate and vortex to mix.
3. Add 3 volumes of ice-cold 100% ethanol and chill the solution at −20 °C for 30 min or longer.
4. Centrifuge for 15 min at maximum speed in a 4 °C centrifuge. Decant the supernatant.
5. Wash the pellet once by adding 500 μl ice-cold 70% ethanol. Centrifuge for 15 min at maximum speed in a 4 °C centrifuge. Decant the supernatant.
6. Air-dry RNA pellet at room temperature or vacuum dry at low heat using DNA Speed Vac (Savant).
7. Resuspend RNA pellet in 50 μl nuclease-free water (*see* **Note 8**). Keep RNA transcripts on ice if proceeding to measuring RNA concentration. Otherwise, store RNA transcripts at −70 °C (*see* **Note 9**).

3.6 Calculation of Total RNA Yield

1. Determine the concentration of RNA transcripts by measuring its absorbance at 260 nm (A260) in a spectrophotometer (*see* **Note 10**).
2. Calculate RNA concentration using the following equation:

$$\text{RNA concentration}\left(\mu g/ml\right) = A_{260} \times 40 \mu g/ml \times \text{dilution factor}$$

3. An RNA sample is considered to be relatively free of protein if the $A_{260/280}$ value ≥2.0 (*see* **Note 11**). An absorbance of 1 U at 260 nm corresponds to 40 μg/ml of RNA per ml.
4. Store RNA transcripts in 2 μl aliquots at −70 °C. Each aliquot is sufficient to construct a standard curve for each PCR run (*see* **Note 12**).

3.7 Determination of RNA Copy Number for In Vitro-Transcribed RNA

1. Calculate the copy number of RNA transcripts by using the following equation:

$$\text{RNA copies}/\mu\text{l} = \frac{6.022\times10^{23}\left[\text{copies}/\text{mol}\right]\times\text{concentration}\left[\text{g}/\mu\text{l RNA}\right]}{\text{MV}\left[\text{g}/\text{mol}\right]}$$

where

$$\text{Molecular weight for single-stranded RNA, MV} = (\text{transcript length in nucleotides, base})\times(340\ \text{Da}/\text{base})$$

2. The copy number of RNA transcripts needs to be determined prior to construction of a standard curve. It is calculated based on RNA concentration measured and its molecular weight.
3. Real-time RT-PCR done on RNA transcripts using CHIKV forward and reverse primers set will result in a single PCR product of 129 bp.

3.8 Construction of RNA Standard for Absolute Quantification

1. Use RNA transcripts prepared in Subheading 3.6. Make a tenfold dilution series of at least five different concentrations (*see* **Notes 13** and **14**).
2. Subject each dilution to real-time RT-PCR. The choice of detection chemistry of real-time PCR depends on the assay design, cost, and feasibility (*see* **Note 15**).

3.9 Absolute Quantification by SYBR Green I-Based Real-Time RT-PCR

1. Perform one-step SYBR Green I-based real-time RT-PCR by using iScript™ One-Step RT-PCR Kit with SYBR® Green premix.
2. Prepare each PCR reaction in a 25 μl reaction volume containing 5 μl of RNA transcripts, 0.25 μl of RNA transcriptase, 12.5 μl of 2× iScript™ SYBR® Green premix, and 0.2 μM of each primer (CHIKV forward primer and CHIKV reverse primer). The concentrations of buffer, dNTPs, *Taq* polymerase, and Mg^{2+} are as recommended by the manufacturer. The thermal profile consists of a 30-min reverse transcription step at 50 °C, 15 min of polymerase activation at 95 °C, followed by 40 cycles of PCR at 95 °C of denaturation for 30 s,

55.8 °C of annealing for 30 s, and 72 °C of extension for 30 s. Collect fluorescence data during the extension step (*see* **Note 16**). The melt curve analysis is carried out from 65 to 95 °C with an increment of 0.5 °C for 5 s. Include a no-template control (NTC) and a positive control (CHIKV) for each PCR run (*see* **Note 17**).

3. After PCR, construct a standard curve by plotting threshold cycle (Cq)/crossing points of different standard dilutions against log of RNA copy number (Fig. 1). A standard curve for quantification assay should have a correlation coefficient of ≥0.95 and a slope between –3.72 and –3.32 (*see* **Note 18**; [15]).

3.10 Absolute Quantification by TaqMan® Real-Time RT-PCR

1. Perform one-step TaqMan® real-time RT-PCR by using QuantiTect® probe RT-PCR Kit.
2. Prepare each PCR reaction in a 25 μl reaction volume containing 5 μl of RNA transcripts, 0.25 μl of RNA transcriptase, 12.5 μl of QuantiTect® Probe premix, 0.2 μM of each primer (CHIKV forward and CHIKV reverse primers), and 0.2 μM of TaqMan® probe. The concentrations of buffer, dNTPs, *Taq* polymerase, and Mg^{2+} are as recommended by the manufacturer. The thermal profile consists of a 30-min reverse transcription step at 50 °C, 15 min of polymerase activation at 95 °C, followed by 40 cycles of PCR at 95 °C of denaturation for 30 s, and 58.7 °C of annealing/extension for 60 s. Collect fluorescence data during the annealing/extension step (*see* **Note 19**). Include a no-template control (NTC) and a positive control (CHIKV) for each PCR run.
3. After PCR, generate a standard curve by plotting a threshold cycle (Cq)/crossing points of different standard dilutions against log of RNA copy number (Fig. 2). A standard curve for quantification assay should have a correlation coefficient of ≥0.95 and a slope between –3.72 and 3.32 (*see* **Note 18**; [15]).

3.11 Quantification of CHIKV Viral Load from an Unknown Sample

1. Extract CHIKV RNA from sample (human sera, monkey sera, adult mosquitoes, etc.) using viral RNA extraction kit following the instructions provided by the manufacturer. Store extracted RNA at –70 °C.
2. Set up PCR reactions for unknown samples, negative control (no-template control), positive control (CHIKV), and serially diluted RNA transcripts. Perform real-time RT-PCR.
3. After PCR, construct a standard curve as described above. The amount of unknown target should fall within the range tested.
4. A sample is diagnosed as CHIKV positive if its cycle number crosses the fluorescence cycle threshold and its *Tm* is similar to that of CHIKV positive control. The viral load of the unknown sample can be extrapolated from the standard curve.

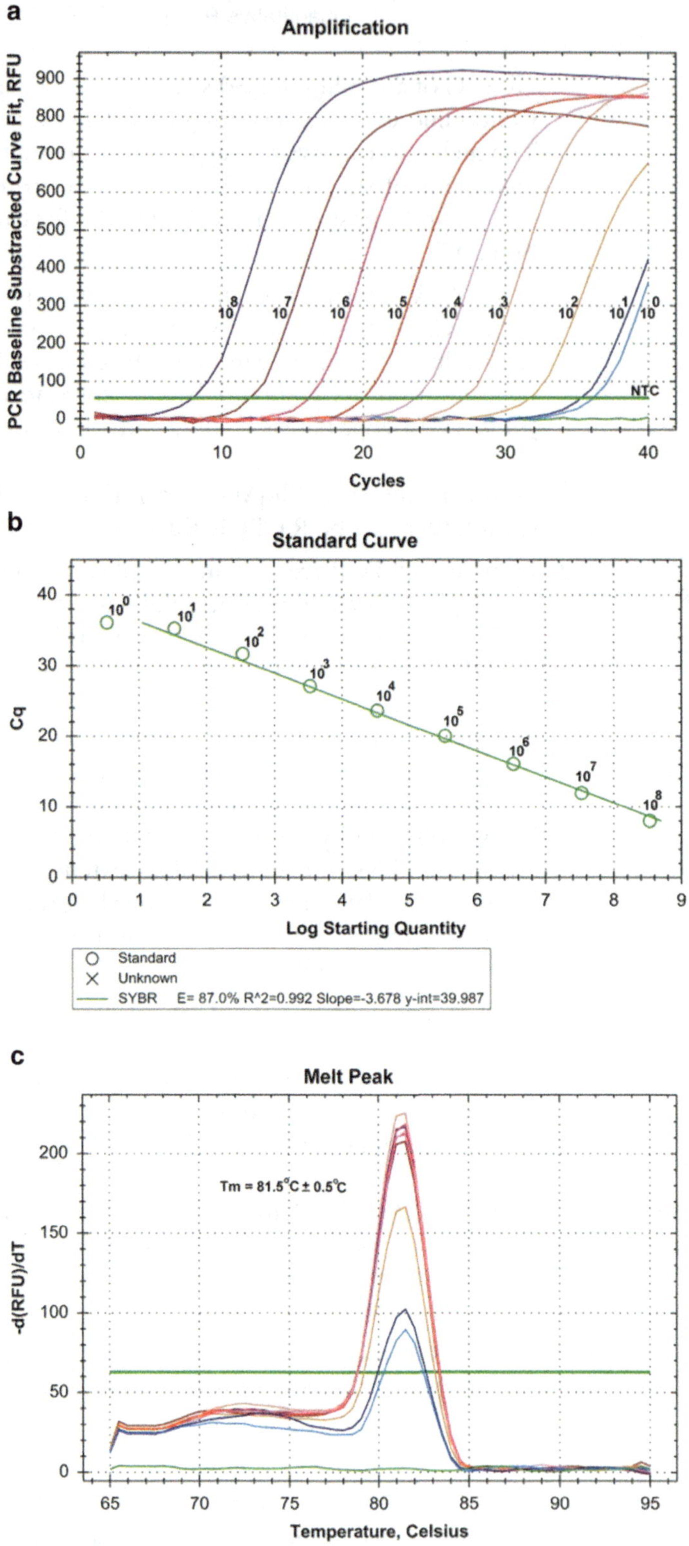

Fig. 1 A representative of SYBR Green I-based real-time RT-PCR assay for detection of CHIKV E1 gene. (**a**) Sensitivity of real-time RT-PCR assay as shown in the amplification plot from left to right is curves of decreasing concentration of CHIKV RNA ranging from 3.5×10^8 to 3.5×10^0 RNA copies/μl in a serial tenfold dilution;

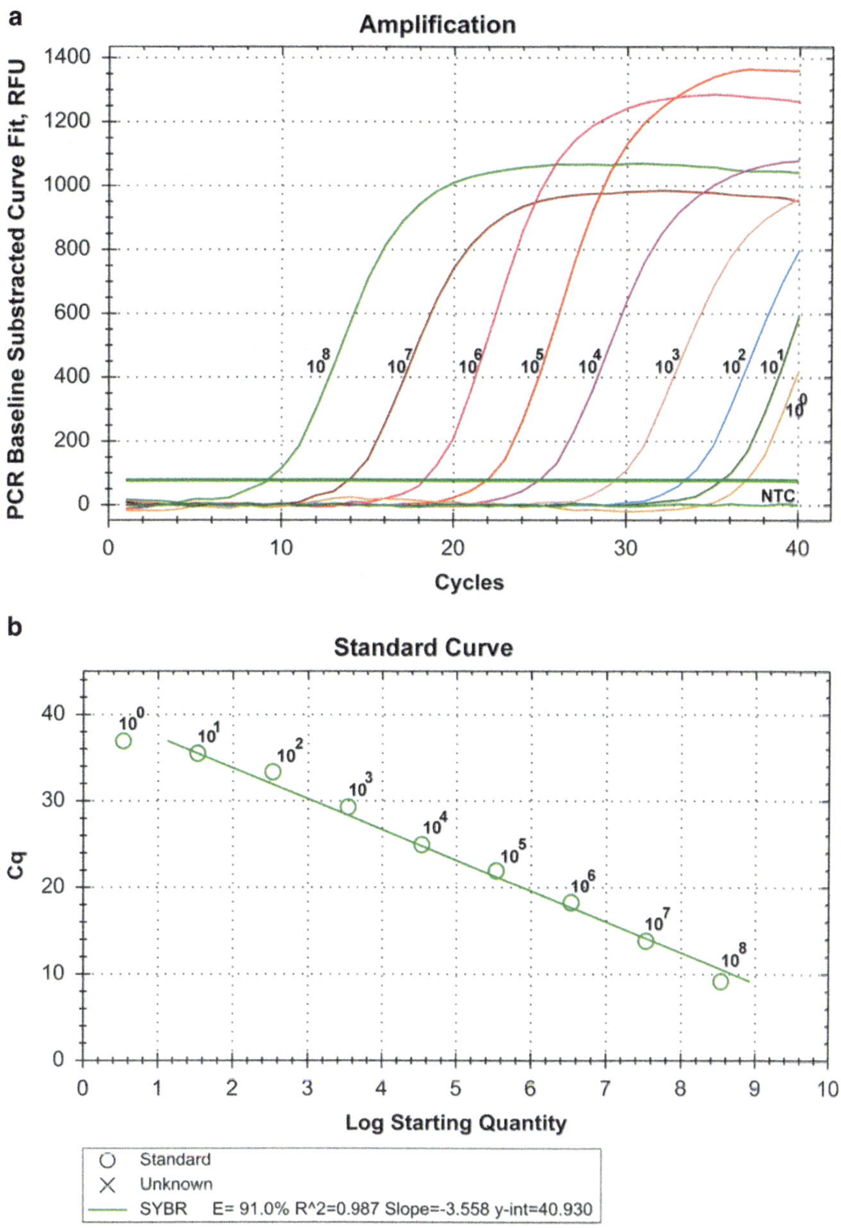

Fig. 2 A representative of TaqMan® real-time RT-PCR assay for detection of CHIKV E1 gene. (**a**) Sensitivity of real-time RT-PCR assay as shown in the amplification plot from left to right is curves of decreasing concentration of CHIKV RNA ranging from 3.5×10^8 to 3.5×10^0 RNA copies/μl in a serial tenfold dilution; NTC: no-template control. (**b**) The standard curve is generated from the Cq values obtained against the known concentration of tenfold serially diluted CHIKV RNA. The standard curve shows a correlation coefficient of 0.987 and detection limit of 3.5 RNA copies/μl

Fig. 1 (continued) NTC: no-template control. (**b**) The standard curve was generated from the Cq values obtained against the known concentration of tenfold serially diluted CHIKV RNA. The standard curve shows a correlation coefficient of 0.992 and detection limit of 3.5 RNA copies/μl. (**c**) Melting curve analysis depicting dissociation plot

4 Notes

1. The E1 gene sequence of MY019IMR/06/BP CHIKV prototype strain (GenBank accession no. EU703761) is used for primer and probe design. The potential primers are selected by identifying highly conserved regions among 25 different CHIKV genotypes by aligning their E1 gene sequences using the ClustalW 1.83 program in FastPCR software (Microsoft Office Software, Biology Software Net). The specificity of selected primers and probe is checked by BLAST analysis against NCBI nucleotide database of CHIKV and related alphaviruses such as O'NyongNyong virus (Gulu strain, GenBank accession no. M20303.1), Ross River virus (8961 strain, GenBank accession no. GQ433357.1), Semliki Forest virus (L10 strain, GenBank accession no. AY112987), and Sindbis virus (GenBank accession no. J02363.1) [13]. Other than the primers and probe sequences listed here, other published primers can also be used in this PCR assay [16–18]. Researchers can opt to design their own primer sequences. Ideally, an amplicon length for real-time PCR should be 50–150 bp for optimal PCR efficiency.
2. To get started, it is helpful to refer to the Minimum Information for Publication of Quantitative Real-time PCR Experiments (MIQE) Guidelines [14].
3. For spin-column-based RNA extraction kit or PCR purification kit, drying the column at 80 °C oven for 5 min prior to elution helps to remove residual alcohol or salts carried over from previous washing steps. It also helps to increase the product yields.
4. The primers are designed to be highly specific and produce a single amplicon only. If there are other bands present, the best option is to excise the 149 bp band and continue with PCR product purification using gel purification protocol which is available from Qiagen.
5. Failure of the IVT reaction might be due to poor-quality DNA template. DNA prepared with standard miniprep procedures should be of a sufficient quality for IVT. However, contaminants such as alcohol or salts carried over from DNA purification process can inhibit the RNA polymerase. Typically, precipitating the DNA template with ethanol followed by resuspension will resolve the contamination issues.
6. To prevent RNase from degrading in vitro-transcribed RNA, use an RNase inhibitor such as RNasin® Ribonuclease Inhibitor (Promega) in the transcription reactions.
7. Lowering the temperature to 16 °C can sometimes improve transcription. Lower reaction temperature slows down the

progression of RNA polymerase, thereby preventing the RNA transcripts from forming secondary structures.

8. RNA transcripts synthesized can be examined for potential DNA contamination by subjecting it to RT-PCR analysis. Perform RT-PCR in the absence of reverse transcriptase enzyme (no-RT control). RNA transcripts with contaminating DNA will yield PCR product in the "no-RT" control, whereas pure RNA transcripts will only yield PCR product in the presence of reverse transcriptase enzyme.
9. It is recommended to determine the concentration of RNA transcripts immediately after IVT procedure. Multiple freeze-thawing cycles may lead to RNA degradation.
10. RNA transcripts must be concentrated in order to measure an accurate A_{260} value.
11. The overall quality of RNA yield may be assessed by running an aliquot of the transcription reaction on a 2% denaturing agarose gel (containing formaldehyde, glyoxal, or 8 M urea). It is important to check the integrity of your RNA preparation because degraded RNA does not perform well in downstream applications.
12. The RNA transcripts should be aliquoted into a suitable volume (e.g., 2 μl per tube), stored at −70 °C, and thawed only once before use. Do not keep the remaining unused RNA transcripts for future applications as multiple freeze-thawing cycles may degrade RNA integrity.
13. RNA transcripts must be diluted 10^6–10^{12}-fold to be at concentrations similar to the target in biological samples. In general, it is best to use an RNA standard series with copy numbers ranging from 10^0 to 10^{10} when constructing the standard curve.
14. Absolute quantification allows quantifying unknown samples by interpolating their quantity from a standard curve. Accurate pipetting is required when diluting the standards over several orders of magnitude.
15. SYBR Green I is cost effective and relatively simple to manipulate although it has a slightly lower sensitivity and specificity compared to TaqMan® probe. TaqMan real-time PCR uses an extra probe that is highly specific for the detection of the chosen region, but it requires being custom-made and is therefore pricier.
16. Ensure that the correct detection channel is activated or the correct filter set is chosen for SYBR Green I.
17. Following amplification, perform a melting curve analysis to verify the authenticity of the amplified product by its specific melting temperature (*T*m) with the melting curve analysis

software of the real-time thermal cycler according to the instructions of the manufacturer. The *Tm* of CHIKV is 81.5 °C ± 0.5 °C.

18. In general, the viral load determined by real-time PCR is approximately 1000–5000-fold higher than the number of infectious particles that is determined by plaque assay [19]. The difference may be due to the production of defective non-infectious particles which are detected by PCR assay but not by plaque assay.
19. Ensure that the correct detection channel is activated or the correct filter set is chosen for the reporter dye.

References

1. Mackay IM, Arden KE, Nitsche A (2002) Real-time PCR in virology. Nucleic Acids Res 30:1292–1305
2. Gibsons UE, Heid CA, William PM (1996) A novel method for real time quantitative RT-PCR. Genome Res 6:995–1001
3. Bronzoni RV, Moreli ML, Cruz AC, Figueiredo LT (2002) Multiplex nested PCR for Brazilian Alphavirus diagnosis. Trans Royal Soc Trop Med Hyg 98:456–461
4. Hasebe F, Parquet MC, Pandey BD, Mathenge EG, Morita K, Balasubramaniam V, Saat Z, Yusop A, Sinniah M, Natkunam S, Igarashi A (2002) Combined detection and genotyping of Chikungunya virus by specific reverse transcription-polymerase chain reaction. J Med Virol 67:370–374
5. Pfeffer M, Linssen B, Parke MD, Kinney RM (2002) Specific detection of Chikungunya virus using a RT-PCR/nested PCR combination. J Vet Med B Infect Dis Vet Public Health 49:49–54
6. Pastorino B, Bessaud M, Grandadam M, Murri S, Tolou HJ, Peyrefitte CN (2005) Development of TaqMan RT-PCR assay without RNA extraction step for detection and quantification of African chikungunya viruses. J Virol Methods 124:65–71
7. Thein S, Linni ML, Aaskov J, Aung MM, Aye M, Zaw A, Myint A (1992) Development of a simple indirect enzyme-linked immunosorbent assay for the detection of immunoglobulin M antibody in serum from patients following an outbreak of Chikungunya virus infection in Yangon, Myanmar. Trans R Soc Trop Med Hyg 86:438–442
8. Van der Velden VHJ, Hochhaus A, Cazzaniga G, Szczepanski T, Gabert J, van Dongen JJM (2003) Detection of minimal residual disease in hematologic malignancies by real-time quantitative PCR: principles, approaches, and laboratory aspects. Leukemia 17:1013–1034
9. Heid CA, Stevens J, Livak KJ, Williams PM (1996) Real-time quantitative PCR. Genome Res 6(10):986–994
10. Abravaya K, Huff J, Marshall R, Merchant B, Mullen C, Schneider G et al (2003) Molecular beacons as diagnostic tools: technology and applications. Clin Chem Lab Med 41(4):468–474
11. Svanvik N, Ståhlberg A, Sehlstedt U, Sjöback R, Kubista M (2000) Detection of PCR products in real time using light-up probes. Anal Biochem 287(1):179–182
12. Solinas A, Brown LJ, McKeen C, Mellor JM, Nicol J, Thelwen N et al (2001) Duplex scorpion primer in SNP analysis and FRET applications. Nucleic Acids Res 29(20):e96–e110
13. Ali UH, Vasan SS, Thayan R, Angamuthu C, Lim LL, Sekaran SD (2010) Development and evaluation of a one-step SYBR-Green I-based real-time RT-PCR assay for the detection and quantification of Chikungunya virus in human, monkey and mosquito samples. Trop Biomed 27(3):611–623
14. Bustin SA, Benes V, Garson JA, Hellemans J, Huggett J, Kubista M, Mueller R, Nolan T, Pfaffl MW, Shipley GL, Vandesompele J, Wittwer CT (2009) The MIQE guidelines: minimum information for publication of quantitative real-time PCR experiments. Clin Chem 55(4), doi: 10.1373/clinchem.2008.112797
15. Argaw T, Ritzhaupt A, Wilson CA (2002) Development of a real-time quantitative PCR assay for detection of porcine endogenous retrovirus. J Virol Methods 106:97–106
16. Arumugam G, Elumalai EK, Jeikarsanthosh D, Jayanthy M, Saroja V, Kamatchiammal S (2011) Isolation and detection of Chikungunya

virus from patients serum samples using RT-PCR and real-time RT-PCR. Int J Pharm BiolSci Arch 2(4):1162–1166

17. Edwards CJ, Welch SR, Chamberlain J, Hewson R, Tolley H, Cane PA, Lloyd G (2007) Molecular diagnosis and analysis of Chikungunya virus. J Clin Virol 39:271–275
18. Santhosh SR, Parida MM, Dash PK, Pateriya A, Pattnaik B, Pradhan HK, Tripathi NK, Ambuj S, Gupta N, Saxena P, Lakshmana Rao PV (2007) Development and evaluation of SYBR Green I-based one-step real-time RT-PCR assay for detection and quantification of Chikungunya virus. J Clin Virol 39:188–193
19. Bae HG, Nitsche A, Teichmann A, Biel SS, Niedrig M (2003) Detection of yellow fever virus: a comparison of quantitative real-time PCR and plaque assay. J Virol Methods 110:185–191

Chapter 11

Chikungunya Virus Infection of *Aedes* Mosquitoes

Hui Vern Wong, Yoke Fun Chan, I-Ching Sam, Wan Yusof Wan Sulaiman, and Indra Vythilingam

Abstract

In vivo infection of mosquitoes is an important method to study and characterize arthropod-borne viruses. Chikungunya virus (CHIKV) is a mosquito-borne alphavirus that is transmitted primarily by *Aedes* mosquitoes. In this chapter, we describe a protocol for infection of CHIKV in two species of *Aedes* mosquitoes, *Aedes aegypti* and *Aedes albopictus*, together with the isolation of CHIKV in different parts of the infected mosquito such as midgut, legs, wings, salivary gland, head, and saliva. This allows the study of viral infection, replication and dissemination within the mosquito vector.

Key words Chikungunya virus, *Aedes aegypti*, *Aedes albopictus*, Oral infection, Blood meal, Alphavirus, Arbovirus

1 Introduction

CHIKV is a mosquito-borne alphavirus transmitted to humans by *Aedes* mosquitoes, primarily *Aedes aegypti* and *Aedes albopictus* [1–3]. CHIKV was first identified in Tanzania in 1952, following isolation from human sera, captive mosquitoes that fed on patients, and wild mosquitoes caught in an epidemic area [4]. Experimental infection of mosquitoes with CHIKV enables study of how the virus disseminates and replicates within their vectors, which may help in development of control measures.

We describe the experimental infection of *Aedes* mosquitoes with CHIKV through an infected blood meal. The midgut and saliva are the key entry and exit points in a mosquito for CHIKV and other arthropod-borne viruses. As a virus-infected blood meal enters the mosquito, it reaches and infects the midgut epithelial cells, the primary replicating site. Dissemination to other secondary organs follows when the virus reaches the hemolymph, as it is flushed throughout the hemocoel. To enable successful onward transmission to a mammalian host, the virus must reach and replicate in the salivary glands, and eventually be shed in the saliva.

Justin Jang Hann Chu and Swee Kim Ang (eds.), *Chikungunya Virus: Methods and Protocols*, Methods in Molecular Biology, vol. 1426,DOI 10.1007/978-1-4939-3618-2_11,

During the infection, the virus will be challenged with numerous stochastic events including bottleneck effects that will significantly reduce the viral titer and population [5, 6].

Isolation or detection of CHIKV can be performed in different parts of the mosquito, including wings, legs, saliva, salivary glands, midgut, head, and ovaries [3, 7–9]. Virus titers vary among different parts of the mosquito. Based on our experience with mosquito infection, the midgut often has the highest titer as it is the initial replicating site, while the saliva usually contains a much lower amount of excreted virus. In most cases, the virus titers range from 10^2 to 10^8 pfu/ml. Infection of mosquitoes is affected by the mosquito species, and strain variation in mosquito and virus. Performing an oral infectious dose (OID_{50}) experiment will give the optimal titer to be used in each experiment for a given mosquito–virus combination [7].

Mosquito dissection to obtain different organs is generally simple although it does require good hand–eye coordination and a stereoscopic microscope. Practicing is recommended before starting an actual study. This chapter describes the infection of mosquitoes with CHIKV, along with procedures for harvesting different organs of the mosquito for virus isolation.

2 Materials

2.1 Blood Meal Containing CHIKV

1. Blood known to be negative for CHIKV RNA and neutralizing antibodies.
2. Virus suspension according to the OID_{50} needed [5], or ranging between 10^4 and 10^7 pfu/ml.
3. Collagen membrane.
4. Scissors.
5. Meal reservoir.
6. Hemotek feeder.

2.2 Mosquito Infection

1. 250 Female 3-day-old *Aedes* mosquitoes (*Ae. aegypti* or *Ae. albopictus*) starved for 24 h prior to blood meal (*see* **Note 1**).
2. Paper cups set-up for mosquito infection: paper cup, nylon netting, glove, masking tape, and rubber band (Fig. 1).
3. Glove box (Fig. 1).
4. Hemotek feeder set.
5. Ice.
6. Forceps.
7. Funnel.

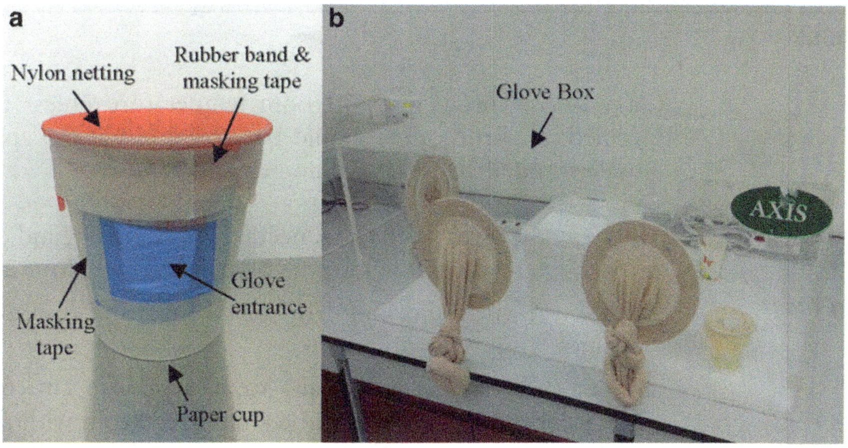

Fig. 1 (a) Paper cup. **(b)** Glove box

8. Cotton.
9. Distilled water.
10. Environmental chamber.
11. 10 % Sucrose supplemented with two tablets of vitamin B complex.

2.3 Harvesting CHIKV from Mosquitoes

1. 70 % Alcohol.
2. Scalpel.
3. Applicator sticks with minute pins attached (dissecting needles).
4. Forceps.
5. Microcentrifuge tubes.
6. 1.5 mm zirconium beads.
7. Serum-free minimal essential medium, MEM.
8. 10 μl non filtered pipette tips.
9. Petri dish.
10. PCR tubes.
11. Phosphate-buffered saline (1× PBS).
12. Stereoscopic microscope.

2.4 Processing of Harvested Parts of Mosquito

1. Homogenizer.

3 Methods

Carry out all procedures at room temperature unless otherwise specified. All procedures should be carried out at the appropriate biosafety and arthropod containment levels.

3.1 Preparation of Blood Meal Containing Virus

1. Thaw the virus suspension needed for infection, and dilute the virus at 1:10 volume into the blood used (*see* **Note 2**).
2. Mix the virus and blood by gently inverting the tubes 7–10 times.
3. Cut the collagen membrane to an area of 6 cm × 6 cm and place it on top of the Hemotek meal reservoir, with the rough surface facing away from the meal reservoir and secure it using the "O" ring provided.
4. Plug the feeder into the PS5 Power Unit.
5. Pipette 1 ml of the blood mixture into each of the two ports.
6. Seal the ports with the plastic plugs provided.
7. Attach the reservoir to the heat transfer plate at the bottom of the feeder.
8. Place the feeder on top of the cup containing the mosquitoes (*see* **Note 3**).

3.2 Mosquito Infection

1. Place 50 female 3-day-old *Aedes* mosquitoes into each of the paper cups prepared.
2. Place the cup into the glove box and carefully put the pre-heated feeder on top of the netting of the cup (*see* **Note 4**).
3. Turn off lighting and leave the feeding process in the dark for 1 h. Maintain the blood meal at 37 °C throughout the feeding period (*see* **Note 5**).
4. After the feeding period, gently remove the feeder and place one cup into a −20 °C freezer for exactly 30 s to snap freeze and immobilize the mosquitoes. Immediately place the cup on ice after snap freezing (*see* **Note 6**).
5. Move everything into another glove box before sorting of mosquitoes.
6. Remove the netting from the paper cup and transfer the blood-fed mosquito (those with engorged abdomens) into another new paper cup using forceps (*see* **Note 7**).
7. Seal the cup's glove entrance with masking tape to prevent mosquitoes from escaping (Fig. 1).
8. Insert ten mosquitoes into each new cup (*see* **Note 8**).

9. Document the amount of all blood-fed mosquitoes and unfed mosquitoes.
10. Kill the remaining unfed mosquitoes by freezing them (*see* **Note 9**).
11. Repeat **steps 4–10** for each cup until all blood-fed mosquitoes are collected.
12. Wet a small piece of cotton with distilled water, and place it on top of the netting of each cup of blood-fed mosquitoes (*see* **Note 10**).
13. Place all the cups into a transparent plastic container and keep them in an environmental chamber. The chamber should be set at 28 ± 1 °C with 88 % relative humidity and a 12:12 h photoperiod.
14. Throughout the infection period, all mosquitoes are to be fed 10 % sucrose supplemented with vitamin B complex. Place a piece of cotton dipped in the sugar solution on each paper cups. Replace the cotton everyday (*see* **Note 11**).
15. Document the number of dead mosquitoes and remove them through the glove entrance every day with forceps in the glove box (*see* **Note 12**).

3.3 Harvesting and Processing of Infected Mosquito Parts

Harvesting CHIKV isolates from different parts of each mosquito requires the use of clean dissecting needles soaked in 70 % alcohol for dissection (*see* **Note 13**). At each planned time point, remove one cup. Remove the dead mosquitoes and snap freeze the remaining live mosquitoes. Transfer each mosquito into an individual microcentrifuge tube and place the tubes on ice.

1. To harvest wings, use a scalpel and forceps, gently remove the wings of the mosquito and place them into individual tubes prefilled with 1.5 mm zirconium beads and 0.5 ml of serum-free medium. Removing the wings first renders the mosquito unable to escape.
2. Keep the tube on ice while processing other samples.
3. After removing the wings, gently pull out the mosquito legs by using two forceps. Use one to hold the mosquito while removing the legs with the other.
4. Keep all the legs in another separate tube containing zirconium beads (*see* **Note 14**).
5. To harvest saliva, pipette 5 μl of MEM solution into a 10 μl non-filtered pipette tip and carefully place it in a Petri dish (*see* **Note 15**).
6. Carefully place the live mosquito's proboscis into the pipette tip (Fig. 2).

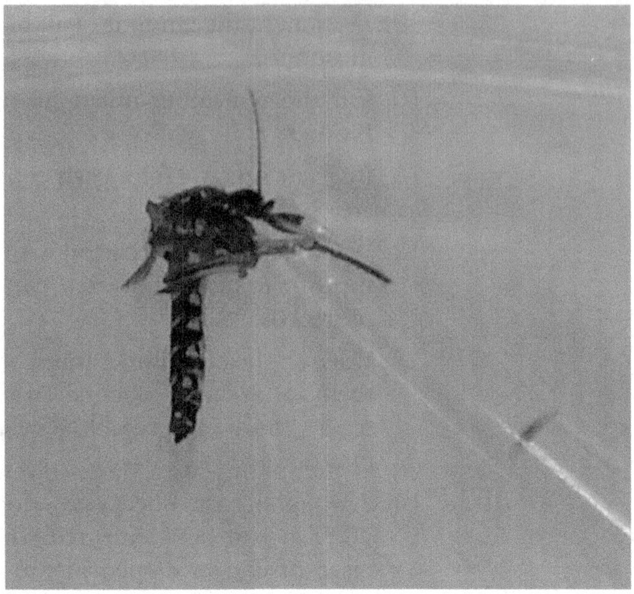

Fig. 2 The proboscis of an *Ae. aegypti* mosquito placed into a 10 μl non-filtered pipette tip for at least 30–45 min before collecting the saliva

7. Cover the Petri dish to prevent the MEM solution from evaporating within the tip.
8. Allow the mosquito to salivate for at least 30–45 min.
9. Carefully remove the mosquito's proboscis from the tip, and place the mosquito in a microcentrifuge tube. If further dissection is needed, place the mosquito onto the middle of a slide. Remove the solution from the pipette tip into a PCR tube, and immediately keep it on ice (*see* **Note 16**).
10. To harvest midgut and ovaries, put two drops of PBS onto the slide (*see* **Note 17**).
11. Using two dissecting needles gently separate the mosquito's abdomen from its thorax. Place each on a separate drop of the PBS (*see* Fig. 3).
12. Hold the cercus (end of the abdomen) with one dissecting needle while using the other to gently squeeze the midgut and ovaries out from the proximal end.
13. If either or both organs cannot be gently squeezed out, use the needle to cut off the cercus. Then pull the organs out from the distal end of the abdomen. Place the midgut and both ovaries into different tubes (*see* **Note 18**).
14. To harvest salivary gland and head, focus the microscope onto the other droplet of PBS solution where the thorax and head of the mosquito are located.

Fig. 3 An *Ae. aegypti* mosquito separated into half and ready to be dissected (5× magnification)

15. Secure the thorax with a dissecting needle, and use the second dissecting needle to gently pull the mosquito's head away from the thorax (*see* **Note 19**; [10]).
16. As the head separates from the thorax, the salivary gland can be found at the anterior portion of the thorax or attached to the head and can be isolated by using the dissecting needle (*see* **Note 20**; Fig. 4).
17. Place the head and salivary gland into different tubes.

3.4 Processing of Harvested Parts of Mosquito

1. Homogenize the mosquito parts in the chilled tubes containing zirconium beads at 4000×g for 15 s using a microtube homogenizer.
2. Store the homogenate samples at −80 °C for downstream work.

4 Notes

1. The number of mosquitoes required should be determined according to experimental needs. Factors such as feeding rate of the mosquito during the blood meal, number of time points needed, and mortality rate at each day post-infection should be considered. We recommend placing no more than 50 mosquitoes in each cup when feeding to prevent overcrowding and to facilitate documenting the numbers of mosquitoes during the experiment. In our experience, it is best to start with two or three times the number of mosquitoes needed for the experiment.

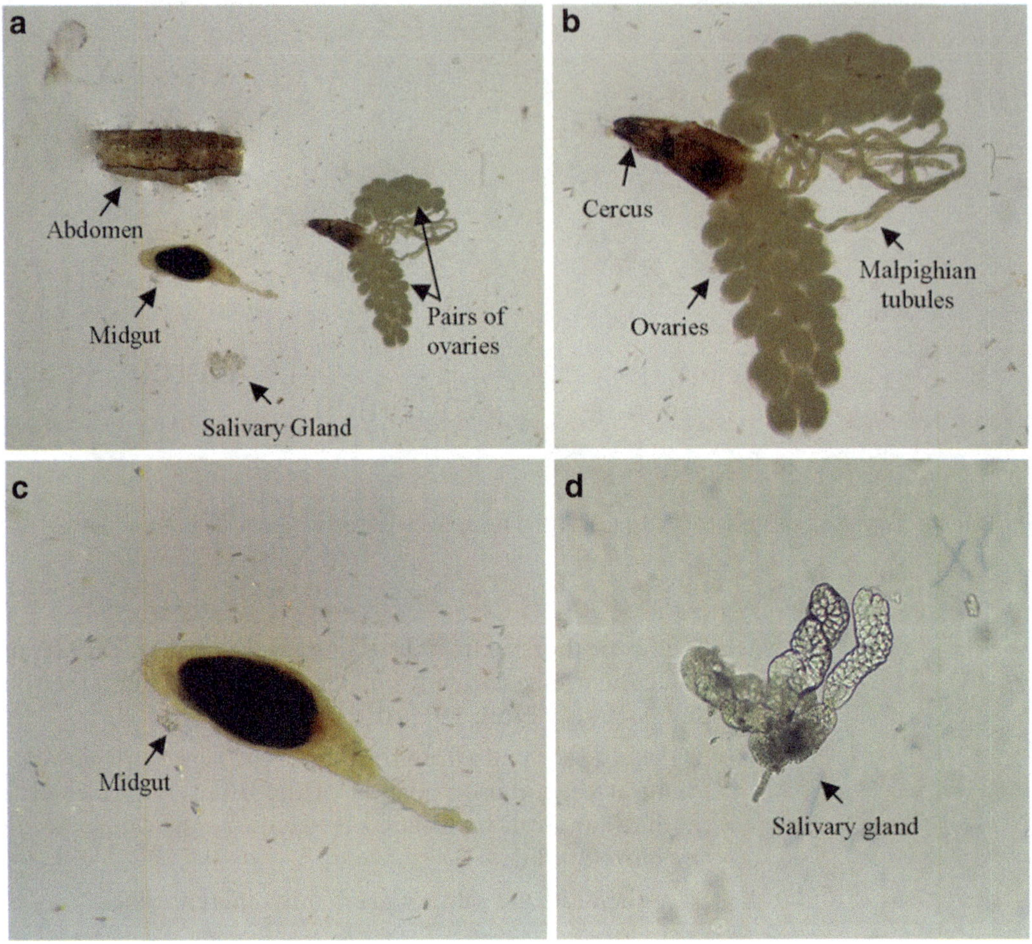

Fig. 4 Individual mosquito organs dissected from an *Ae. aegypti.* **(a)** Overview of mosquito organs dissected (2× magnification). **(b)** A pair of ovaries with developed eggs obtained through the cercus (5× magnification), **(c)** Midgut with undigested blood (5× magnification). **(d)** Salivary gland (10× magnification)

2. Blood used in the experiment should be obtained from the same source to ensure standardization.
3. It is important to warm up the blood mixture to 37 °C prior to feeding, as unsuitable temperatures will adversely affect the feeding rate of the mosquitoes.
4. Make sure the netting of the cup is strong enough to hold the feeder. Overstretching of the netting may allow mosquitoes to escape through the tiny holes.
5. Dark conditions enhance the feeding rate. It is also crucial to maintain the temperature of the surrounding room temperature at 24–28 °C. Lower temperatures may reduce feeding rate of the mosquitoes.
6. Avoid snap freezing the mosquitoes for more than 45 s as this will weaken or kill them. Placing the mosquitoes on

ice will continue to immobilize them and prevent escape during sorting.

7. A funnel can be placed into the new cup through the glove entrance to aid the transfer of mosquitoes.
8. Placing the same amount of mosquitoes into each cup makes it easier to count and document the number of mosquitoes.
9. Make sure all mosquitoes are dead before disposing of them.
10. Do not wet the cotton with too much distilled water as this may drown the mosquitoes inside the cup. Squeeze to remove excess water from the cotton.
11. Change the cotton every day to prevent growth of other microorganisms within the cup or on the mosquitoes.
12. It is important to remove dead mosquitoes to prevent growth of other microorganisms. It will also prevent the researcher from inadvertently selecting a dead mosquito during a particular harvesting time point.
13. The dissecting needles should be washed in 70 % alcohol between dissection of each mosquito to prevent cross-contamination.
14. The mosquito will remain alive for up to 1 h if the legs and wings are carefully removed to minimize damage to the body. Proceed to saliva collection immediately, as it will not salivate when it is dead.
15. Make sure there are no air bubbles between the edge of the tip and the MEM.
16. Dilute the solution into more MEM if needed.
17. Do not mix the two drops of PBS to prevent cross-contamination.
18. The midgut and ovary will be connected by the Malpighian tubules. Do not include the Malpighian tubules with the midgut. The Malpighian tubules can be discarded.
19. Do not cut to separate the head from the thorax, as the salivary glands attached to the head may be damaged.
20. The salivary glands comprise of six lobes: four lateral lobes and two medial lobes.

Acknowledgements

This work was supported by the European Union's Seventh Framework Program (Integrated Chikungunya Research, grant agreement no. 261202), the University of Malaya (UMRG grant RG526-13HTM), and the Ministry of Higher Education, Malaysia (FRGS grant FP036/2013A).

References

1. Jupp PG, McIntosh BM (1988) Chikungunya virus disease. In: Monath TP (ed) The arboviruses: epidemiology and ecology, vol II. CRC Press, Boca Raton, FL, pp 137–157
2. Powers AM, Brault AC, Tesh RB et al (2000) Re-emergence of chikungunya and o'nyong-nyong viruses: evidence for distinct geographical lineages and distant evolutionary relationships. J Gen Virol 81:471–479
3. Sam IC, Loong SK, Michael JC et al (2012) Genotypic and phenotypic characterization of chikungunya virus of different genotypes from Malaysia. PLoS One 7:e50476
4. Ross RW (1956) The Newala epidemic. III. The virus: isolation, pathogenic properties and relationship to the epidemic. J Hyg (Lond) 54:177–191
5. Forrester NL, Coffey LL, Weaver SC (2014) Arboviral bottlenecks and challenges to maintaining diversity and fitness during mosquito transmission. Viruses 6:3991–4004
6. Coleman J, Juhn J, James AA (2007) Dissection of midgut and salivary glands from *Aedes aegypti* mosquitoes. J Vis Exp 5:e228
7. Tsetsarkin KA, Vanlandingham DL, McGee CE et al (2007) A single mutation in Chikungunya virus affects vector specificity and epidemic potential. PLoS Pathog 3:e201
8. Chen R, Wang E, Tsetsarkin KA et al (2013) Chikungunya virus 3′ untranslated region: adaptation to mosquitoes and a population bottleneck as major evolutionary forces. PLoS Pathog 9:e1003591
9. Tsetsarkin KA, Weaver SC (2011) Sequential adaptive mutations enhance efficient vector switching by chikungunya virus and its epidemic emergence. PLoS Pathogen 7:e1002412
10. World Health Organization (1975) Manual on practical entomology in malaria, parts I and II. World Health Organization Part II. pp 88–111

Chapter 12

Analysis of CHIKV in Mosquitoes Infected via Artificial Blood Meal

Jeremy P. Ledermann and Ann M. Powers

Abstract

Having a mechanism to assess the transmission dynamics of a vector-borne virus is one critical component of understanding the life cycle of these viruses. Laboratory infection systems using artificial blood meals is one valuable approach for monitoring the progress of virus in its mosquito host and evaluating potential points for interruption of the cycle for control purposes. Here, we describe an artificial blood meal system with Chikungunya virus (CHIKV) and the processing of mosquito tissues and saliva to understand the movement and time course of virus infection in the invertebrate host.

Key words Chikungunya virus, Artificial blood meal, Mosquito infection, Viral transmission dynamics, Saliva collection, Mosquito processing

1 Introduction

Arthropod-borne viruses (arboviruses) have a complex life cycle consisting of replication in and transmission from both vertebrate and invertebrate hosts; therefore, it is critical to understand both halves of the cycle to comprehend how the viruses are maintained and how these cycles can be interrupted to prevent and control disease. One method for assessing the dynamics of the virus in the mosquito vector is to perform an artificial blood meal infection (mimicking a naturally acquired blood meal) to evaluate the movement of the virus throughout the mosquito over time. During the process, the mosquitoes ingest virus that is present in the mixture of washed red blood cells and virus suspension. The virus initially infects the epithelial cells of the midgut which is an indication of the infection rate of the virus in that mosquito species. After replication in the midgut, virus escapes and spreads through the mosquito via the hemocoel infecting secondary organs such as the neural tissue, ovarian tissue, flight muscle, and salivary glands. When the salivary glands become infected, the virus can then be transmitted when the mosquito takes another blood meal. This

Justin Jang Hann Chu and Swee Kim Ang (eds.), *Chikungunya Virus: Methods and Protocols*, Methods in Molecular Biology, vol. 1426,DOI 10.1007/978-1-4939-3618-2_12, © Springer Science+Business Media New York 2016

transmission can be measured by collecting the saliva from the mosquitoes and assaying for the presence of virus. The methods below demonstrate these laboratory methods for measuring the infection, dissemination, and transmission rates of Chikungunya virus within the vector mosquito.

This chapter will describe the generation of mosquitoes for laboratory virus infections and the process of infecting the mosquitoes using an artificial blood meal system. To assess the percentage of mosquitoes infected and with disseminated infections, a cell-culture CPE procedure is used on the heads and bodies of fed mosquitoes respectively. Finally, to determine the transmission rates, a procedure to collect saliva and assay for the presence of virus is described.

2 Materials

A risk assessment should be performed prior to initiating any new procedures and appropriate biological safety measures should be adhered to when preparing these materials or performing these procedures. The Biosafety in Microbiological and Biomedical Laboratories and the Arthropod Containment Guidelines [1, 2] are useful resources when initiating mosquito experiments such as those in this chapter.

2.1 Mosquito Rearing

1. 0.4 % Liver powder solution. Dissolve 4 g of liver powder in 1 l of water. Store at 4 °C.
2. 5 % Sucrose solution. Dissolve 50 g of sucrose in 1 l of water. Store at 4 °C. Fill glass baby food jar (or similar) and secure a gauze pad on the top with a rubber band.
3. Rabbit pellets (Harlan Laboratories Inc.)
4. Hemotek membrane feeding system (Hemotek, Accrington; Fig. 1).
5. Transfer pipets, 5 ml.
6. D-cell mosquito aspirator (Clarke).
7. Webbed scoop. Cut a mesh fabric, such as organdy, to size and fold to make a scoop. Fasten end with a small clip.
8. Emergence Cup. Plastic cup, ~100 ml. The mosquito breeder (BioQuip Inc.) with the top chamber removed works effectively as an emergence cup.
9. Larval rearing tray (BioQuip Inc.). A clear Plexiglas lid can be obtained from most hardware stores. Cut to completely cover rearing tray.
10. Adult rearing cages (BioQuip Inc.)
11. Oviposition container. Fill a 50 ml glass baby food jar (or similar) with 20 ml of water. Roll an approximately sized 2″ × 6″ piece of seed paper (Anchor Paper Co.) and place into the jar.

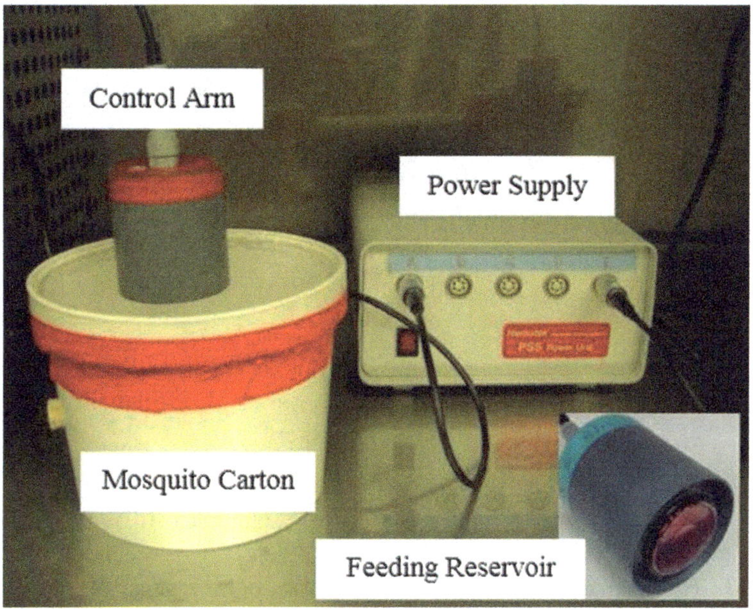

Fig. 1 Hemotek artificial blood feeding system (inset: filled reservoir attached to control arm). (Photo: J.P. Ledermann)

12. Defibrinated sheep or calf blood (Colorado Serum Company).
13. Mosquito cartons: pint, half-gallon, and gallon cartons (Huhtamaki).
14. Tupperware® or airtight container.
15. ZipLoc® bag.

2.2 CHIKV Infectious Blood-Meal Preparation

1. Mosquito cartons: remove the inside portion of the half gallon (64 oz.) container lid with a scalpel or other sharp instrument keeping the outer ring. With a hole puncher, make a ½ in. hole in the side of the container and plug the hole with a rubber or cork stopper. Cut a circle from the mesh lining material that is approximately 2 in. larger than the top opening of the container. Position the mesh lining atop the ½ gal container, secure with the cut out lid ring, and pull the mesh taught. Apply tape to secure the mesh and lid to the container base. Repeat procedure with the pint (16 oz.) containers (*see* **Note 1**).
2. 10 % Sucrose in fetal bovine serum (FBS). Dissolve 0.1 g of sucrose in 1 ml of FBS in a 15 ml conical tube.
3. Compacted Sheep Blood. Pour ~30 ml of sheep blood into a 50 ml conical tube. Bring sheep blood to 50 ml volume with PBS and mix by inversion. Spin sheep blood/PBS mixture in a swing bucket rotor for 10 min at 3000×*g*. Remove approximately 25 ml of the light red or clear layer from the top of the tube with a transfer pipette. Repeat washing two more times (until upper layer is

clear or nearly clear) bringing the total volume of compacted red blood cells (RBCs) to around 10 ml after the last wash.

4. D-cell mosquito aspirator (Clarke).
5. Glass petri dish with lid on a tray of wet ice (*see* **Note 2**).
6. Conical centrifuge tubes (15 and 50 ml).
7. Mosquito media: Dulbecco's minimal essential medium (DMEM) (Gibco) supplemented with 10 % FBS, 100 U/ml of penicillin and streptomycin, 1 U/ml of fungizone, and gentamycin.
8. Dulbecco's phosphate-buffered saline (DBPS).
9. Collection tubes: 1.7 ml centrifuge and 2.0 ml tubes with O-ring gaskets.
10. Glove box for mosquito manipulations (Bel-Art).
11. Chikungunya virus stock (6–7 $\log_{10}$ pfu/ml).

2.3 Post-exposure Mosquito Processing

1. Glass microscope slides (*see* **Note 3**).
2. No. 2 scalpel.
3. Forceps.
4. Kimwipes.
5. 70 % Ethanol.
6. 0.05 % Trypsin with 1× ETDA (0.53 mM) solution.
7. Microfuge pestles.
8. 1 ml syringe attached to a filter, 0.2 μM, 13 mm.
9. Tissue culture plate, flat bottomed, 96-well.
10. Tissue culture basins, sterile.
11. Vero cells (*see* **Note 4**).
12. Saliva collection tube. Fill a glass capillary tube (Chase Scientific Glass) with 5 μl of type B immersion oil (Cargill). Plug the opposite end using Critoseal putty (McCormick Scientific, *see* **Note 5**).
13. Scalpel.
14. Forceps.
15. 70 % Ethanol.
16. Gauze pads (4 ×4).
17. Mosquito media.
18. Cryovial tubes.
19. Blood-fed mosquitoes.

2.4 Nucleic Acid Detection and Quantification

A variety of molecular kits are available. The following are only recommendations that have been successfully used with these procedures.

1. QIAamp Viral RNA kit (Qiagen).
2. OneStep RT-PCR kit (Qiagen).
3. QuantiTect Probe real-time RT-PCR kit (Qiagen).
4. 1 % Agarose gel. Prepare gel using a low electroendosmosis agarose (Fisher Scientific).
5. Mosquito samples.
6. Primers: *see* Tables 1 and 2 for sequences.

Table 1
CHIKV oligonucleotides for conventional RT-PCR

Name	Sequence (5′–3′)	T_m (°C)
CHIKV 17 FWD	CACGTAGCCTACCAGTTTCTTAC	68
CHIKV 301 FWD	CAGGAAGTACCACTGCGTCTGCC	70
CHIKV 1184 FWD	GGCAGAACGCAACGGAATATG	64
CHIKV 1303 REV	CCCCAGGAGTTTTTCATCTTCCATGTC	72
CHIKV 1814 FWD	GCGGAGCAAGTGAAGACGTG	64
CHIKV 2450 REV	ACGCAAACGCCTCGTCTACGTACA	74
CHIKV 2523 FWD	GCGGCTTCTTCAATATGATGCAG	68
CHIKV 3217 FWD	CTCACCTGAAGTAGCCCTGAATG	70
CHIKV 3918 FWD	GATCGTCTAGAGCGTTGAAACCACC	76
CHIKV 4758 FWD	AGCAAGTCTGCCTATATGCCCT	66
CHIKV 4923 REV	CTTGTGACGTGGTTCATGCGAAG	70
CHIKV 5331 FWD	GGAATATAACACCCATGGCTAGC	68
CHIKV 6061 FWD	CTATCCAACTGTCTCATCATACC	66
CHIKV 6340 REV	CACGTTGAATACTGCTGAGTCC	66
CHIKV 6822 FWD	CTGCTTTGATGCTGTTAGAGG	62
CHIKV 7417 FWD	GTCCATGGCCACCTTTGCAAGCTC	76
CHIKV 7750 REV	GCTTCTTATTCTTCCGATTCCTGCG	74
CHIKV 8071 FWD	GTGCACATGAAGTCCGACGC	64
CHIKV 8712 FWD	GACGGATGACAGCCACGATTG	66
CHIKV 9230 REV	GTGACCGCGGCATGACATTGATC	72
CHIKV 9396 FWD	GCTACTGTATCCTGACCACCC	66
CHIKV 10110 FWD	GCCAACACTATCGCTTGATTAC	64
CHIKV 10657 REV	GCAGTACCAGTTGTGTATTAGC	64
CHIKV 10818 FWD	CATGCCCATCTCCATCGACATAC	70
T25V REV	TTTTTTTTTTTTTTTTTTTTTTTTT TV	54

Table 2
CHIKV real-time qRT-PCR oligonucleotide sets

Name	Sequence (5′-3′)
CHIKV 6856 FWD	TCACTCCCTGTTGGACTTGATAGA
CHIKV 6981 REV	TTGACGAACAGAGTTAGGAACATACC
CHIKV 6919 FWD-FAM (probe)	AGGTACGCGCTTCAAGTTCGGCG
CHIKV 243 FWD	GAYCCCGACTCAACCATCCT
CHIKV 330 REV	CATMGGGCARACGCAGTGGTA
CHIK 273 FWD-FAM (probe)	AGYGCGCCAGCAAGGAGGAKGATGT

3 Methods

3.1 Aedes aegypti *and* Aedes albopictus *Colony Rearing*

1. Day 1: To hatch eggs, add egg liner to a rearing tray with water (approximately ½ full) and 2 ml of liver powder solution. Leave overnight in an incubation chamber or room stabilized at 28 °C with ≥80 % humidity and light:dark cycle of 14:10 or 12:12 (*see* **Note 6**).
2. Day 2: Split the first instar larvae into multiple rearing trays with water so each tray density reaches approximately 150 larvae/tray. Add one or two rabbit pellets to each tray and cover with Plexiglas lid.
3. Days 3–6: Monitor the larvae development daily, assuring that food is present. Add more food (rabbit pellets) when necessary (*see* **Note 7**).
4. Days 7–10: If optimal larval growth conditions are present, pupation will begin on day 6 or 7 and continue till approximately day 10. Pick pupae with a transfer pipet or webbed scoop and place into an emergence cup with 5–10 ml of water. Place cup into the adult rearing cage. Place baby food jar with 5 % sugar solution on top of the mosquito cage.
5. Adult mosquitoes will take a blood meal as early as 24 h post-emergence. For colony maintenance, provide calf or sheep blood via the Hemotek blood feeding system (*see* **Note 8**). Allow the mosquitoes to feed for 45–60 min. Place oviposition container into the adult rearing cage. For infectious blood meal, transfer 50–100, 3- to 5-day-old female adults (never previously blood-fed) from the rearing cage to a 1 gal prepared mosquito carton using a D-cell mosquito aspirator. Cork the carton with the mosquitoes inside.
6. For colony maintenance, remove egg liner and place on a paper towel to dry until still slightly damp. Once damp, place egg

liner into a Ziploc bag. Place Ziploc bag into a Tupperware container or other air tight vessel at 28 °C (*see* **Note 9**) for storage until use.

3.2 Mosquito Infectious Blood-Meal Protocol

1. Prepare virus by thawing the virus stock at room temperature in biosafety cabinet.
2. Dilute the virus in cell culture medium (or PBS). Final volume of virus should be 1 ml or volume equal to 1/3 of the entire meal.
3. Prepare the complete blood meal by mixing the packed RBCs, the sucrose/FBS, and the diluted virus in a 1:1:1 ratio in a 15 ml conical tube (*see* **Note 10**).
4. Pre-warm the blood meal preparation in a 37 °C water bath 5 min prior to filling a feeder.
5. Prepare a 1.7 ml centrifuge tube for each meal to use to back-titer the sample.
6. In a secure container, transfer 3- to 5-day-old uninfected female mosquitoes (from Subheading 3.1) in a 64 oz. mosquito carton to the BSL-3 insectary (*see* **Note 11**).
7. Set-up the Hemotek blood feeding system in the reach-in incubator set at the appropriate temperature and humidity for the species being used (*see* **Note 12**).
8. Affix the feeding membrane onto the empty feeding reservoirs (Fig. 1) and move into the biosafety cabinet (*see* **Note 13**).
9. Using a transfer pipet, fill the reservoirs through the port, being careful not to puncture the membrane and cap the fill port.
10. Transfer ≥150 μl of each blood meal to a labeled back titer tube and place the tube in a 37 °C water bath for the duration of the feed.
11. Affix the prepped and filled feeding reservoirs onto the Hemotek heating arms and place onto the top of mosquito cartons to allow the mosquitoes to take a blood meal. A period of 60 min is adequate time for most mosquitoes to feed. Monitor the feeding periodically to assure the quality of the feed (*see* **Note 14**).
12. Upon completion of the feed, remove blood feeding reservoirs from the control arms and place the reservoirs into disinfectant. Remove the blood meal back-titer sample from the water bath and place into −80 °C freezer.
13. Remove each carton containing mosquitoes exposed to an infectious blood meal and place at 4 °C to render the mosquitoes unconscious.

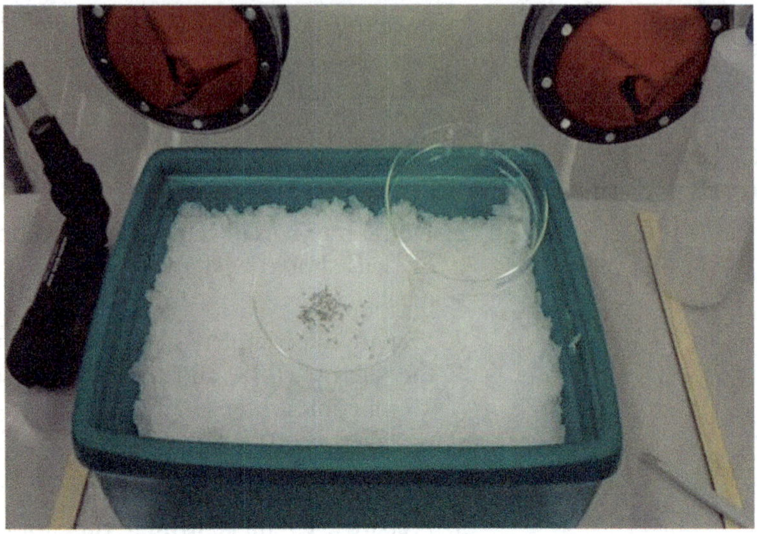

Fig. 2 Glove box set up showing tray of ice, anesthetized mosquitoes in a petri plate, ethanol bottle, and D-cell aspirator. (Photo J.P. Ledermann)

14. Once all the mosquitoes are knocked down (approximately 5–8 min), transfer the carton from 4 °C to the glove box which contains a glass petri dish on a tray of ice (Fig. 2; *see* **Note 15**). Immediately transfer the mosquitoes from the carton to the petri dish and separate engorged mosquitoes into easily countable groups.
15. Count the blood-fed mosquitoes as they are moved into a new 16 oz. mosquito carton (through the hole in the side of the carton) using forceps. Place corked cartons containing the exposed mosquitoes into the secondary containment cage inside the reach-in incubator. Place sugar water onto the top of each carton. Discard any unfed or partially engorged mosquitoes by saturating the mosquito bodies in disinfectant and then autoclaving.
16. Upon completion of the intrinsic incubation period (typically 7 days), observe the number of dead mosquitoes inside each container and record. Do not process the dead mosquitoes. Alternatively, mosquitoes can be removed from each carton on multiple days to assess virus infection patterns over time.

3.3 Post-exposure Mosquito Processing: Detection of Virus in Mosquitoes by CPE Analysis

1. Process knocked-out mosquitoes as in Subheading 3.2. Place all mosquitoes in the petri plate on wet ice, contained in the glove box. Count the number of live mosquitoes and record.
2. Remove a single knocked-out mosquito and place onto a glass microscope slide for dissection (*see* **Note 16**).
3. Separate the head from the abdomen by cutting with the scalpel and place each individual part into a labeled 1.7 ml tube or

similar. Place each tube into the tray of ice to keep cool. Dip scalpel and forceps into a small cup of 70 % ethanol to sterilize. Using a Kimwipe or gauze pad, wipe off any excess liquid before dissecting the next sample.

4. Repeat **steps 2–4** until all mosquitoes have been dissected.
5. Remove samples from the ice and place into a labeled freezer box and place into −70 °C freezer until processing.
6. Remove mosquitoes to be assayed from −70 °C and place in the ice tray inside BSC (*see* **Note 17**).
7. Add 200 μl of mosquito media to the sample in a 1.5 ml microfuge tube and grind the sample with a microfuge pestle until the sample is completely homogenized. Add an additional 200 μl, slowly rinse the pestle of any sample into the tube.
8. Draw the sample into a 1 ml syringe by slowly drawing up the plunger. Affix a 0.2 μM syringe filter to the syringe tip and slowly filter the homogenized sample through the filter (by depressing the plunger) into a new cryovial tube (*see* **Note 18**).
9. If not processing for CPE immediately, remove samples from the ice, place into a labeled freezer box, and place into −70 °C freezer until processing.
10. To process, remove homogenized and filtered samples from the freezer and allow to thaw at 4 °C. Keep samples on ice during processing.
11. Add 100 μl of sample to a single well of the 96-well culture plate (*see* **Note 19**).
12. Prepare Vero cell suspension by trypsinizing a stock culture flask and transferring cells to a sterile tissue culture basin (*see* **Note 20**). Add 50 μl of cell suspension to each well of the plate using a multi-channel pipet.
13. Incubate plates at 37 °C and 5 % CO_2 in a cell culture incubator (*see* **Note 21**). Check for cytopathic effects (CPE) daily for up to 5 days for alphaviruses (Fig. 3).

3.4 Post-exposure Mosquito Processing: Forced Salivation

1. To measure virus transmission rates, prepare saliva collection capillary tubes with Critoseal putty and 1.7 ml tube with 500 μl of DMEM for each saliva sample.
2. Knock out mosquitoes that have been previously exposed to virus-infected blood meals as in Subheading 3.2. Place all mosquitoes in the petri plate on wet ice, contained in the glove box. Count the number of live mosquitoes and record.
3. While gently holding the mosquito by the halteres or the base of one wing with a pair of forceps, remove the wings and legs from each mosquito with forceps and insert proboscis into the saliva collection tube (Fig. 4).

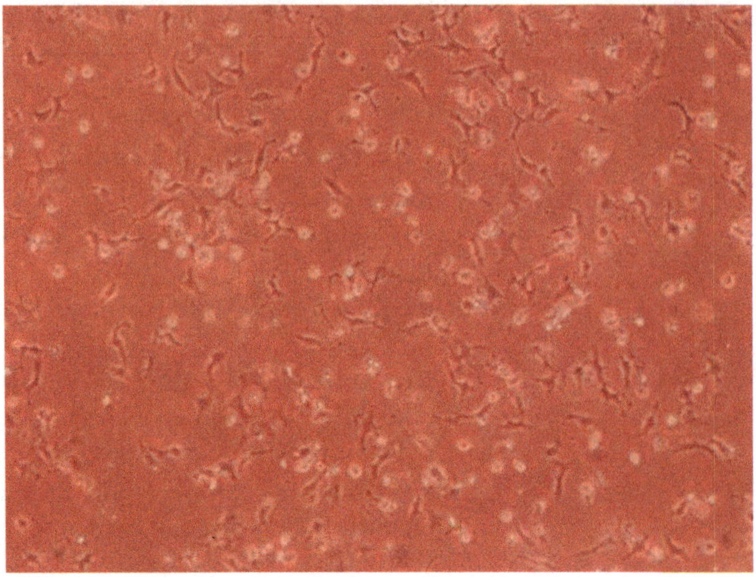

Fig. 3 Typical alphaviral cytopathic effects (CPE) in Vero cells, 48 h post inoculation (40× mag). (Photo K. S. Shaw)

Fig. 4 Mosquito saliva collection showing wingless/legless mosquitoes with proboscis inserted into oil in capillary tubes. (Photo J. Ledermann)

4. After 60 min, remove mosquito and process through the CPE procedure (Subheading 3.3), or freeze at −70 °C (*see* **Note 22**).
5. Remove each capillary collection tube and place into the labeled 1.7 ml tube with the oil side in the bottom of the tube. Carefully snap the collection tube in half, discarding the empty portion. Close the tube lid.
6. Centrifuge the tube for 1 min at 10,000 × *g* to expel the oil from the capillary into the tissue culture medium. Place samples into −70 °C freezer until analysis.

3.5 CHIKV Nucleic Acid Detection and Quantification

1. Remove saliva, or ground and filtered mosquito head/body samples to be processed from freezer.
2. Process the sample through an RNA extraction kit or similar process to obtain clean viral nucleic acid.
3. For RT-PCR, use 5–10 μl of RNA with any set of the CHIKV-specific primer sets listed (Table 1; *see* **Note 23**) and reagent kit per the manufacturer's instructions. Briefly, final concentrations and thermocycling conditions are as follows: 1× PCR buffer, 400 μM dNTP, 0.6 μM each primer at 50 °C for 30 min, 95 °C for 15 min, 40 cycles of 94 °C for 0.5 min, 50–68 °C (5 °C below T_m of primers) for 1 min and 72 °C for 1 min, followed by 72 °C for 10 min. Analyze the PCR product on a 1 % agarose gel.
4. For CHIKV nucleic acid quantitation and detection, analyze 5–10 μl of RNA combined with any set of primer and probe oligonucleotides (Table 2). Follow the instructions from the real-time qRT-PCR reagent kit and the referenced publications ([3, 4]; *see* **Note 24**). Briefly, final concentrations and thermocycling conditions are as follows: 1× PCR buffer, 0.4 μM each primer, 0.2 μM probe at 50 °C for 30 min, 95 °C for 15 min and 40–45 cycles of 94 °C for 15 s and 60 °C for 1 min (*see* **Note 25**).

4 Notes

1. The mosquito-holding containers should be carefully constructed with emphasis on insect containment. A chiffon or organdyfabric or other fine mesh material is appropriate for the cover and can be found at any craft or fabric store. Make a hole half way up the side of the container so that it is not covered by any tape, but allows for easy access with the aspirator tube. The 64 oz. container is temporary and is only utilized for holding the mosquitoes before and during the blood feed. The 16 oz. container is used to hold the mosquitoes after the blood feed and during the incubation period.

2. Assure that petri dish is dry before placing mosquitoes into it.
3. With an acrylic ink pen or similar, divide the glass slide into 8–10 squares. Each square can be used for placement of a single mosquito for dissection. The slide can be reused after disinfection/sterilization with ethanol and wipe.
4. Vero cells should be of low passage (<50) and grown to confluent monolayer in a T-75 or T-150 tissue culture flask.
5. To add putty, simply press the capillary tube into the putty tray and remove. More putty will need to be added as the oil level falls. The oil level will stabilize after just a few putty additions.
6. The egg liner in water can be placed into a chamber under vacuum for 1 h to increase the hatch rate. In addition, a light:dark cycle is typically 12:12 or 14:10 for *Aedes* spp. but adjust to the mosquito species being studied.
7. The addition of rabbit pellet food should only occur on days 1 and day 3 or 4. Too much food will promote bacteria growth. If this occurs, sieve larvae and replace the dirty water with clean water.
8. Generally, field-collected mosquitoes take several generations to adapt to feeding using artificial blood feeding systems. In these cases, a mouse or hamster should be provided (under appropriate animal protocol regulations) in order to maintain the colony. Removing the sugar water solution from the mosquitoes for 12–16 h prior to blood meal will increase the likelihood of feeding.
9. The egg liner must remain in a humidified environment to preserve the eggs. Under optimal conditions, egg liners can last ≥6 months.
10. Normally use 1 ml of each component. Feeders have a max capacity of 3 ml.
11. Adult mosquitoes that are used in the infectious blood meal should be 3–5 days old as well as starved of sugar water for 12–16 h before the blood meal. In addition, only fill the 64 oz. container with a maximum of 300 adults.
12. For *Aedes* spp., 80 % humidity and 28 °C is typical.
13. The collagen membrane which is supplied with the feeding system is adequate, but it has been observed that newly colonized field strains prefer a pig gut membrane or a mouse skin.
14. Expelling breath over the carton during feeding can enhance the feeding rate. However, wear proper PPE as CHIKV is easily aerosolized.

15. Place a 70 % ethanol bottle in the glove box and use to kill any escaped mosquitoes (as they cannot fly when their wings are wet).
16. The mosquito will remain "anesthetized" during this step. However, if the mosquito starts to move, place it back into the petri dish on ice. Also, since CHIKV can aerosolize, take appropriate PPE precautions such as wearing an N-95 mask and safety glasses.
17. Remove only 10–15 mosquitoes at any one time from the freezer to limit the time your samples are at room temperature.
18. An alternate method for clarification is to centrifuge the ground sample at $16,000 \times g$ for 1 min to pellet sample. Transfer supernatant to a new cryovial being careful to avoid transferring any of the debris.
19. Two wells per plate should contain a tissue culture medium and cells only as negative control wells. Positive control wells should not be added as this could lead to false positives.
20. A confluent T-150 contains sufficient cells (3×10^7 cells) for 5, 96-well plates. Trypsinize your cells and resuspend in 5 ml of tissue culture medium, making sure to break up any cell clumps by pipetting up and down. Add an additional 20 ml of media to bring the cells to the correct density.
21. For containment, place plates in a Nalgene container or similar with a wet paper towel. Crack the lid slightly to allow for gas exchange.
22. Saliva droplets will exude from the proboscis into the oil and may appear as tiny bubbles [5, 6]. It is also recommended to process the mosquito head and bodies through the CPE procedure on the same day as the saliva collections.
23. Many primers are listed in the table and each could be used for genome amplification in RT-PCR or in sequencing. However, amplicons should not exceed 3–4 K in nucleotide length as both polymerase activity and read quality will decrease with increasing product size.
24. Many CHIKV real-time PCR sets are described in the literature. Those listed in the table are broadly reactive, detecting both Asian and the ECSA genotypes.
25. Each primer and probe set should be optimized as described in the manufactures' guidelines. Those conditions listed are only recommendations.

References

1. U.S. Department of Health and Human Services (2009) Biosafety in microbiological and biomedical laboratories (BMBL), 5th edn. HHS Publication No. (CDC) 21–1112
2. Arthropod Containment Guidelines (2003) Vector-Borne Zoonot Dis 3(2):75–90
3. Lanciotti RS, Kosoy OL, Laven JJ, Panella AJ, Velez JO et al (2007) Chikungunya virus in US travelers returning from India, 2006. Emerg Infect Dis 13:764–767
4. Partidos CD, Weger J, Brewoo J, Seymour R, Borland EM et al (2011) Probing the attenuation and protective efficacy of a candidate chikungunya virus vaccine in mice with compromised interferon (IFN) signaling. Vaccine 29(16):3067–3073
5. Aitken THG (1977) An in vitro feeding technique for artificially demonstrating virus transmission by mosquitoes. Mosquito News 37:130–133
6. Aitken TH, Tesh RB, Beaty BJ, Rosen L (1979) Transovarial transmission of yellow fever virus by mosquitoes (*Aedes aegypti*). Am J Trop Med Hyg 28:119–121

Chapter 13

Chikungunya Virus Growth and Fluorescent Labeling: Detection of Chikungunya Virus by Immunofluorescence Assay

Meng Ling Moi and Tomohiko Takasaki

Abstract

Immunofluorescence assay (IFA) is a highly versatile and sensitive assay for detection and titration of chikungunya virus (CHIKV). The IFA technique requires virus-infected cells (viral antigen) and antibodies specific to the viral antigens for detection. Suitable antibodies for detection include monoclonal antibodies specific to CHIKV structural and nonstructural proteins, polyclonal antibodies, and convalescent serum samples. Here, the details of virus antigen preparation, detection by IFA method, and applications are described. The described IFA method is potentially useful in a wide range of studies including virus growth kinetics and virus infection mechanism studies. Additionally, the described IFA method can be modified for applications in arbovirus diagnosis, including CHIKV.

Key words Chikungunya virus, Immunofluorescence assay, Fluorescence microscopy, Flow cytometry

1 Introduction

Chikungunya is a re-emerging disease of global significance. It is estimated that a few million CHIKV infections had occurred worldwide since the re-emergence of the disease in 2005 [1]. Common symptoms of Chikungunya virus (CHIKV) infection includes fever, rash, and debilitating joint pain that may persist for several months [2]. Current virus detection tools include reverse-transcriptase PCR (RT-PCR) for detection of viral genome, immunofluorescence assay (IFA) for detection of viral antigen, and cell culture for infectious viral particle titration [3]. RT-PCR provides a rapid method for viral genome titration. Virus titration using cell culture is time-consuming and laborious, but is the most specific tool for providing information on viral infectivity. IFA is vital in the identification and detection of viral antigen, and because the amount of infected cells containing the viral antigen is proportionate to viral growth, the IFA method has also been adopted to detect and quantify virus concentration [4].

Justin Jang Hann Chu and Swee Kim Ang (eds.), *Chikungunya Virus: Methods and Protocols*, Methods in Molecular Biology, vol. 1426,DOI 10.1007/978-1-4939-3618-2_13, © Springer Science+Business Media New York 2016

IFA is versatile, because it can be used to detect virus antigen, virus growth, presence of virus-specific immunoglobulins and antibodies with neutralizing activity in a sample. Virus antigen (samples) is required as a starting point for IFA assay. The next step would require antibodies that cross-react with the viral antigens. Antibodies that are generally used to detect CHIKV antigens are antibodies that detect the envelop (E) protein. Because an array of cross-reactive antibodies can be utilized in the IFA, the assay can be adapted to detect secretion of other structural and nonstructural proteins that are involved in virus growth. Alternatively, IFA offers a rapid method in the detection of changes in levels of viral protein expression, characterization of spatial preferences within different cellular compartments and characterization of the events of virus life cycle [5–7].

In this chapter, we will describe the detection of CHIKV in mammalian cell lines (Vero, BHK) by using IFA. We will further describe the applications of IFA in determination of virus growth using fluorescent microscopy and flow cytometry. The IFA assay can also be modified to detect CHIKV-specific antibodies in a sample. Here, we further discuss a modified IFA method to detect virus-specific antibodies sera from a suspected case of arbovirus infection with chronic joint pain [8]. Because presence of viral antigens is proportional to virus growth, the immunofluorescence staining method also offers a rapid and simple alternative in detecting virus neutralization activity of a sample [9, 10]. Additionally, IFA is suitable in studies using nonadherent cell lines, because IFA method can be adapted for both adherent and nonadherent cell lines. However, further characterization and validation of the IFA assay against the standard plaque assay would be needed [11]. The described IFA method is potentially useful in a wide range of studies including virus growth kinetics studies, CHIKV infection mechanism studies, and for diagnostics.

2 Materials

2.1 Virus Growth

1. CHIKV stock. Virus stock collected from cell culture supernatant could be stored at −80 °C for long-term storage. Avoid repeated freeze–thaw cycles.
2. Cell lines for virus culture: Vero cells (African green monkey kidney-derived cell lines), BHK cells (baby hamster kidney-derived cell lines), or C6/36 cells (*Aedes albopictus*-derived cell lines). Other cell lines can be used, however cell susceptibility to CHIKV infection should be predetermined.
3. Growth and maintenance medium: Eagles' Minimum Essential Medium (EMEM) supplemented with 10 % heat-inactivated fetal bovine serum (FBS). Optional addition: 20 U/ml of

penicillin and 20 μg/ml of streptomycin into growth medium to prevent potential bacterial contamination.

4. 1× Trypsin-EDTA solution: 0.05 % Trypsin-EDTA solution (v/v).
5. 1× Dulbecco's Phosphate Buffered Saline (DPBS), calcium and magnesium free.

2.2 Fluorescent Microscopy

Prepare and keep reagents at 4 °C or on ice during the experiment unless indicated otherwise.

1. Primary antibody: CHIKV primary antibodies (monoclonal antibodies or polyclonal antibody) or CHIKV convalescent serum sample.
2. Blocking reagent: 4 % Block ace (AbDSerotac) or 1 % skim milk in DPBS. Filter with 0.45 μM syringe filter. Blocking reagent can also be used as a diluent for antibodies.
3. Secondary antibody: Fluorescence conjugated cross-reactive secondary antibody.
4. Wash solution: 1× DPBS, pH 7.
5. Growth and maintenance medium: 10 % heat-inactivated FBS in EMEM.
6. Ice-cold fixation solution: acetone:methanol (1:1).
7. Chemical-resistant glass slides with coverslip: Glass slide printed with water-repellent mark, 10–12 holes
8. Mounting solution: 20 % glycerol in 1× DPBS or Vecta Shield Mounting Media for Fluorescence.

2.3 Serological Diagnosis Using IFA

Prepare and keep all reagents at 4 °C or on ice during the experiment unless indicated otherwise.

1. Virus stock: CHIKV, Dengue virus (DENV), Ross River virus (RRV) stock. Virus stock collected from cell culture supernatant must be stored at −80 °C for long-term storage. Avoid repeated freeze–thaw cycles.
2. Patient serum sample. Heat-inactivate the samples prior to use.
3. Control serum sample: serum sample collected from healthy individuals that are negative for anti-DENV, anti-RRV, and anti-CHIKV IgM and IgG antibodies
4. Blocking reagent: 4 % Block ace (AbD Serotac) or 1 % skim milk in DPBS. Filter with 0.45 μM syringe filter. Blocking reagent can also be used as a diluent for antibodies.
5. Secondary antibody: fluorescence-conjugated anti-human IgM or IgG secondary antibody.
6. Wash solution: 1× DPBS, pH 7.

7. Growth and maintenance medium: 10 % heat-inactivated FBS in EMEM.
8. Fixation solution: Ice-cold acetone and methanol prepared at a ratio of 1:1.
9. Chemical-resistant glass slides for cell adhesion, coverslips: glass slide printed with water-repellent mark, 10–12 holes.
10. Mounting solution: 20 % glycerol in 1× DPBS or Vecta Shield Mounting Media for fluorescence.

2.4 Flow Cytometry

Prepare and keep all reagents at 4 °C or on ice during the experiment unless indicated otherwise.

1. Primary antibody: CHIKV cross-reactive primary antibodies (monoclonal antibodies or polyclonal antibody) or CHIKV convalescent patient serum sample. For intracellular antigen staining, dilute antibody to optimal concentration in 1× BD Perm/Wash solution. For cell surface staining, dilute antibody in 1× DPBS.
2. Secondary antibody: Fluorescence-conjugated cross-reactive secondary antibody. For intracellular antigen staining, dilute antibody to optimal concentration in 1× BD Perm/Wash solution. For cell surface staining, dilute antibody in 1× DPBS.
3. Blocking solution: 4 % Block ace (AbD Serotac) or 1 % skim milk in PBS. Filter with 0.45 μM syringe filter.
4. Growth and maintenance medium: 10 % heat-inactivated FBS in EMEM.
5. Wash solution for cell harvest: DPBS 1×, pH 7.
6. Fixation and permeabilization buffer (BD Cytofix/Cytoperm solution).
7. Wash solution for intracellular antigen staining. 1× BD Perm/Wash buffer.
8. 1× DPBS for cell surface staining.

3 Methods

3.1 Virus Growth

1. Prepare a monolayer of BHK or Vero cells at a confluency of 70–80 %. The cells can be prepared using either 12-well plate or 24-well plate. For 12-well plate, seed between 1×10^5 to 5×10^5 cells per well (cell density is dependent on cell line and cell growth conditions). Incubate the plate overnight in a CO_2 incubator at 37 °C.
2. Discard the old cell culture supernatant and infect the cells at a 1:10 serial dilution of multiplicity of infection (MOI) of 1.0, 0.1, 0.01, and 0.001 [12]. Determine the virus titers using the

same cell lines for infection assay by standard plaque assay [13] or endpoint dilution assay (tissue culture infectious dose, $TCID_{50}$) before starting this step (*see* **Note 1**). For 12-well plate, prepare 50 μl of virus solution containing the corresponding MOI for each well. After virus inoculation, gently tilt the plate every 10 min for a total of 6 times (total of 60 min). Incubate the plate in a CO_2 incubator at 37 °C during the infection assay. Add 1 ml of growth and maintenance medium after the 1 h incubation period (*see* **Note 2**).

3. Harvest the cells each day from day 0 (day of infection) up to day 5 after infection.
4. For cell harvest, after wash with 1 ml 1× DPBS twice, trypsinize the cells with 0.1 ml of 1× Trypsin-EDTA solution for 5–10 min.
5. Add 1 ml of growth and maintenance medium. Gently detach the cells by repeated pipetting and transfer the solution to a 1.5 ml centrifuge tube.
6. Spin the microfuge tubes at 800 × *g* for 5 min and discard the supernatant.
7. Resuspend the cells in 1× DPBS at concentrations optimal for fluorescent microscopy and flow cytometry.
8. Incubation and harvest period can be shortened or prolonged according to study design. In CHIKV susceptible cells, cytopathic effect (cell death) could be observed using a light microscope or by naked-eye under fluorescent light. Alternatively, harvest the cell culture supernatant to determine virus titers released into the culture fluids (*see* **Note 3**).

3.2 Determination of Virus Growth by Fluorescent Microscopy

1. After the centrifugation step, prepare the cells in 1× DPBS, at a concentration of 5×10^5 cells/100 μl to 1×10^6 cells/100 μl (*see* **Note 4**).
2. Add 10 μl of cells in 1× DPBS suspension to each wettable surface on the glass slide and allow the slide to dry under the clean bench airflow for at least 10 min (*see* **Note 5**).
3. Fix and permeabilize the cells with 50 ml of ice-cold fixation solution at 4 °C for 1 min. If no staining chambers are available, a chemical-resistant 50 ml polyethylene centrifuge tube could be used to fix up to two slide glasses at a time.
4. Wash the slide glass in 50 ml of wash solution twice. Remove excess wash solution by gently knocking the corner of the slide glass on paper towels (*see* **Note 6**).
5. Block the cells with 10 μl of blocking solution for 10 min, wash the slides with 50 ml of wash solution once.
6. Apply 10 μl of unlabelled CHIKV cross-reactive primary antibody to the well surface of the slide glass (*see* **Notes 7** and **8**).

Optimal concentration of primary and secondary antibodies for IFA should be determined prior to this step (*see* **Note 9**). Incubate the slide glass in a moist chamber (*see* **Note 10**) for 60 min at 37 °C. After incubation, wash the slide glass in wash solution twice. Volume and incubation times for washing can be optimized according to the type of antibody used.

7. Remove excess wash solution by gently knocking the corner of the slide glass on paper towels. Blot excess wash solution on the water-resistant surface gently and avoid touching the wells with the paper towel.
8. Apply evenly 5–10 μl of fluorescence-conjugated secondary antibody to the well surface (*see* **Note 11**).
9. Incubate the slide glass in a humid chamber for 30 min at 37 °C, in dark.
10. After incubation, wash the slide glass with 50 ml of wash solution twice. Remove excess wash solution by gently knocking the corner of the slide glass on paper towels.
11. Add 100 μl of mounting solution to the slide glass and cover with a cover slip. Observe and determine the number of infected cells under confocal microscope (*see* **Note 12**).
12. Count the number of infected and uninfected cells in 3–6 random fields using a square grid under a light microscope at 200× magnification. Alternatively, count 100 cells in a random field and determine the percentage of cells that were infected within the specified 100 cells. Endpoint dilution assay ($TCID_{50}$) can be alternatively used to determine CHIKV titers.

3.3 Serological Diagnosis Using IFA

1. Prepare 1×10^6 cells/100 μl of Vero cells, infected for 3 days with either CHIKV, Ross River virus (RRV), or dengue virus (DENV) at MOI = 0.1. Use noninfected cells as negative control.
2. Add 10 μl of cell suspension in 1× DPBS in each of the wells on the multi-well slide, and allow the wells to dry for 10–20 min under the laminar flow.
3. Fix and permeabilize the cells with 50 ml of ice-cold fixation solution at 4 °C for 1 min.
4. Apply 10 μl of serially diluted patient serum sample (start from undiluted serum sample, and with tenfolds serially diluted samples, i.e. 1:10–1:1000). Sera from healthy individuals can be used for antibody negative control. Incubate slides for 60 min at 37 °C. After incubation, wash the slide glass in 50 ml of wash solution twice.
5. Apply evenly 10 μl of fluorescence-conjugated anti-human IgG secondary antibody or anti-human IgM secondary antibody to each of the wells. Incubate the slide glass in a humid

chamber for 30 min at 37 °C, in dark. After incubation, wash the slide glass in 1× DPBS twice (*see* **Note 13**).

6. Add 100 μl of mounting solution to the slide glass and cover with a cover slip. Observe and determine the number of infected cells under confocal microscope [5].

3.4 Determination of Virus Growth by Flow Cytometry

1. Wash the virus-infected cells in 1× DPBS. Prepare 250 μl of cell suspension in 1× PBS, at a concentration of 5×10^5 to 1×10^6 cells/100 μl. Cell concentration can be adjusted according to requirements of the flow cytometry instrument in use.
2. Fix and permeabilize the cells with fixation and permeabilization solution at 4 °C or on ice for 20 min. Alternatively, to stain viral proteins on cell surface, the permeabilization step can be omitted.
3. Wash cells twice in fixation and permeabilization solution and resuspend the cells in the same buffer. To stain viral protein on cell surface, replace the solution with 1× PBS. Block the cells with 10 μl of blocking solution (Block Ace) for 10 min, on ice.
4. Add 10 μl of diluted primary antibody to the cell solution. Incubate the cells on ice for 60 min. Optimal staining concentration and incubation periods should be determined prior to this step using flow cytometry.
5. After incubation, vortex or gently flick the microfuge tubes a few times to avoid cell aggregation. Wash the cells with Perm/Wash buffer twice and pellet the cells by centrifuging at 800 × *g* for 5 min (or optimal centrifugation speed). Resuspend the cells in the same buffer.
6. Add 10 μl of diluted secondary antibody to the cell suspension. Incubate the cells on ice for 30 min, in dark. Wash the cells with wash solution. Resuspend the cells in 1× DPBS prior to flow cytometric analysis at a concentration between 5×10^5 and 1×10^6 cells/100 μl.

4 Notes

1. Because Vero and BHK cells are susceptible to CHIKV infection, these cell lines can be used as positive controls and to determine virus titers. The cells can be replaced with other adherent or nonadherent cells, however the optimal multiplicity of infection (MOI) and incubation period for each of the cell lines should be predetermined with a positive control (Vero or BHK cell lines).
2. When using nonadherent cells, add the virus suspension directly to the cells. After the incubation step, the old medium can be exchanged with fresh medium (EMEM supplemented with 10 % heat-inactivated FBS).

3. Cell supernatant culture could be used to determine virus titers in the supernatant. Methods including RT-PCR and plaque titration can be adopted to determine virus concentration in the cell supernatant.
4. Adjust the optimal cell concentration for each cell line. Overly dense concentration will result in difficulty in determining the number of infected cells. Low cell concentration will result in sparse cell seeding, leading to false-negative results, particularly in cells that require a high MOI for infection.
5. The slides can be dried under the laminar flow for 10 min and stored at −20 °C for a few months. The staining step can be continued at a later period; the slides should be allowed to reach room temperature by incubation under the laminar flow for 10 min.
6. Cell surface should be dry after 10 min of incubation under the air-flow of the clean-bench. Incubate a further 10 min if the surface does not appear to be dry. Until the drying and fixation steps are completed, the slide glasses should be treated as materials containing infectious agents according to institutional regulations.
7. Adding antibodies to the slide surface can be tricky. Gently add the diluted antibodies to avoid spill-over to the next well. Do not leave the slide surface dry after this step. Excess DPBS on hydrophobic surface should also be removed to avoid cross-contamination of antibodies. Also, do not touch the well surface as this will result in the removal of cells from the slide glass.
8. Either fluorescent-labeled, unlabeled primary antibodies or convalescent sera can be used for staining. Choose a fluorescent secondary antibody that is raised against the host of the primary antibody. Generally, Alexa Fluor fluorescence-conjugated antibodies are suitable for staining.
9. The optimal concentration of primary and secondary antibody for fluorescent detection is determined by using a series of serially diluted antibody (i.e. 1:100, 1:500, 1:1000, 1:5000). Most purified monoclonal antibodies (0.1–1 mg/ml) work at an optimal concentration with dilutions ranging from 1:100 to 1:1000 for immunofluorescent staining. However, for convalescent sera, polyclonal antibodies or ascites, higher antibody concentration (i.e. 1:10–1:100) may be needed for optimal staining.
10. Moist chambers can be obtained commercially or prepared by using polystyrene storage box or bread box. For bread box, attach two tubular cylinders (serological pipette or plastic straw) to the bottom of the container, at widths suitable for horizontal placing of slide glasses. Place paper towels moistened with

ddH$_2$O on the bottom of the container and place the slide glasses on top of the cylinders. Cover the container with aluminum foil and proceed with the incubation step.

11. In the event that the antibody solution does not spread evenly across the well, use a 1–10 μl pipette tip positioned horizontally to spread the antibody evenly across the well. Avoid touching the pipette tip on the surface of the well as this will result in removal of the cells from the well.
12. Divide 100 μl of the mounting solution evenly across the slide glass. Gently lay the cover slip over the slide glass with a pair of tweezers to avoid bubble formation on the slide glass. Alternatively, add a few drops of mounting solution to the edge of the slide, and lay the cover slip (at an angle of 30–40°) starting from the edge of the slide. Blot off excess amounts of mounting solution with paper towel.
13. Detection of anti-IgM antibodies (serological diagnosis) can be used to identify acute viral infection after viremia clearance (typically 1 week after infection for CHIKV patients and up to a few months after acute infection; [14]). Thus, the modified IFA method to detect virus-specific antibodies is particularly useful in differential diagnoses of CHIKV patients with persistent arthralgia and convalescent CHIKV cases, in the event where ELISA kits are not available.

Acknowledgements

This work was supported in part by the research grant, Research on Emerging and Re-emerging Infectious Diseases (H26-shinkou-jitsuyouka-007), from the Ministry of Health, Labour and Welfare, Japan, the Environment Research and Technology Development Fund (S-8) of the Ministry of the Environment, and a Grant-in-Aid for Young Scientists (B) from JSPS (26870872).

References

1. Weaver SC, Forrester NL (2015) Chikungunya: evolutionary history and recent epidemic spread. Antiviral Res 120:32–39
2. Borgherini G, Poubeau P, Jossaume A, Gouix A, Cotte L, Michault A, Arvin-Berod C, Paganin F (2008) Persistent arthralgia associated with chikungunya virus: a study of 88 adult patients on reunion island. Clin Infect Dis 47(4):469–475
3. Yap G, Pok KY, Lai YL, Hapuarachchi HC, Chow A, Leo YS, Tan LK, Ng LC (2010) Evaluation of Chikungunya diagnostic assays: differences in sensitivity of serology assays in two independent outbreaks. PLoS Negl Trop Dis 4(7):e753
4. Raquin V, Wannagat M, Zouache K, Legras-Lachuer C, Moro CV, Mavingui P (2012) Detection of dengue group viruses by

fluorescence in situ hybridization. Parasit Vectors 5:243

5. Moi ML, Lim CK, Takasaki T, Kurane I (2010) Involvement of the Fc gamma receptor IIA cytoplasmic domain in antibody-dependent enhancement of dengue virus infection. J Gen Virol 91(Pt 1):103–111
6. O'Brien CA, Hobson-Peters J, Yam AW, Colmant AM, McLean BJ, Prow NA, Watterson D, Hall-Mendelin S, Warrilow D, Ng ML, Khromykh AA, Hall RA (2015) Viral RNA intermediates as targets for detection and discovery of novel and emerging mosquito-borne viruses. PLoS Negl Trop Dis 9(3):e0003629
7. Teo CS, Chu JJ (2014) Cellular vimentin regulates construction of dengue virus replication complexes through interaction with NS4A protein. J Virol 88(4):1897–1913
8. Tochitani K, Shimizu T, Shinohara K, Tsuchido Y, Moi ML, Takasaki T (2014) The first case report of Ross River virus disease in a Japanese patient who returned from Australia (In Japanese). Kansenshogaku Zasshi 88(2):155–159
9. Chawla T, Chan KR, Zhang SL, Tan HC, Lim AP, Hanson BJ, Ooi EE (2013) Dengue virus neutralization in cells expressing Fc gamma receptors. PLoS One 8(5):e65231
10. deAlwis R, de Silva AM (2014) Measuring antibody neutralization of dengue virus (DENV) using a flow cytometry-based technique. Methods Mol Biol 1138:27–39
11. World Health Organization (2007) Guidelines for plaque-reduction neutralization testing of human antibodies to dengue viruses. http://www.who.int/immunization/documents/date/en/index.html
12. Moi ML, Lim CK, Tajima S, Kotaki A, Saijo M, Takasaki T, Kurane I (2011) Dengue virus isolation relying on antibody-dependent enhancement mechanism using FcγR-expressing BHK cells and a monoclonal antibody with infection-enhancing capacity. J Clin Virol 52(3):225–230
13. Lim CK, Nishibori T, Watanabe K, Ito M, Kotaki A, Tanaka K, Kurane I, Takasaki T (2009) Chikungunya virus isolated from a returnee to Japan from Sri Lanka: isolation of two sub-strains with different characteristics. Am J Trop Med Hyg 81(5):865–868
14. Aoyama I, Uno K, Yumisashi T, Takasaki T, Lim CK, Kurane I, Kase T, Takahashi K (2010) A case of Chikungunya fever imported from India to Japan, follow-up of specific IgM and IgG antibodies over a 6-month period. Jpn J Infect Dis 63(1):65–66

Chapter 14

Virus Isolation and Preparation of Sucrose-Banded Chikungunya Virus Samples for Transmission Electron Microscopy

Chang-Kweng Lim

Abstract

Virus isolation and purification is an invaluable technique in virology to detect and characterize viruses. This chapter describes a large-scale Chikungunya virus (CHIKV) propagation and purification methods by using discontinuous sucrose gradient, and sample preparation for transmission electron microscopy. Sucrose-banding yields large quantities of high-titer (10^{10} pfu/ml) CHIKV stocks. Such stocks are stable for years when stored at −70 °C.

Key words Virus propagation, Sucrose gradient, Negative staining, Immunogold labeling

1 Introduction

Clinical symptoms of Chikungunya virus (CHIKV) infection are acute febrile illnesses accompanied by chills, headache, nausea, vomiting, joint pain, myalgia, and rash [1]. Most patients have high and relatively persistent levels of viremia (>10^7 RNA copies/ml) during the first 48 h of illness, and this may persist for up to 6 days in some patients [2]. Virus isolation by inoculation on mosquitoes or intracerebral inoculation of suckling mice is possible. In vitro cell culture methods using mosquito cell lines (C6/36) and other mammalian cell lines (Vero and BHK-21) have comparable sensitivity to in vivo methods and are quicker.

Experimental reproducibility also requires CHIKV stocks to be prepared in a consistent manner. One simple approach is to prepare a cytopathic effect (CPE) stock. After CPE appears, the culture supernatant is collected and then the CHIKV is titered by plaque assay. The CPE stocks, which are typically 10^7–10^8 pfu/ml, are adequate for exploratory studies. Large-scale virus stocks are usually prepared by banding the virus in discontinuous sucrose gradients, and this procedure will be described.

Justin Jang Hann Chu and Swee Kim Ang (eds.), *Chikungunya Virus: Methods and Protocols*, Methods in Molecular Biology, vol. 1426, DOI 10.1007/978-1-4939-3618-2_14,

Negative staining for electron microscope (EM) was introduced in 1959 and revolutionized virology in generating virus images of unprecedented clarity [3, 4]. It is indispensable for structural studies and virus identification [5]. Morphologically, CHIKV is spherical, approximately 70 nm in diameter, with a lipid bilayer membrane which is composed of lipids derived from the plasma membrane of the host cell. Mature virions sediment at 280S, and have a buoyant density of 1.22 g/cm^3 in sucrose. The virion relative molecular mass (Mr) is about 52×10^6. Lipid comprises 30 % of the dry weight of virions and is enriched in cholesterol and sphingolipid. Because of the lipid envelope, alphaviruses are readily inactivated by organic solvents and detergents. Virus infectivity is stable in the pH range of 7–8, but it is rapidly inactivated by low pH, or by 58 °C [1, 6].

Virus isolation is the most reliable evidence of infection and transmission electron microscopy (TEM) allows rapid morphological identification and differential diagnosis of different agents contained in the specimen [7]. Thus, this chapter describes the CHIKV isolation from acute patent serum followed by virus purification and sample preparation for TEM.

2 Materials

2.1 Isolation and Propagation of CHIKV

1. Acute patient serum. Collect serum specimens 2 days after onset of symptoms from Chikungunya fever patient and store at −70 °C until use [2].
2. Vero cell (ATCC: CCL-81).
3. Growth medium: 10 % heat-inactivated FCS, Eagle's MEM (EMEM), 1 % penicillin-streptomycin.
4. Tissue culture flasks (150 cm^2).
5. Tissue culture flasks (75 cm^2).
6. Tissue culture multilayer flask (1720 cm^2).
7. 1× Dulbecco's Phosphate-Buffered Saline (DPBS) (−).
8. 30× staining dye. Add 2.25 g of methylene blue and 0.375 ml of 1 N (1 mol/l) NaOH in 200 ml of double-distilled water (DW$_2$). Filter through Whatman filter paper, and store at room temperature (RT).
9. Methylcellulose (MC) overlay medium: 1 % Methylcellulose in EMEM, 1× Glutamax, 0.22 % $NaHCO_3$.
10. 10 % formalin in DPBS (−): Add 50 ml of formalin in 450 ml of DPBS (−).

2.2 Virus Purification

1. 5× PEG-it virus precipitation solution (System Bioscience).
2. 1 M Tris–HCl, pH 8.0. Add 60.55 g of Tris in 400 ml of DW$_2$, adjust to pH 8.0 with HCl and fill up to 500 ml with DW$_2$.

3. TNE buffer: 50 mM Tris–HCl, pH 8.0, 150 mM NaCl, 5 mM EDTA.
4. 30 % (w/v) sucrose in TNE buffer. Add 30 g of sucrose, 5 ml of 1 M Tris–HCl, pH 8.0, 5 ml of 0.1 M EDTA, 3 ml of 5 M NaCl and fill up to 100 ml with DW_2.
5. 60 % (w/v) sucrose in TNE buffer. Add 60 g of sucrose, 5 ml of 1 M Tris–HCl, pH 8.0, 5 ml of 0.1 M EDTA, 3 ml of 5 M NaCl and fill up to 100 ml with DW_2.
6. Centrifuge clear tube for SW28 bucket rotor (38-ml, 25 × 89 mm).
7. Centrifuge clear tube for SW41Ti bucket rotor (12-ml, 14 × 89 mm).
8. SW28 swinging bucket rotor.
9. SW41Ti swinging bucket rotor.
10. P200 micropipettor.

2.3 Negative Staining and Immunogold Labeling

1. EM forceps, reverse N4-style (DUMONT).
2. P20 and P200 micropipettors.
3. Carbon-coated copper 400 mesh EM grids (VECO B. V.).
4. DW_2 or other extremely clean water.
5. 0.2 M phosphate buffer, pH 7.4. Prepare 27.6 g/l of sodium dihydrogen phosphate monohydrate (0.2 M NaH_2PO_4-H_2O; [Solution A]) and 53.6 g/l of disodium hydrogen phosphate 7-water (0.2 M Na_2HPO_4-$7H_2O$; [Solution B]). To make 0.2 M phosphate buffer, pH 7.4, use 22.6 parts of Solution A, 77.4 parts of Solution B in a total of 100 parts.
6. 2.5 % (w/v) glutaraldehyde in 0.1 M phosphate buffer, pH 7.4. Add 40 ml of DW_2 to 10 ml of 25 % glutaraldehyde solution. Then add 50 ml of 0.2 M phosphate buffer, pH 7.4.
7. 4 % uranyl acetate solution in DW_2. Dissolve 0.08 g uranyl acetate in 2 ml DW_2, filter through a 0.2 μm syringe filter to remove precipitates.
8. Whatman filter paper cut into wedges.
9. 1× DPBS (−).
10. Blocking solution: 0.1 % (w/v) bovine serum albumin (BSA) in DPBS (−). Add 10 mg of BSA in 10 ml of DPBS (−). Filter through a 0.45 μm filter, aliquot, and store at −70 °C.
11. First antibody: Anti-CHIKV antibody in ascitic fluid (AF).
12. Gold particles (diameter, 5 nm) conjugated to goat anti-mouse IgG and IgM antibody (British Biocell International).

3 Methods

Carry out all procedures at room temperature unless otherwise specified.

3.1 CHIKV Isolation

1. Plate the Vero cells on a 12-well plate (final concentration; 2×10^5 cells/well) and culture at 37 °C in 5 % CO_2 for o/n.
2. Inoculate 50 μl of acute patient serum to each well of 12-well plates containing a confluent Vero cell monolayer (*see* **Note 1**).
3. Incubate the plates at 37 °C for 90 min in a 5 % CO_2 incubator after virus inoculation. Tilt the plates gently every 15 min.
4. Add 2 ml of growth medium per well and incubate the plates at 37 °C in 5 % CO_2 incubator for 7 days.
5. After approx. 2–4 days, the cells in the monolayer will show typical CHIKV CPE, i.e., they will round up and detach from the dishes into individual floating (Fig. 1). After CPE appears, collect the culture supernatant, stock at −80 °C until ready for use.
6. Perform virus titration of stock CHIKV by using a plaque assay.

3.2 CHIKV Titration

1. To perform virus titration of stock virus by plaque assay, dilute the virus stock by tenfold from 1:10 to $1{:}10^8$ in a growth medium (*see* **Note 2**).
2. Inoculate 50 μl of the serially diluted stock virus to each well of 12-well plates containing a confluent Vero cell monolayer.
3. Incubate the plates at 37 °C for 90 min in a 5 % CO_2 incubator after virus inoculation. Tilt the plates gently every 15 min.
4. Add 1 ml of MC overlay medium to each well, and then incubate the plates at 37 °C in a 5 % CO_2 incubator for 3 days.
5. Fix with 0.5 ml of 10 % formalin in DPBS (−) at room temperature for 30 min, discard formalin and rinse plates with tap water.
6. Add 0.5 ml of 1× staining dye in DW_2 to each well at room temperature for 30 min, discard staining dye, rinse plates with tap water and air-dry.
7. Count the number of plaques in each well against blue-stained cell background, and select wells with 50–150 plaque for each specimen.
8. Calculate the average of plaque number for each specimen in the wells.
9. Multiply the average by dilution factor and divide by inoculum volume (0.05).
10. Resulting value is stock virus infectivity titer by pfu/ml.

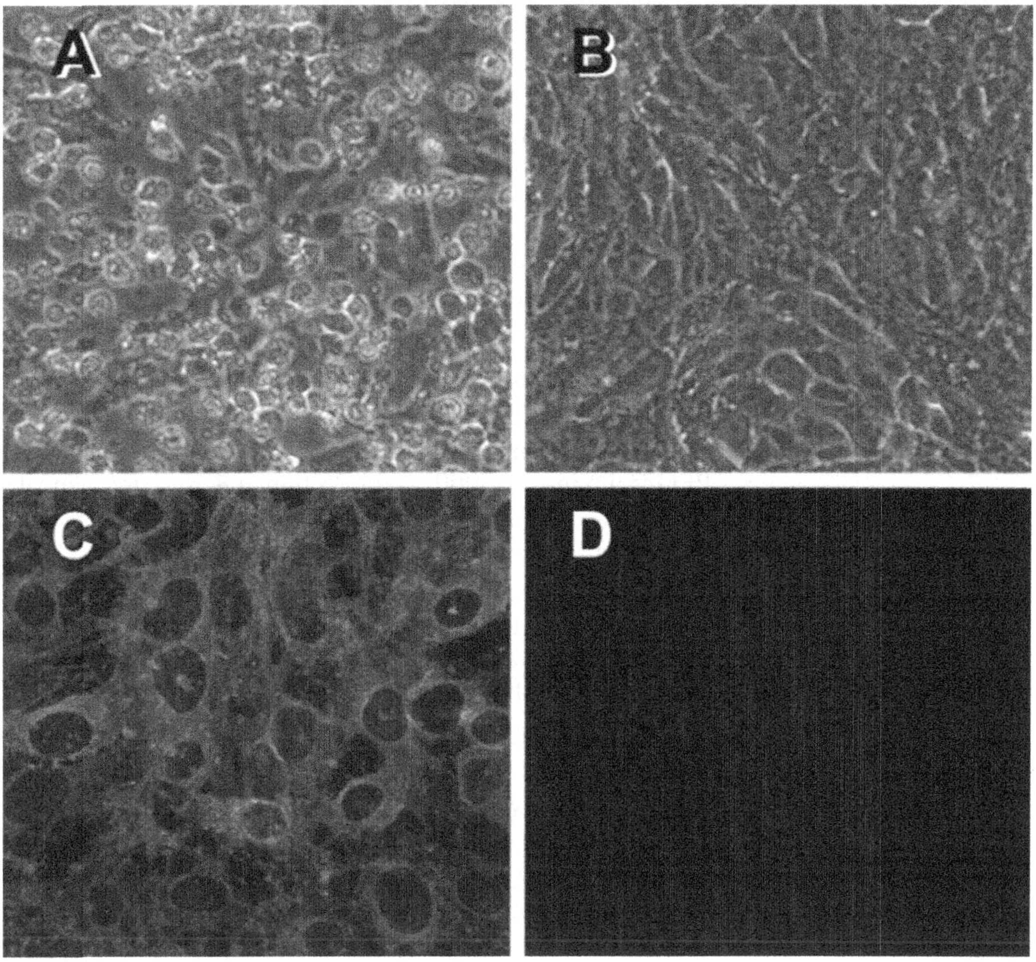

Fig. 1 Virus identification and isolation. The serum specimens obtained on day 2 and day 8 were examined for virus detection and isolation. The cytopathic effect (CPE) was detected 4 days after the inoculation with the day 2 acute serum (**a**), but not with the day 8 convalescent serum (**b**). Indirect immunofluorescence staining revealed that the cells infected with the CPE culture supernatant contained the viral antigen (**c**), but not those infected with the day 8 serum culture supernatant (**d**)

3.3 Large-Scale CHIKV Preparation for Virus Purification

1. Plate the cells at 560 ml/flask (final concentration; 1.1×10^8 cells/flask) on a multilayer flask (1720 cm^2) and culture at 37 °C in 5 % CO_2 for overnight.
2. Dilute the virus stock to 1×10^7 pfu/ml. Remove 5 ml of culture medium and inoculate 1 ml of the diluted virus to multilayer flask containing a confluent Vero cell monolayer (MOI 0.1; *see* **Note 3**).
3. Incubate the plates at 37 °C for 60 min in a 5 % CO_2 incubator after virus inoculation. Tilt the plates gently every 15 min.
4. Add 4 ml of growth medium to a multilayer flask, and then incubate the flask at 37 °C in a 5 % CO_2 incubator for 2–3 days.

5. Harvest the culture medium when weak cytopathic effects (CPE) appeared and spin down cell debris with low-speed centrifugation (15 min at 4 °C and 15,000 × *g*).
6. Aliquot the CHIKV-containing culture supernatant into cryogenic vials and store at −80 °C until ready use.
7. Perform virus titration of stock CHIKV by using a plaque assay (*see* **Note 2**).

3.4 Virus Purification by Polyethylene Glycol (PEG Precipitation)

1. Transfer the CHIKV-containing culture supernatant to a sterile vessel and add 1 volume of cold 5× PEG-it virus precipitation solution (4 °C) to every 4 volumes of the culture supernatant.
2. Refrigerate overnight (at least 12 h). Centrifuge supernatant/PEG-it mixture at 3000 × *g* for 30 min at 4 °C. After centrifugation, the CHIKV particles appear as a white pellet at the bottom of the vessel (*see* **Note 4**).
3. Transfer supernatant to a fresh tube. Spin down residual PEG-it solution by centrifugation at 1500 × *g* for 5 min. Remove all traces of fluid by aspiration, taking great care not to disturb the precipitated CHIKV particles in pellet.
4. Resuspend/combine viral pellets in 1/10 of original volume using cold, TNE buffer at 4 °C.
5. Aliquot into cryogenic vials and store at −80 °C until ready for purification with a sucrose step gradient (*see* **Note 2**).

3.5 Virus Purification by Discontinuous Sucrose Gradient

1. To prepare the 60–30 % (w/v) sucrose step gradient, add 5 ml of 60 % (w/v) sucrose in TNE buffer to a 38.5-ml ultracentrifuge tube.
2. Slowly layer the 5 ml of 30 % (w/v) sucrose in TNE buffer above the 60 % by dripping it down the side of the tube, holding the pipette close to the existing liquid level and keeping the tube at eye level to observe the formation of a layer between the two solutions. Continue dripping slowly until the gradient is complete.
3. Slowly layer up to 25 ml of virus stock from the PEG precipitation step above the 30 % (w/v) sucrose. If necessary add TNE buffer, so that the tube is filled to within a few millimeters of the top.
4. Centrifuge the samples in an SW28 swinging bucket rotor at 121,000 × *g* for 3 h at 4 °C. The density of alphavirus is 1.22 g/cm^3 in sucrose.
5. After centrifugation, inspect the samples for white band. The virus is generally observed above the 60 % (w/v) sucrose layer. Collect the band using a Pasteur pipette. A black piece of paper behind the tubes will help visualize the bands present after centrifugation.

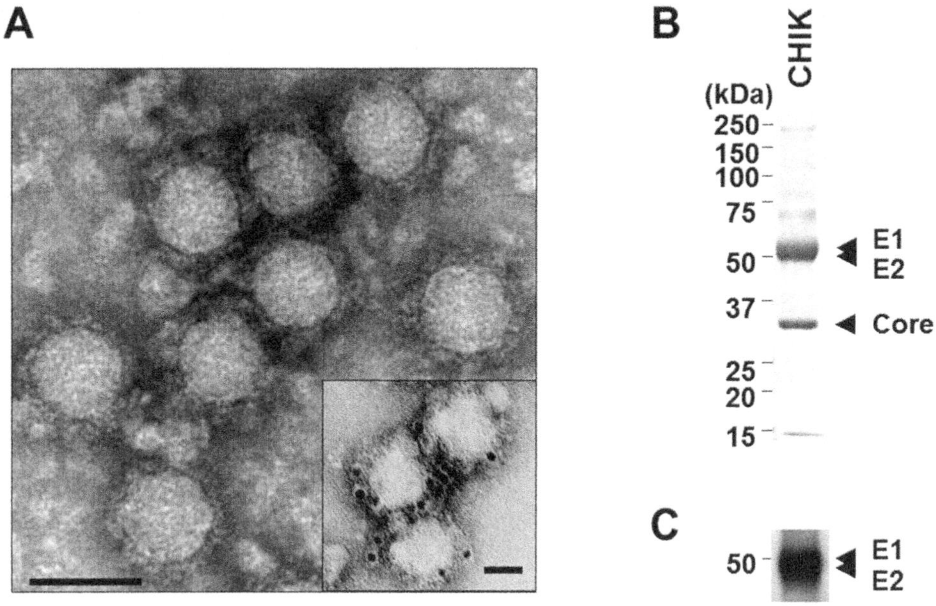

Fig. 2 Morphological and biochemical characteristics of the purified CHIKV. The purified virus (SL10571, GenBank accession AB455494) was fixed and visualized by TEM (bar = 100 nm). The *inset* shows the immunogold labeling of partially purified particles with anti-CHIKV monoclonal antibody and gold particles (diameter, 10 nm) conjugated to goat anti-mouse IgG and IgM antibody (bar = 50 nm) (**a**). The purified virus was electrophoresed on a 12.5 % gel, stained with Coomassie Brilliant Blue (**b**) and analyzed by Western blot (**c**)

6. Dilute the collected band with TNE buffer to get density less than 1.1 (g/cm^3; *see* **Note 5**).
7. Then, layer the diluted sample onto 5 ml of 30 % (w/v) and 5 ml of 60 % (w/v) discontinuous sucrose gradient and repeat centrifugation with the SW28 swinging bucket rotor at 121,000 × *g* for 3 h at 4 °C.
8. Collect the band and dilute sample with TNE buffer to get density less than 1.1 g/cm^3 to load onto a 30 % (w/v) sucrose gradient bed to pellet down the virus.
9. To prepare the sucrose gradient bed, add 2 ml of 30 % (w/v) sucrose in TNE buffer to a 12-ml ultracentrifuge tube.
10. Add up to 10 ml of virus above the 30 % sucrose by dripping it down the side of the tube. Continue dripping slowly until the gradient is complete. If necessary add TNE, so that the tube is filled to within a few millimeters of the top.
11. Centrifuge the samples in an SW41Ti swinging bucket rotor at 281,000 × *g* for 1 h at 4 °C.
12. After centrifugation, discard the supernatant carefully. Resuspend the pellet with 200 μl of TNE.
13. Place the resuspended pellet into 1.5 ml tube and spin for 15,000 × *g* for 15 min at 4 °C.
14. Collect the supernatant and observe under TEM (Fig. 2; *see* **Note 6**).

3.6 Negative Staining

1. Use the reverse grip forceps to pick up a grid. Set the forceps with the carbon-side up on the EM grid.
2. Apply 10 μl of the purified CHIKV onto a carbon-coated copper 400 mesh EM grid for 1 min and blot onto filer paper by holding the grid perpendicular to the paper. Use a Whatman paper wedge to soak up the stain on the grid by only touching the outer copper ring of the grid.
3. Wash the grid with 10 μl of DW_2 for three times to wash away all the sucrose. In between each step, carefully blot onto filter paper. Try not to get the water on the back of the grid.
4. Immediately fix the sample with 10 μl of the 2.5 % (w/v) glutaraldehyde in 0.1 M phosphate buffer for 5 min at RT and blot onto filter paper, followed by three quick washes in DW_2.
5. Apply 10 μl of the 4 % uranyl acetate onto the carbon/sample side of grid for 1 min and blot onto filter paper and allow to air-dry. Keep drying the grid until the surface looks like a shiny oil slick. The stain is a fixative and will preserve the structures. Leave grids along for another few minutes to finish air-drying, then place in a grid box (*see* **Notes 7** and **8**).
6. Examine the grid under the TEM. Grids should be looked at immediately—water absorbed from the air can ruin samples (Fig. 2). If it is not possible to look at samples immediately, place grid box in a container with desiccant.

3.7 Immunogold Labeling

1. Put a 50 μl of droplet blocking solution on a plastic 10 cm petri dish.
2. Float the sample-absorbed grids sample side down, onto the blocking solution surface for 15 min. In between each step, carefully blot onto filter paper.
3. Then, float the grids for 15 min on a droplet of the appropriate anti-CHIKV monoclonal antibody (AF diluted in 1:500 blocking solution), followed by three quick washes in DPBS (−).
4. Float the grids for 15 min on a droplet of gold particles (diameter, 5 nm) conjugated to goat anti-mouse IgG and IgM antibody diluted 1:20 in blocking solution, followed by three quick washes in DPBS (−).
5. Immediately fix the sample with 15 μl of the 2.5 % (w/v) glutaraldehyde in a 0.1 M phosphate buffer for 5 min at RT and blot onto filter paper, followed by three quick washes in DW_2.
6. Stain 1 min in 10 μl of 4 % uranyl acetate and blot onto filter paper and allow to air-dry. Leave grids along for another few minutes to finish air-drying, then place in a grid box.
7. Examine the grid under the TEM (Fig. 2).
8. In each experiment, prepare negative control grids with no CHIKV antibody and a nonspecific antibody.

4 Notes

1. Collect the serum samples 2 days after onset of symptoms from the acute Chikungunya fever patient and store at −80 °C until ready for use [2]. Because viremia of CHIKV is brief, success of virus-based tests depends on early admission and clinical recognition of infection [2].
2. Perform virus titration on all the harvested supernatant fluids by a plaque assay to confirm the presence of infectious CHIKV and to determine viral titers.
3. If virus stock is not enough for preparation of large-scale stock virus, perform a small-scale stock preparation. For a small-scale stock CHIKV preparation, plate the cells on T-75 flask (final concentration; 2.0×10^6 cells/flask) and culture at 37 °C in 5 % CO_2 for overnight. Then, remove growth medium, inoculate 1 ml of the diluted virus to the flask (MOI 0.1), and incubate for 90 min at 37 °C in 5 % CO_2. Add 10 ml of growth medium to the flask and incubate at 37 °C in 5 % CO_2 for 7 days. Harvest the culture medium when weak cytopathic effects (CPE) appear and spin down cell debris with low-speed centrifugation (15 min at 4 °C and $15{,}000 \times g$). Take the supernatant and store at −80 °C until use. Perform virus titration of stock virus by a plaque assay.
4. Precipitation of viral particles from large volumes can be achieved by using the Nalgene 250 ml polypropylene centrifuge tube, following manufacturer's instructions.
5. According to sucrose conversion table by USDA (file code 135-A-50; 1981), the apparent specific gravity at 20/20 °C of 60 and 30 % (w/v) sucrose are 1.28908 and 1.12913, respectively.
6. Analyze the viral proteins by 12.5 % sodium dodecyl sulfate polyacrylamide gel electrophoresis (SDS PAGE) and Western blotting with anti-CHIKV monoclonal antibody and sheep anti-mouse IgG antibody HRP conjugate.
7. Aspirate 10 μl of stain into a P20 micropipettor. Take stain from the top of the tube without mixing the tube. Uranyl acetate makes precipitates that will sink to the bottom of the tube and will show up as big black marks on grid that should be avoided.
8. For a diluted sample, longer waiting times may be necessary. If there is too much stain left on the grid, that is all you will see in the TEM is stain. Conversely, if there is too little stain left on the grid, the viruses will not be negatively stained.

Acknowledgments

The author would like to express his deep sense of gratitude to Dr. Tomohiko Takasaki for countless discussions, and for critical review of this manuscript. I also want to thank Keiko Tanaka for her great help with TEM. I also would like to thank Dr. Polly Roy, in whose laboratory I learned the fundamentals of viral purification and TEM operation.

References

1. Lim CK, Kurane I, Takasaki T (2010) Re-emergence of chikungunya virus. In: Maeda A (ed) Animal viruses. Transworld Research Network, Kerala, India, pp 1–22
2. Lim CK, Nishibori T, Watanabe K, Ito M, Kotaki A, Tanaka K, Kurane I, Takasaki T (2009) Chikungunya virus isolated from a returnee to Japan from Sri Lanka: Isolation of two sub-strains with different characteristics. Am J Trop Med Hyg 81(5): 865–868
3. Brenner S, Horne RW (1959) A negative staining method for high resolution electron microscopy of viruses. Biochim Biophys Acta 34:103–110
4. Ackermann HW, Heldal M (2010) Basic electron microscopy of aquatic viruses. In: Wilhelm SW, Weinbauer MG, Suttle CA (eds) Manual of aquatic viral ecology. ASLO, Waco, TX, pp 182–192
5. Limn CK, Roy P (2003) Intermolecular interactions in a two-layered viral capsid that requires a complex symmetry mismatch. J Virol 77: 11114–11124
6. Strauss JH, Strauss EG (1994) The alphaviruses: gene expression, replication, and evolution. Microbiol Rev 58:491–562
7. Hazelton PR, Gelderblom HR (2003) Electron microscopy for rapid diagnosis of infectious agents in emergent situations. Emerg Infect Dis 9(3):294–303

Chapter 15

Viral–Host Protein Interaction Studies Using Yeast Two-Hybrid Screening Method

Namrata Dudha and Sanjay Gupta

Abstract

Yeast two-hybrid (Y2H) assay is one of the earliest methods developed to study protein–protein interactions. In the proteomics era, Y2H has created a niche of its own by providing protein interaction maps for various organisms. Owing to limited coding capacities of their genomes, viruses are dependent on their host cellular machinery for successful infection. Identification of the key players orchestrating the survival of virus in their host is essential for understanding viral life cycle and devising strategies to prevent interactions resulting in pathogenesis. In this chapter, Y2H assay will be explained in detail for studying viral–host protein interactions of Chikungunya virus (CHIKV).

Key words Chikungunya virus, Human cDNA library, Protein–protein interactions, Virus–host interactions, Yeast two-hybrid

1 Introduction

Yeast two-hybrid (Y2H) assay, one of the most widely used techniques to study protein–protein interactions, was first described by Fields and Song [1, 2]. This assay takes advantage of the fact that distinct DNA-binding domain (DNA-BD) and transcription activation domain (AD) of many eukaryotic transcription factors are capable of activating transcription when in close proximity. Genes of the proteins to be tested for interaction are fused to either Gal4 DNA-BD (bait) or Gal4 AD (prey), and co-expressed in budding yeast cells. In case the two candidate proteins interact, they are capable of initiating transcription of reporter genes that have been engineered into the yeast genome (Fig. 1). As a genetic technique, Y2H offers a sensitive means to test the direct interaction between two targeted proteins, or to use one protein as a bait to screen cDNA libraries from desired cell types, tissues, or entire organisms. The identity of the interacting partners is then obtained by sequencing the corresponding plasmids from the selected yeast colonies. Screening of libraries has been made easy by exploiting recent advances in

Justin Jang Hann Chu and Swee Kim Ang (eds.), *Chikungunya Virus: Methods and Protocols*, Methods in Molecular Biology, vol. 1426,DOI 10.1007/978-1-4939-3618-2_15,

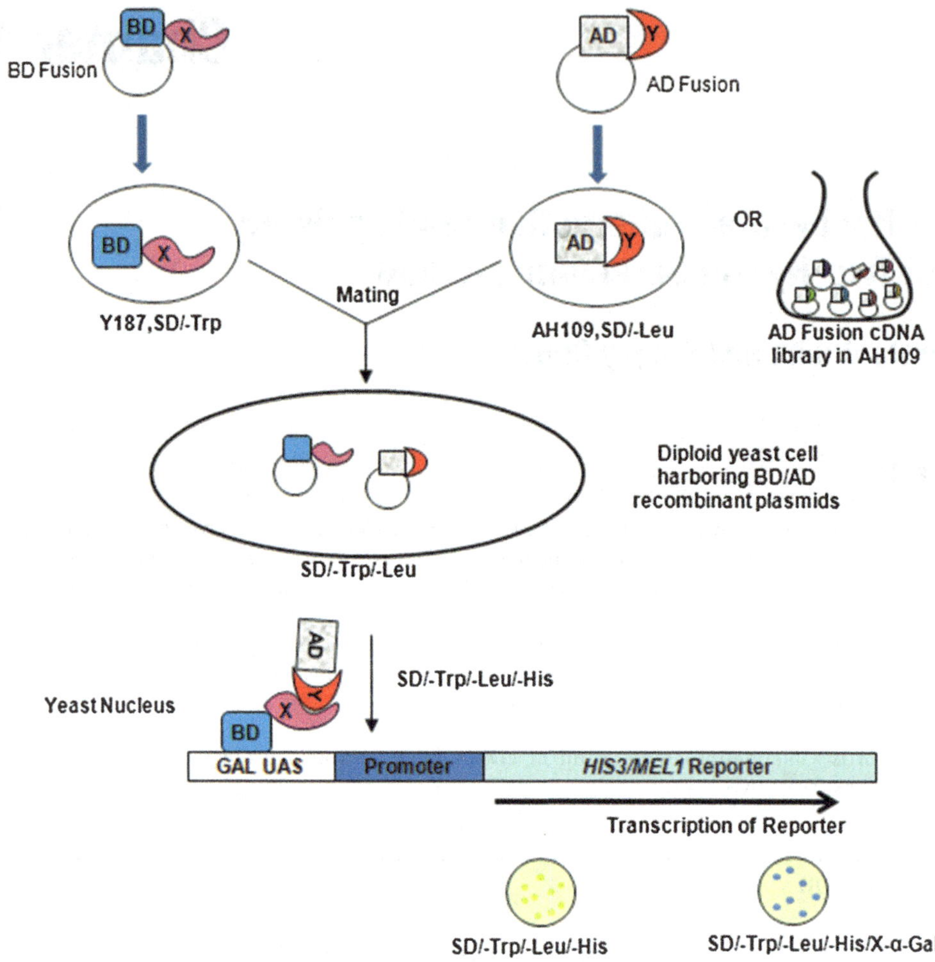

Fig. 1 The principle of the yeast two-hybrid system: The bait plasmid encodes protein X fused to the C-terminus of a transcription factor DNA-binding domain (BD) and the prey plasmid encodes protein Y fused to an activation domain (AD). Alternatively, the prey can consist of proteins encoded by an expression library. Each plasmid is introduced into an appropriate yeast strain (BD-X in Y187 and AD-Y in AH109) for mating. Following mating, the diploid clones containing both the plasmids were selected on minimal synthetic dropout medium containing all amino acids except tryptophan, leucine, and histidine (SD/-Trp/-Leu/-His). Only if proteins X and Y physically interact with one another, the BD and AD brought together to reconstitute a functionally active transcription factor that binds to upstream specific activation sequences (UAS) of the promoter and activate the expression of the reporter genes (*HIS3/MEL1*)

"interaction mating," a method that uses yeast cell conjugation to combine two plasmids in a yeast cell. Currently, two transcription factor systems based on either Gal4 or LexA are used in Y2H studies. In addition, sensitivity of the assay has been enhanced by utilizing multiple nutritional markers and enzymatic reporters (e.g. *HIS*, *MEL1*, and *Lac Z*) in tandem for interaction analysis.

Y2H assay has been previously employed to study viral–viral [3–9] and virus–host interactions in a global fashion [10–16]. This chapter describes a Gal4-based Y2H system to study the viral–host interactions of CHIKV envelope proteins. Ectodomains of envelope proteins E1 and E2 expressed as BD fusions in AH109 yeast cells will be used to screen human fetal brain cDNA library encoding host proteins as AD-fusion in Y187 yeast cells by mating [17]. Briefly, the method involves transformation of yeast AH109 cells using viral baits as BD fusion, followed by nutritional selection of transformants on a synthetic medium lacking tryptophan. Expression of the envelope proteins in yeast cells will be detected by Western blot of the supernatant obtained on yeast cell lysis, using an anti-c-Myc antibody. Yeast AH109 cells expressing the viral envelope protein as BD fusion will be selected on synthetic medium in the absence of tryptophan and histidine to test autoactivation of reporter gene (*HIS*) by the viral envelope proteins. If none of the envelope proteins (E1/E2) activate the reporter; mate each viral BD fusion with the human fetal brain cDNA library encoding host proteins as AD-fusion in Y187 yeast cells. Select the diploid cells harboring viral (bait) and host (prey) proteins on a triple dropout medium (TDO), devoid of tryptophan, leucine, and histidine; supplemented with X-α-Gal for blue-white screening of the interacting partners. The blue-colored colonies obtained on TDO medium are to be further selected on a high stringency medium in the absence of tryptophan, leucine, histidine, and adenine; supplemented with X-α-Gal, to eliminate false positives. The positive transformants will be screened for presence of the prey construct by yeast colony PCR followed by elimination of duplicate and multiple prey plasmids from the screened diploid cells. The prey plasmids will then be isolated from yeast and transformed in *E. coli* DH5α cells. They will then be rescued from bacterial cells and the identity of each prey protein will be analyzed by sequencing. In this analysis, host binders of E1 and E2 proteins identified may be involved in facilitating viral entry, translocation of viral proteins across the cell, maturation of viral spike at the time of budding and in suppression of immune response by the virus.

2 Materials

Prepare all the solutions in deionized water unless stated otherwise.

2.1 Yeast Strains and Plasmids Used in GAL4 Y2H System

1. AH109 (*MATa, trp1-901, leu2-3,112, ura3-52, his3-200, gal4D, gal80D, LYS2:: GAL1UAS GAL1TATA HIS3, GAL2UAS GAL2TATA ADE2, URA3:: MEL1UAS MEL1TATA lacZ*; Clontech).
2. Y187 (*MATα, ura3-52, his3-200, ade2-101, trp1-901, leu2-3,112, gal4D, met-, gal80D, URA3:: GAL1UAS GAL1TATA-lacZ*; Clontech).
3. Plasmids: GAL4-system bait—pGBKT7 (BD; selectable yeast marker-*TRP1*), GAL4-system prey—pGADT7 (AD; selectable yeast marker-*LEU2*).
4. Control vectors of Y2H system: pGBKT7-53 (*TRP1*), pGBKT7-Lam (*TRP1*), and pGADT7-T (*LEU2*; Clontech).

2.2 Media

1. 40 % glucose solution. For 100 ml of 40 % glucose solution, dissolve 40 g glucose in 50 ml of water. Make up the volume to 100 ml and sterilize using a 0.22 μm syringe filter into an autoclaved reagent bottle. Store at 4 °C till further use.
2. YPDA medium, pH 6.5. 2 % glucose, 2 % peptone, 1 % yeast extract, 0.003 % adenine hemisulfate. Weigh 20 g of peptone, 10 g yeast extract, 40 mg adenine hemisulfate and dissolve in 900 ml of water and adjust pH to 6.5, make up the volume to 950 ml. Autoclave the media and let it cool to room temperature (RT). Add 50 ml of 40 % stock glucose solution (*see* **Note 1**). For making solid media, add 20 g agar to the medium before autoclaving.
3. 10× dropout supplement/-Leu (DO/-Leu). Add 0.62 g of triple dropout supplement (DO/-Trp/-Leu/-His; Clontech), 20 mg tryptophan (Trp), 20 mg of histidine (His) to 100 ml of water and autoclave. Store at 4 °C till further use.
4. 10× DO/-Trp. Dissolve 0.62 g of DO/-Trp/-Leu/-His, 100 mg of Leu, 20 mg of His and autoclave. Store at 4 °C till further use.
5. 10× DO/-Trp/-Leu. Add 0.62 g of DO/-Trp/-Leu/-His, 20 mg of histidine to 100 ml of water and autoclave. Store at 4 °C till further use.
6. 10× DO/-Trp/-Leu/-His. Dissolve 0.62 g of DO/-Trp/-Leu/-His in 100 ml of water and autoclave. Store at 4 °C till further use.
7. 10× DO/-Trp/-Leu/-His/-Ade. Add 0.60 g of DO/-Trp/-Leu/-His/-Ade in 100 ml water and autoclave. Store at 4 °C till further use.
8. Minimal Synthetic Dropout (SD) medium. For 500 ml media, take 13.35 g minimal SD base (Clontech) and dissolve in 450 ml of water and autoclave. When the media cools to approximately 55 °C, add 50 ml of appropriate autoclaved dropout supplement (e.g. for 500 ml of SD/-Trp broth add 50 ml of 10× DO/-Trp to 450 ml of SD medium). For making solid media, add 10 g agar to the medium before autoclaving.

2.3 Yeast Transformation

1. 50 % PEG-3350. Dissolve 50 g PEG-3350 in 40 ml of water and make up volume to 100 ml. Autoclave and store at RT.
2. 10× lithium acetate (LiAc). Add 10.2 g of lithium acetate to 60 ml of water. Adjust pH to 7.5 with acetic acid and make up the volume to 100 ml. Autoclave and store at RT.
3. 0.5 M EDTA. Add 18.6 g disodium EDTA·$2H_2O$ in 70 ml water. Mix on magnetic stirrer while adjusting the pH to 8.0 with 5 N NaOH. Make up the volume to 100 ml, autoclave and store at room temperature (RT).

4. 1 M Tris–HCl. To make 100 ml solution, add 12.1 g of Tris salt to 60 ml of water. Set pH to 7.5 with 2 N HCl and make up the volume to 100 ml. Autoclave and store at RT.
5. 10× TE buffer. 100 mM Tris–HCl, 10 mM EDTA, pH 7.5. For 100 ml stock solution, take 10 ml of 1 M Tris–HCl (pH 7.5) and 2 ml of 0.5 M EDTA (pH 8.0) and make up the volume to 100 ml. Autoclave and store at RT.
6. TE-LiAc-PEG solution. 10 mM Tris–HCl, 1 mM EDTA, 100 mM LiAc, 40 % PEG. For 10 ml of solution, add 1 ml of 10× TE, 1 ml of 10× LiAc and 8 ml of 50 % PEG in a 50 ml sterile centrifuge tubes and make up the volume to 10 ml (*see* **Note 2**).
7. TE-LiAc solution. 10 mM Tris–HCl, 1 mM EDTA, 100 mM LiAc. Take a sterile 50 ml centrifuge tubes and add 1 ml each of 10× LiAc and 10× TE. Make up the volume to 10 ml.
8. 10 mg/ml salmon sperm DNA/Herring testis carrier DNA. For each transformation use 10 μl of 10 mg/ml stock.
9. Ectodomains of viral gene E1 and E2 as BD fusions (BD-TrE1 and BD-TrE2, respectively) cloned in plasmid pGBKT7 (BD).
10. Dimethylsulfoxide (DMSO).

2.4 Preparation of Yeast Protein Extracts

1. 100× phenylmethyl sulfonyl fluoride. Dissolve 174 mg of phenylmethyl sulfonyl fluoride (PMSF) in 1 ml isopropanol to make 100× solution. Cover it with foil and store at −20 °C.
2. Cracking buffer stock solution. 8 M Urea, 5 % SDS (w/v), 40 mM Tris–HCl, 0.1 mM EDTA, 0.4 mg/ml bromophenol blue. In 20 ml of water mix 48 g urea, 5 g SDS (*see* **Note 3**), 4 ml of 1 M Tris-Cl (pH 6.8), 20 μl of 0.5 mM EDTA (pH 8.0), 40 mg bromophenol blue. Make up the volume to 100 ml and store at RT.
3. Cracking buffer complete solution. For 1.13 ml of complete cracking buffer, mix 1 ml stock cracking buffer solution, 10 μl β-mercaptoethanol, 50 μl of 100× PMSF and 70 μl protease inhibitor cocktail for yeast.
4. Phosphate-buffer saline (PBS). 137 mM NaCl, 2.7 mM KCl, 1.8 mM KH_2PO_4, 10 mM Na_2HPO_4. Dissolve 8 g NaCl, 0.2 g KCl, 0.24 g KH_2PO_4 and 1.44 g Na_2HPO_4 in 900 ml water to make 1× PBS. Adjust the pH to 7.4 with 1 N HCl. Make up the volume to 1 l with water. Autoclave and store at RT.
5. PBST. PBS, 0.05 % Tween 20. Add 50 μl of Tween-20 in 100 ml of 1× PBS (*see* **Note 4**).
6. Anti-c-myc antibody. In 10 ml of PBST, add 5 μl anti-c-myc antibody to obtain a dilution of 1:2000.
7. Anti-mouse antibody. In 10 ml of PBST, add 5 μl anti-mouse antibody to obtain a dilution of 1:2000.

8. Blocking buffer. 5 % bovine serum albumin (BSA) in PBS. To make 5 % BSA solution, add 1 g BSA 10 ml of 1× PBS and invert mix gently to avoid frothing. Make up volume to 20 ml and store at 4 °C.
9. Diaminobenzidine. In 10 ml of PBS dissolve 0.05 % diaminobenzidine (DAB). Add 60 μl of 30 % H_2O_2 just before developing the blot.
10. SDS-PAGE and Western blotting equipment and consumables.

2.5 Testing Autoactivation of BD Viral Fusion Constructs

1. 1 M 3-amino-1,2,4-triazole (3-AT). Dissolve 0.841 g of 3-AT to 10 ml water and filter sterilize (*see* **Note 5**). Store at 4 °C.
2. SD/-Trp/-His/3-AT agar plates. Prepare agar plates as described in Subheading 2.2. The SD plates prepared with 3-AT can be stored for 2 months at 4 °C.

2.6 Screening of Library for Viral–Host Interaction Study

1. 100 mg/ml kanamycin stock solution. Dissolve 500 mg of kanamycin to 5 ml of water to make stock solution of 100 mg/ml. Filter sterilize using a 0.22 μm filter. Make aliquots of 500 μl and store at −20 °C.
2. 0.5× YPDA broth. To make 1 l of 0.5× YPDA broth, dissolve 10 g of peptone, 5 g of yeast extract, 0.1 % adenine hemisulfate and dissolve in 900 ml of water. Set pH to 6.5 and make up volume to 975 ml. Autoclave and when media cools down make up the volume to 1 l by adding 25 ml of 40 % stock glucose solution.
3. YPDA broth as described in Subheading 2.2.
4. 2× YPDA broth. To make 1 l of 2× YPDA broth, dissolve 40 g peptone, 20 g yeast extract and 0.4 % adenine hemisulfate in 850 ml of water. Adjust pH to 6.5, make up the volume to 900 ml and autoclave. When media cools down, add 100 ml of 40 % stock glucose solution.
5. 20 mg/ml X-α-Gal solution. Dissolve 20 mg/ml of 5-Bromo-4-chloro-3-indoyl-α-d-galactopyranoside (X-α-Gal) in *N,N*-dimethylformamide (DMF). Store at −20 °C wrapped in aluminum foil.
6. SD/-Trp/-Leu/X-α-Gal media plates. Prepare SD/-Trp/-Leu agar media as described in Subheading 2.2 and add X-α-Gal solution to a final concentration of 0.02 mg/l.
7. SD/-Trp/-Leu/-His/X-α-gal plates. Prepare SD/-Trp/-Leu/-His agar as described in Subheading 2.2 and add X-α-Gal solution to a final concentration of 0.02 mg/l.
8. SD/-Trp/-Leu/-His/-Ade/X-α-gal plates. Prepare SD/-Trp/-Leu/-His/-Ade agar plates as described in Subheading 2.2 and add X-α-Gal solution to a final concentration of 0.02 mg/l.

2.7 Yeast Colony PCR

1. 10 KU/ml lyticase. Resuspend 10 KU lyophilized lyticase in 1 ml of 1× TE.
2. PCR reagents: Taq DNA polymerase, 10 mM dNTP mix, nuclease-free water, T7 sequencing (forward) and AD sequencing (reverse) primers.
3. 20 % sodium dodecyl sulfate. Dissolve 20 g of sodium dodecyl sulfate (SDS) in 100 ml of water to make 20 % solution. Store at room temperature (*see* **Note 6**).
4. Agarose gel electrophoresis equipment and consumables.

2.8 Elimination of Multiple Library Plasmids

1. SD/-Trp/-Leu/X-α-gal plates as described in Subheading 2.6.

2.9 Restriction Digestion to Eliminate Library Plasmids in Duplicates

1. *Hae* III restriction enzyme.

2.10 Isolation of Prey Plasmid DNA from Yeast

1. SD/-Leu broth as described in Subheading 2.2.
2. Lyticase as Described in Subheading 2.7.
3. 20 % sodium dodecyl sulfate.
4. Plasmid DNA isolation kit.
5. Agarose gel electrophoresis equipment and consumables.

2.11 Transformation of Yeast Plasmids in E. coli Cells

1. Electrocompetent *E. coli* (DH5α) cells.
2. Electroporator and 0.1 cm cuvettes.
3. Prey plasmid DNA isolated from yeast cells.
4. LB Broth. Dissolve 10 g tryptone, 5 g yeast extract and 5 g NaCl in 950 ml of water. Adjust pH to 7.0 with 5 N NaOH and make up the volume to 1 l. Autoclave and store at 4 °C until further use.
5. 100 mg/ml ampicillin stock. Dissolve 500 mg of ampicillin to 5 ml of water to make a stock solution of 100 mg/ml. Filter sterilize using a 0.22 μm disposable filter. Make 500 μl aliquots and store at −20 °C.
6. LB/amp plates. Prepare LB broth as above, add agar (15 g) and autoclave. Cool to 50 °C and add ampicillin to a final concentration of 50 μg/ml. Pour plates and store at 4 °C.

2.12 Bacterial Colony PCR

1. PCR Reagents as described in Subheading 2.7.

2.13 Isolation of Plasmid DNA from E. coli DH5α Cells

1. LB Broth and plates as described in Subheading 2.11.
2. Plasmid DNA isolation kit.

3 Methods

3.1 Small-Scale Transformation of Yeast Cells

1. Prepare primary culture by inoculating 5 ml of YPDA or SD broth with 1–2 colonies of yeast (AH109 or Y187 strain; each of 2–3 mm diameter) and incubate at 30 °C for 16–18 h with shaking at 220 rpm (*see* **Note 7**).
2. Transfer appropriate amount of primary culture to a flask containing 100 ml of YPDA broth such that the OD_{600} ranges between 0.2 and 0.3 (*see* **Note 8**).
3. Incubate secondary culture at 30 °C for ~3 h with shaking at 220 rpm until its OD_{600} reaches 0.4–0.6.
4. Following incubation, centrifuge the culture at RT for 5 min at 3800 × *g*. Discard the supernatant and resuspend the cell pellet in 50 ml (half the culture volume) of autoclaved water.
5. Pellet the cells again at RT for 5 min at 3800 × *g*. Discard the supernatant and resuspend cell pellet in 0.5 ml (for 100 ml culture) of freshly prepared sterile 1× TE/LiAc solution (*see* **Note 9**).
6. Add 0.1 μg of plasmid DNA (gene of interest as BD fusion; control plasmids) and 0.1 mg of herring testes carrier DNA to a fresh 1.5 ml centrifuge tube and mix them (*see* **Notes 10** and **11**).
7. After mixing, add 0.1 ml of competent yeast cells (prepared in **step 5**) and 0.6 ml of sterile PEG/LiAc sequentially to the tube and mix intermittently by vortexing (*see* **Note 12**).
8. Incubate the transformation mix at 30 °C for 30 min with shaking at 220 rpm (*see* **Note 13**).
9. Following incubation, add 70 μl DMSO and mix the contents of the tube well by gentle inversion.
10. Give heat shock to cells at 42 °C for 15 min in a water bath followed by chilling on ice for 1–2 min.
11. Pellet cells at RT for 5 s at 13,400 × *g*. Discard the supernatant and resuspend cells in 0.5 ml of sterile 1× TE buffer.
12. Plate 0.1 ml transformation mixture on SD agar plate to select the desired transformants and incubate at 30 °C for 3–5 days, until the colonies appear (*see* **Note 14**).

3.2 Preparation of Yeast Protein Extracts

1. Inoculate 10 ml of SD broth having appropriate dropout, with 1–2 colonies of yeast cells transformed with gene of interest as BD fusion (each of 2–3 mm diameter) and incubate at 30 °C overnight with shaking at 220 rpm (*see* **Note 15**).
2. For secondary culture, inoculate 100 ml of YPDA broth with appropriate amount of primary culture such that the initial OD_{600} of culture ranges between 0.2 and 0.3.

3. Incubate the culture at 30 °C for ~3 h with shaking at 220 rpm until OD_{600} reaches 0.4–0.6.
4. After incubation, quickly pour the cultures into pre-chilled falcons and centrifuge at 4 °C for 5 min at 3800 × *g*. Discard the supernatant and resuspend cells in 50 ml (half the culture volume) of ice-cold autoclaved water.
5. Pellet the cells by centrifugation at 4 °C for 5 min at 3800 × *g*. Discard the supernatant and use the cell pellet for protein extraction (*see* **Note 16**).
6. Resuspend the cell pellet swiftly in pre-warmed (60 °C) complete cracking buffer. Per 7.5 OD_{600} units of cells 100 μl of complete cracking buffer is used (1 OD_{600} units = OD_{600} × volume of secondary culture; *see* **Note 17**).
7. Transfer each cell suspension to a 1.5 ml microcentrifuge tube containing 80 μl of glass beads per 7.5 OD_{600} units of cells and heat at 70 °C for 10 min followed by vigorous vortexing for 1 min.
8. Centrifuge the cell suspension at 4 °C for 5 min at 13,400 × *g* and collect the supernatant (first supernatant) in a fresh 1.5 ml microcentrifuge tube kept on ice.
9. Add same amount of cracking buffer to the centrifuge tube with cell pellet and place at 100 °C for 3–5 min in a boiling water bath. Vortex vigorously for 1 min.
10. Repeat **step 8** and pool the supernatant with the corresponding first supernatant.
11. Boil the samples at 100 °C and load the samples on SDS-PAGE (10 %) for immuno-blotting (*see* **Note 18**).
12. Once the dye front reaches end of the gel, stop the electrophoresis. Remove the gel carefully from the cassette and wash it in transfer buffer.
13. Charge the PVDF membrane in methanol for 15 s, followed by washing for 2 min with water and then incubate in transfer buffer for 5 min.
14. Arrange the transfer cassette by placing the sponge on negative side of the cassette. Then, place a pre-wet 3 MM Whatman filter on it followed by gel, charged membrane, another Whatman filter and sponge.
15. Electroblot the proteins onto PVDF membrane for 2 h at 100 mA.
16. Disassemble the apparatus after transfer and remove the PVDF membrane. Place it in blocking buffer (5 % BSA in PBS) for 1 h at RT to prevent nonspecific binding of antibodies to the membrane.
17. Wash the membrane three times with PBST for 5 min each.

18. Dilute primary antibody in PBST (1:2000 for anti-c-myc for BD fusions) and incubate the membrane for 1.5 h at RT on a rocker. Remove excess of antibody by washing the membrane three times with PBST after incubation as in **step 17**.
19. Incubate the membrane with secondary antibody conjugated with horseradish peroxidase (HRP) diluted in PBST (1:2000 dilution) for 1.5 h at RT with shaking.
20. Repeat **step 17** and finally wash with PBS. Develop the blot with 10 ml of DAB, 0.05 % as substrate for HRP and hydrogen peroxide (0.1 %) as reaction catalyst. Stop the reaction with distilled water.

3.3 Autoactivation Analysis

1. To test the activation of reporter gene by BD viral constructs in the absence of prey plasmid (AD fusion), select the BD fusion transformants on SD/-Trp/-His media. Take empty BD and AD vectors as negative controls (*see* **Note 19**).
2. If viral proteins activate the yeast reporter then plate the bait (BD fusion) protein responsible for autoactivation on SD/-Trp/-His media containing variable concentrations of 3-AT (*see* **Note 20**). Select the concentration, at which no growth is observed on SD medium lacking histidine, for screening of interaction. Interactions of viral proteins autoactivating the reporter can be selected on SD/-Trp/-Leu/-His/3-AT.

3.4 Yeast Two-Hybrid Library Screening

1. For primary culture inoculate 50 ml of SD/-Trp broth with 1–2 colonies (each of 2–3 diameter) of yeast strain AH109 transformed with recombinant BD plasmid (bait strain; BD-TrE1 or BD-TrE2) and incubate at 30 °C for 16–18 h at 220 rpm.
2. Pellet the culture next day for 5 min at 3800 × *g*. Discard the supernatant and resuspend the cell pellet in a 4 ml SD/-Trp broth (cell density >1×10^8 cells/ml).
3. Thaw a vial of cDNA library (cDNA cloned in pGADT7-Rec plasmid) pre-transformed in yeast strain Y187 (library strain) at RT in a water bath.
4. Before proceeding for mating, use 10 μl of the library for Library Titration. Mix library vial by gentle vortexing and transfer 10 μl of library to 1 ml of 1× YPDA broth in a 1.5 ml microcentrifuge tube. Mix by gentle vortexing. This is Dilution A (1:10^2). Remove 10 μl from Dilution A and add to 1 ml of 1× YPDA broth. This is Dilution B (1:10^4). Plate 100 μl aliquots of Dilution A and B on SD/-Leu plates. And incubate at 30 °C until colonies appear (3–5 days).
5. Count the number of colonies to determine the titer (cfu/ml). The library titer can then be calculated by the following formula: No. colonies/plating volume (ml) × dilution factor = cfu/ml.

6. Transfer the contents of the vial to a 2 l flask containing 45 ml of 2× YPDA broth supplemented with 50 μg/ml kanamycin (*see* **Note 21**) along with 4 ml of bait strain (from **step 2**) and incubate at 30 °C for 20–24 h with shaking at 50 rpm.
7. After 24 h, harvest the cells by centrifugation at RT for 10 min at 3800 × *g*. Discard the supernatant and resuspend pellet in 10 ml of 0.5× YPDA broth supplemented with 50 μg/ml kanamycin.
8. After resuspension, plate 250 μl of mated culture on SD/-Trp/-Leu/-His plates (150 mm) supplemented with X-α-Gal (*see* **Note 22**) and incubate at 30 °C for 3–5 days, until appearance of blue-colored colonies (*see* **Notes 23** and **24**).
9. Amplify the positive clones (blue-colored colonies) on SD/-Trp/-Leu/-His plates.
10. Transfer well isolated colonies to SD/-Trp/-Leu/-His/-Ade plates supplemented with X-α-Gal and incubate at 30 °C for a week for high stringency selection (*see* **Notes 25** and **26**).

3.5 Yeast Colony PCR

1. For generation of template for colony PCR, resuspend a 2–4 mm yeast colony in 50 μl of nuclease-free water in a 1.5 ml microcentrifuge tube by vortexing.
2. Add 10 μl of lyticase in the cell suspension to be tested and incubate at 37 °C for 45 min with shaking at 220 rpm.
3. Add 10 μl of 20 % SDS to each tube and vortex for 1 min to mix.
4. Put the samples through one freeze/thaw cycle (at −20 °C) and vortex again to ensure complete lysis of cells. Samples can be stored frozen at −20 °C. If samples have been frozen, vortex them again before using them.
5. At this point template is ready for use.
6. Prepare a PCR reaction mixture of 10–20 μl with the conditions indicated in Table 1.

Table 1

Components	Final CONCENTRATION
10× PCR buffer	1×
dNTPs (10 mM each)	0.25 mM each
Forward primer	0.5 pmol
Reverse primer	0.5 pmol
Taq DNA polymerase	6 U/reaction
Template DNA	1–10 ng/reaction
Nuclease free water	Variable

Table 2

Cycle	Temperature	Time
1. Initial denaturation	95 °C	10 min
2. 25 cycles		
Denaturation	95 °C	1 min
Annealing	56 °C	1 min
Extension	72 °C	1 min/kb (*see* **Note 27**)
3. Final extension	72 °C	10 min

7. Follow the thermal cycler program outlines in Table 2.
8. Analyze the PCR products by agarose gel electrophoresis.

3.6 Elimination of Multiple Library Plasmids

1. Re-streak the positive clones exhibiting multiple bands after colony PCR on SD/-Trp/-Leu plates supplemented with X-α-Gal, 2–3 times (*see* **Note 28**). This will allow the segregation of the AD plasmids while maintaining selective pressure on both the DNA-BD and AD vectors.
2. Incubate the plates at 30 °C for 3–5 days (until blue- or white-colored colonies appear). A mixture of blue and white colonies indicates segregation.
3. Perform colony PCR of blue colonies after each substreak to confirm segregation of prey plasmids.

3.7 Restriction Digestion to Eliminate Library Plasmids in Duplicates

1. Amplify the AD/library inserts by PCR using vector-specific primers (T7 forward and AD reverse).
2. Digest these PCR products with *Hae* III, a frequent cutter restriction enzyme.
3. Analyze the fragment sizes by electrophoresis on a 0.8 % TAE agarose gel.
4. Plasmids having same restriction profile are regarded as duplicates and only one of them is considered for subsequent studies.

3.8 Isolation of Prey Plasmid DNA from Yeast

1. For isolation of pGADT7-Rec (prey) plasmid, prepare a primary culture by inoculating 5 ml of SD/-Leu broth with 1–2 colonies (each of 2–3 mm diameter, positive interactor) and incubate it at 30 °C for 16–18 h with shaking at 220 rpm.
2. Pellet the cells by centrifugation at RT for 5 min at 3800 × *g*. Discard the supernatant and resuspend the pellet in the residual liquid (~50 μl).
3. Mix the cells thoroughly by vortexing after the addition of 10 μl of lyticase solution.

4. Incubate the cell suspension at 37 °C for 45 min with shaking at 220 rpm.
5. Add 10 μl of 20 % SDS to each tube and vortex vigorously for 1 min to mix.
6. Subsequently, freeze/thaw the samples 2–3 times and vortex again to ensure complete lysis of the cells.
7. The plasmid DNA can then be isolated from lysed cells using a plasmid DNA isolation kit.
8. Analyze the eluted plasmids on 1 % agarose gel.

3.9 Transformation of Yeast Plasmids in E. coli (DH5α)

1. Thaw electrocompetent *E. coli* (DH5α) cells on ice. Add 5 μl of yeast plasmid DNA to 40 μl cells on ice.
2. Transfer samples to a prechilled cuvette having a 0.1 cm gap and pulse for 1 s.
3. Quickly add 200 μl of LB and transfer the cell suspension to a 1.5 ml microcentrifuge tube.
4. Incubate at 37 °C for 1 h with shaking (220 rpm).
5. Plate cell on LB agar supplemented with ampicillin (100 μg/ml). Incubate plates at 37 °C overnight.

3.10 Bacterial Colony PCR and Plasmid DNA Isolation

1. Confirm the presence of insert in isolated prey plasmids by bacterial colony PCR. Program the Thermal cycler as discussed in Subheading 3.5.
2. Isolate prey plasmids from *E. coli* DH5α cells (showing amplification of inserts) using a commercially available plasmid DNA isolation kit.

3.11 Sequencing and Analysis

1. Sequence the positive clones identified by Y2H using T7 forward and 3′ AD-specific reverse sequencing primers.
2. To identify the host proteins interacting with the viral bait, subject the sequenced clones to nucleotide BLAST and check for sequences for which full-length coding sequence is available.

4 Notes

1. If sugar solution must be added before autoclaving set the autoclaving cycle at 121 °C for 15 min. Autoclaving for a longer duration or at a higher temperature may result in charring of sugar solution which may lead to decreased performance of the medium. In case of solid media, sugar solution must be added to media at ~55 °C.
2. Prepare TE-LiAc-PEG solution freshly at the time of experiment.

3. Add SDS last and mix gently to avoid frothing.
4. Use cut tip for aspiration of Tween-20 as it is very viscous.
5. 3-AT is light-sensitive and should be covered in foil.
6. Do not mix vigorously or autoclave SDS solution.
7. Vortex the primary culture to resuspend yeast colonies before incubating for overnight growth at 30 °C.
8. Five ml of primary culture is sufficient for inoculating 50 ml of secondary culture. If overnight grown culture is visibly clumped then disperse the clumps by vortexing.
9. At this stage our yeast competent cells are ready for transformation. Competent cells have highest transformation efficiency at this point and should be used within 1 h.
10. Herring testes carrier DNA is activated by heating at 100 °C for 10 min prior to its usage.
11. Gene of interest (TrEl/TrE2) is to be cloned in pGBKT7. Then transform control plasmids pGBKT7, pGBKT7-53 and pGBKT7-Lam in AH109 cells, and pGADT7 and pGADT7-T in Y187 cells.
12. Gentle vortexing is important for mixing of DNA and cells for efficient transformation of yeast cells.
13. At the time of incubation centrifuge tubes containing the mix should be kept at a slanting position to allow better mixing by providing more surface area.
14. The transformants containing BD fusion are to be selected on SD/-Trp whereas yeast cells transformed with AD fusion are to be selected on SD/-Leu agar plates.
15. Inoculate SD/-Trp broth with cells containing bait (BD fusion) plasmids.
16. The pellets are either used directly for protein extraction following 2–3 freeze–thaw cycles or frozen immediately on dry ice or in liquid nitrogen and stored at −80 °C until further use.
17. Since the initial excess PMSF in the complete cracking buffer degrades quickly, add an additional aliquot of the PMSF stock solution (100×) to the samples after 15 min and approximately every 7 min thereafter until when they are placed on dry ice or are safely stored at −80 °C.
18. At this point samples can be stored at −80 °C until further use.
19. Autoactivation refers to activation of transcription factor by BD fusion in the absence of AD-fusion. Therefore, it is crucial to test for the autoactivation capacity of target protein. For this purpose the BD-viral fusion (bait) transformant is to be assayed for *HIS* (reporter gene) activation by selecting them on the SD/-Trp/-His medium. On selection of BD fusion on this

medium, if growth is observed, then the viral protein is autoactivating the reporter gene (i.e. viral protein as BD fusion in the absence of prey forms a functional transcription factor).

20. 3-AT is a competitive inhibitor of *HIS3* protein in yeast and is used to inhibit low levels of *HIS3* leaky expression. The selection of bait plasmids autoactivating the reporter gene on SD/-Trp/-Leu/-His/3-AT may be used to titrate the minimum level of *HIS3* expression required for growth on histidine-deficient media. Further, deletion mutants of the viral protein as BD fusion may be used to eliminate autoactivation.
21. Kanamycin added in the medium is to prevent any bacterial contamination.
22. X-α-Gal is light-sensitive; therefore, SD/Dropout plates containing X-α-Gal should be covered with foil.
23. It is important to use appropriate controls at the time of interaction analysis. Viral proteins as BD (bait) fusion are mated with empty vectors containing AD domain (pGADT7). Similarly, plasmids pGBKT7-p53 and pGADT7-T encoding the known interacting proteins tumor suppressor protein p53 and Simian Virus 40 (SV40) large T-antigen fused with BD and AD domains, respectively, are mated with each other and can be used as positive control. The noninteracting proteins Lamin and SV40 large T-antigen encoded by pGBKT7-Lam and pGADT7-T as BD and AD fusions, respectively, are mated with each other to serve as negative control for interaction studies.
24. Plate the mated culture (100 μl) on SD/-Trp, SD/-Leu and SD/-Trp/-Leu plates for determining the mating efficiency. Mating efficiency (percentage of diploids) = (No. of cfu/ml of diploid/No. of cfu/ml of limiting partner) × 100. In this case limiting partner is prey library.
25. *ADE2* and *HIS* gene are part of adenine and histidine biosynthesis pathway of yeast, respectively. Since the yeast strains AH109 and Y187 have deletion for both these genes, they have the ability to grow in the histidine and adenine-deficient medium if the GAL promoter has been activated. This activation can only occur if the bait and prey proteins come together reconstituting the DNA-binding and activation domains of the GAL4 transcription, i.e. if interaction is taking place between the bait and the prey protein. Further, *ADE2* reporter provides a stronger nutritional selection compared to *HIS* reporter.
26. Selection of interacting proteins on the SD/-Trp/-Leu/-His/-Ade/X-α-Gal medium allows more stringent screening of the interactions by testing expression of all three reporter genes, *HIS3*, *ADE2*, and *MEL1*, respectively. This virtually eliminates false-positive interactions; however less stringent interactions may be lost.

27. Adjust the time in accordance with the size of inserts in the cDNA library employed for screening host interactors.
28. Library co-transformant can harbor more than one prey plasmid. This means, in addition to containing a prey vector that expresses a protein responsible for activating the reporters, yeast cell may also contain one or more prey plasmids that do not express an interacting protein. Appearance of multiple bands on yeast colony PCR indicates the presence of multiple prey constructs in the yeast cell.

References

1. Fields S, Song O (1989) A novel genetic system to detect protein–protein interactions. Nature 340:245–246
2. Brent R, Finley R (1997) Understanding gene and allele function with two-hybrid methods. Annu Rev Genet 31:663–704
3. Rozen R, Sathish N, Li Y et al (2008) Virion wide protein interactions of Kaposi's Sarcoma associated Herpesvirus. J Virol 82(10):4742–4750
4. Calderwood MA, Venkatesan K, Xing L et al (2007) Epstein-Barr virus and virus human protein interaction maps. Proc Natl Acad Sci U S A 104:7606–7611
5. von Brunn A, Teepe C, Simpson JC et al (2007) Analysis of intraviral protein–protein interactions of the SARS coronavirus ORFeome. PLoS One 2:e459
6. Chen M, Cortay JC, Gerlier D (2003) Measles virus protein interactions in yeast: new findings and caveats. Virus Res 98:123–129
7. Mccraith S, Holtzman T, Moss B et al (2000) Genome wide analysis of Vaccinia virus protein–protein interactions. Proc Natl Acad Sci U S A 97:4879–4884
8. Zhang L, Villa NY, Rahman MM et al (2009) Analysis of vaccinia virus–host protein–protein interactions: validations of yeast two-hybrid screenings. J Proteome Res 8(9):4311–4318
9. Kumar K, Rana J, Sreejith R et al (2012) Intraviral protein interactions of Chandipura virus. Arch Virol 157:1949–1957
10. Uetz P, Dong YA, Zeretzke C et al (2005) Herpesviral protein networks and their interaction with the human proteome. Science 311:239–242
11. Vliet KV, Mohamed MR, Zhang L et al (2009) Poxvirus proteomics and virus–host protein interactions. Microbiol Mol Biol Rev 73(4): 730–749
12. Mairiang D, Zhang H, Sodja A et al (2013) Identification of new protein interactions between dengue fever virus and its hosts, human and mosquito. PLoS One 8(1), e53535
13. Geng Y, Yang J, Huang W et al (2013) Virus host protein interaction network analysis reveals that the HEV ORF3 protein may interrupt the blood coagulation process. PLoS One 8(2):e56320
14. Dolan PT, Zhang C, Khadka S et al (2013) Identification and comparative analysis of Hepatitis C virus–host cell protein interactions. Mol Biosyst 9:3199–3209
15. de Chassey B, Aublin-gex A, Ruggieri A et al (2013) The interactomes of Influenza virus NS1 and NS2 proteins identify new host factors and provide insights for ADAR1 playing a supportive role in virus replication. PLoS Pathog 9(7):e1003440
16. Sreejith R, Rana J, Dudha N et al (2012) Mapping of interactions among Chikungunya virus nonstructural proteins. Virus Res 169(1):231–236
17. Dudha N, Rana J, Rajasekharan S et al (2015) Host–pathogen interactome analysis of Chikungunya virus envelope proteins E1 and E2. Virus Genes 50(2):200–209

Chapter 16

Application of GelC-MS/MS to Proteomic Profiling of Chikungunya Virus Infection: Preparation of Peptides for Analysis

Atchara Paemanee, Nitwara Wikan, Sittiruk Roytrakul, and Duncan R. Smith

Abstract

Gel-enhanced liquid chromatography coupled with tandem mass spectrometry (GeLC-MS/MS) is a labor intensive, but relatively straightforward methodology that generates high proteome coverage which can be applied to the proteome analysis of a range of starting materials such as cells or patient specimens. Sample proteins are resolved electrophoretically in one dimension through a sodium dodecyl sulfate (SDS) polyacrylamide gel after which the lanes are sliced into sections. The sections are further diced and the gel cubes generated are subjected to in-gel tryptic digestion. The resultant peptides can then be analyzed by tandem mass spectroscopy to identify the proteins by database searching. The methodology can routinely detect several thousand proteins in one analysis. The protocol we describe here has been used with both cells in culture that have been infected with chikungunya virus and specimens from Chikungunya fever patients. This protocol details the process for generating peptides for subsequent mass spectroscopic and bioinformatic analysis.

Key words Chikungunya, Proteome, Mass spectrometry, Peptides, GeLC-MS/MS

1 Introduction

Chikungunya virus (CHIKV) is a mosquito borne *Alphavirus* of the family *Togaviridae* that causes a disease in humans similar to dengue fever, but is typically characterized by an arthralgic syndrome that can persist long after the fever period [1]. Since the sudden explosive reemergence of the virus from 2004 [2], a number of studies have sought to understand the pathobiology of this virus through the application of global protein profiling as reviewed elsewhere [3]. Methodologies employed previously were either gel based, with separation through polyacrylamide gels in either 1- or 2-dimensions, or solution based. Gel-enhanced liquid chromatography tandem mass spectroscopy (GeLC-MS/MS) is based on the

Justin Jang Hann Chu and Swee Kim Ang (eds.), *Chikungunya Virus: Methods and Protocols*, Methods in Molecular Biology, vol. 1426,DOI 10.1007/978-1-4939-3618-2_16, © Springer Science+Business Media New York 2016

pre-separation of proteins through one dimension by polyacrylamide gel electrophoresis before proteins are subjected to in-gel tryptic digestion and analysis of peptides by coupled nanocapillary liquid chromatography and tandem mass spectroscopy (LC-MS/MS). The pre-separation of proteins by electrophoresis through a SDS-polyacrylamide gel in one dimension serves to both reduce protein complexity in the subsequent mass spectroscopy analytical runs as well as to provide a degree of filtering to remove low molecular weight impurities such as detergent and buffer components which may affect the subsequent mass spectroscopy analysis. The method pioneered by Schirle and colleagues [4] generates more complete proteome coverage than standard two-dimensional gel electrophoresis methodologies particularly with regards to the ability to detect less abundant and transmembrane proteins. The methodology is highly labor intensive, and, with appropriate replication, several thousand samples must be subjected to preparative handling and analysis.

However, the method generates a high degree of proteome coverage and can be interpreted to identify alterations in protein expression between samples from different sources. The protocol described here has been previously used to analyze samples originating from patients with different severities as a consequence of CHIKV infection [5] and from cells infected in vitro with CHIKV [6]. In both cases large numbers of differentially expressed proteins were identified and subsequently validated by different methodologies [5, 6].

Here, we present a GelC-MS/MS protocol using protein lysate prepared from both cell-cultured material and clinical specimens. We describe the step-by-step protocols for cell lysate preparation, protein quantification, protein separation by SDS-PAGE, gel staining, and subsequent sample processing for mass spectrometry analysis.

2 Materials

2.1 Protein Preparation and Quantitation

1. 0.5 M EDTA. Weigh 186.12 g of ethylenediaminetetraacetic acid (EDTA) disodium salt, dihydrate ($EDTA{\cdot}Na_2{\cdot}2H_2O$, molecular weight 372.24 g/mol) into a 2 l beaker. Add 800 ml deionized water. While stirring vigorously on a magnetic stirrer, add NaOH (*see* **Note 1**) to adjust the solution pH to 8.0. Transfer to a volumetric flask and adjust the volume to 1000 ml with deionized water and sterilize by autoclaving.
2. 10× Phosphate-buffered saline (PBS): 1370 mM NaCl, 27 mM KCl, 100 mM Na_2HPO_4, 20 mM KH_2PO_4. Measure 80 g of sodium chloride (NaCl), 2 g of potassium chloride (KCl),

14.4 g of disodium hydrogen phosphate (Na_2HPO_4), and 2.4 g of potassium dihydrogen phosphate (KH_2PO_4) into a glass beaker and add 800 ml of deionized water and then dissolve using a magnetic stirrer. When the solution has dissolved completely, adjust the pH to 7.4 using concentrated HCl (*see* **Note 2**) and transfer to a volumetric flask, adjust the volume to 1000 ml with deionized water and sterilize by autoclaving.

3. 1× PBS. Measure 10 ml of 10× PBS into a 100 ml sterile volumetric flask and add 90 ml of sterile water and mix thoroughly.
4. 0.22 μm filter.
5. 0.25 % (w/v) trypsin–EDTA. Measure 1.25 g of trypsin powder into a glass beaker and add 400 ml of 1× PBS and then dissolve using a magnetic stirrer. When a clear solution has been obtained, add 1 ml of 0.5 M EDTA, pH 8.0 and continue stirring for 1–5 min. Adjust the volume to 500 ml with 1× PBS and sterilize by passing through a 0.22 μm filter in a laminar flow hood. Aliquot into 50 ml conical tubes and store at −20 °C. Before use, thaw a tube from storage by either allowing the tube to sit in a cold room overnight or by incubating the tube in a 37 °C water bath until the solution has melted, after which it should be kept at 4 °C.
6. 10 % SDS (w/v) stock solution. Measure 10 g of SDS (*see* **Note 3**) into a glass beaker and add 80 ml of distilled water and stir with a magnetic bar until dissolved. The solution can be heated to 68 °C if the SDS does not go into solution easily. Adjust the volume to 100 ml using a volumetric flask. The solution is stable at room temperature for up to 6 months.
7. 5 % sodium dodecyl sulfate (SDS). Measure 20 ml of 10 % SDS into 50 ml tube and add 20 ml of deionized water then mix thoroughly.
8. 0.5 % sodium dodecyl sulfate (SDS). Measure 2 ml of 10 % SDS stock solution into 50 ml tube and add 38 ml of distilled water then mix thoroughly.
9. Acetone (analytical grade).
10. CTC stock solution: 0.2 % $CuSO_4$, 0.4 % tartaric acid (w/v). Weigh 0.1 g of copper sulfate ($CuSO_4$) and 0.2 g of tartaric acid into a 50 ml tube. Add 50 ml of deionized water and mix thoroughly.
11. 20 % sodium carbonate (Na_2CO_3). Weigh 10 g of Na_2CO_3 into a 100 ml glass beaker and add 50 ml of deionized water and dissolve using a magnetic stirrer. Transfer the solution to a volumetric flask and adjust the volume to 50 ml with deionized water.

12. 0.8 N sodium hydroxide (NaOH). Weigh 1.6 g of NaOH into a 50 ml tube and add 50 ml of deionized water and vigorously shake to mix.
13. Lowry reagent solution A: 0.025 % $CuSO_4$, 0.05 % tartaric acid, 2.5 % Na_2CO_3, 0.2 N NaOH, 0.5 % SDS. Measure 5 ml of CTC stock solution, 5 ml of 20 % sodium carbonate, 10 ml of 0.8 N sodium hydroxide and 20 ml of 5 % SDS into a 50 ml tube and mix by inverting.
14. Lowry reagent solution B. Measure 2 ml of Folin-Ciocalteu's phenol reagent and 10 ml of deionized water into a falcon tube and mix by inverting.
15. Bovine serum albumin (BSA) protein standards. Weigh 2 mg of BSA into a 1.5 ml microcentrifuge tube and add 1 ml of 0.5 % SDS and dissolve the protein by vortexing for 2 min to generate a stock 2 mg/ml BSA solution. Serially dilute the stock BSA solution with 0.5 % SDS to generate standards of 0.4, 0.8, 1.2, 1.6, and 2 μg/μl (giving final concentrations of 2, 4, 6, 8, and 10 μg BSA per 5 μl, respectively).
16. MicroWell 96-well plates.
17. Microplate reader.

2.2 SDS Polyacrylamide Gel

1. Resolving gel buffer: 1.5 M Tris–HCl buffer pH 8.8. Weigh 181.65 g of Tris base into a glass beaker and dissolve in 800 ml of deionized water using a magnetic stirrer. When the solution has dissolved completely, adjust the pH to 8.8 using concentrated HCl (*see* **Note 2**). Adjust the volume to 1000 ml with a volumetric flask using deionized water.
2. Stacking gel buffer: 0.5 M Tris–HCl buffer pH 6.8. Weigh 60.55 g of Tris base into a glass beaker and dissolve in 800 ml of deionized water using a magnetic stirrer. When the solution has dissolved completely, adjust the pH to 8.8 using concentrated HCl (*see* **Note 2**). Adjust the volume to 1000 ml with a volumetric flask using deionized water.
3. 40 % acrylamide–bisacrylamide solution (29:1).
4. 10 % SDS (w/v).
5. 10 % (w/v) ammonium persulfate solution. Weigh 1 g of ammonium persulfate $(NH_4)_2S_2O_8$ into a plastic tube then add 10 ml of distilled water then mix thoroughly. Aliquot 500 μl portions into microcentrifuge tubes, and use immediately or store at 4 °C or at −20 °C (*see* **Note 4**).
6. *N*,*N*,*N*′,*N*′-Tetramethyl-ethylenediamine (TEMED).
7. 10× SDS-PAGE running buffer: 0.25 M Tris–HCl pH 8.3, 1.92 M glycine, 1 % SDS. Weigh 30.3 g of Tris base, 144 g of glycine, and 10 g of SDS (*see* **Note 3**) into a beaker and dissolve in 800 ml of deionized water using a magnetic stirrer.

The pH of the buffer should be pH 8.3 and pH adjustment is normally not required. Transfer the solution to a volumetric flask and adjust volume to 1000 ml using deionized water. The running buffer can be stored at room temperature.

8. 1× SDS-PAGE running buffer: 25 mM Tris–HCl pH 8.3, 192 mM glycine, 0.1 % SDS. Measure 100 ml of 10× SDS-PAGE running buffer into a 1 l measuring cylinder and add 900 ml of deionized water and mix thoroughly.
9. Mini-slab size polyacrylamide gel electrophoresis system with glass plates and a power pack.
10. 3 M Tris–HCl buffer pH 6.8. Weigh 36.33 g of Tris base into a glass beaker and dissolve in 80 ml of deionized water using a magnetic stirrer. Mix thoroughly and adjust the pH to 6.8 using concentrated HCl (*see* **Note 2**). Transfer the solution to a volumetric flask and adjust the volume to 100 ml with deionized water.
11. 5× SDS loading buffer: 0.5 M dithiothreitol (DTT), 10 % SDS, 0.1 mg/ml bromophenol blue, 0.4 M Tris–HCl pH 6.8, 50 % glycerol. Weigh 1.925 g of DTT, 2.5 g of SDS (*see* **Note 3**) and 2.5 mg of bromophenol blue into a 50 ml tube. Add 3.25 ml of 3 M Tris-HCl pH 6.8, 12.5 ml glycerol and sterile water to 25 ml then mix vigorously. Aliquot and store at −20 °C.
12. Low molecular weight protein marker.

2.3 Gel Staining and Destaining (Coomassie Blue)

1. Coomassie blue staining solution: 50 % methanol, 10 % acetic acid, 2.5 g/l Coomassie Brilliant Blue R-250. Weigh 2.5 g of Coomassie brilliant blue R-250 into a beaker then add 400 ml of deionized water, 500 ml of methanol and 100 ml of glacial acetic acid. Mix the solution using a magnetic stirrer. Filter the solution through Whatman No. 1 filter paper.
2. Destaining solution: 16.5 % ethanol, 5 % acetic acid. Measure 785 ml of deionized water, 165 ml of absolute ethanol and 50 ml glacial acetic acid into a glass beaker and add deionized water to 1000 ml and mix thoroughly.

2.4 Gel Staining and Destaining (Silver Staining)

1. Fixing solution: 50 % methanol, 12 % acetic acid, 0.0185 % formaldehyde. Measure 500 ml of absolute methanol, 120 ml of glacial acetic acid, and 500 μl of 37 % formaldehyde into a glass beaker and add deionized water to 1000 ml and mix thoroughly.
2. Washing solution: 35 % ethanol. Measure 350 ml of absolute ethanol and 650 ml of deionized water into a glass bottle and mix thoroughly.

3. Sensitizing solution: 0.02 % sodium thiosulfate ($Na_2S_2O_3$). Weigh 0.2 g of $Na_2S_2O_3$ into a volumetric flask and add deionized water to 1000 ml and mix thoroughly.
4. Staining solution: 0.2 % silver nitrate ($AgNO_3$). Weigh 2 g of $AgNO_3$ into plastic container with a lid and add deionized water to 1000 ml and then shake vigorously.
5. Developing solution: 6 % Na_2CO_3, 0.0004 % $Na_2S_2O_3$, 0.0185 % formaldehyde. Weigh 60 g of Na_2CO_3 into plastic container with a lid, add 20 ml of sensitizing solution, 500 μl of 37 % formaldehyde, and deionized water to 1000 ml and mix thoroughly.
6. Stopping solution: 1.46 % EDTA disodium salt. Weigh 14.6 g of EDTA disodium salt into a plastic container with a lid and add deionized water to 1000 ml and mix thoroughly.
7. Gel storage solution: 0.1 % acetic acid. Add 1 ml of glacial acetic acid to 999 ml of deionized water and mix.

2.5 Gel Processing

1. Clear glass plate.
2. 95 % ethanol.
3. Medical scalpel and blades.
4. 96-well polypropylene microplates, V-bottom.
5. Sequencing grade modified trypsin.
6. 100 mM ammonium bicarbonate (NH_4HCO_3). Weigh 79.06 mg of NH_4HCO_3 then add 80 ml of deionized water and mix in a volumetric flask. Adjust the solution volume to 100 ml with deionized water.
7. 50 mM NH_4HCO_3. Mix 50 ml of 100 mM NH_4HCO_3 and 50 ml of deionized water and then mix thoroughly.
8. 10 mM NH_4HCO_3. Measure 10 ml of 100 mM NH_4HCO_3 and add 90 ml of deionized water and then mix thoroughly.
9. 25 mM ammonium bicarbonate (NH_4HCO_3)/50 % methanol. Mix 50 ml of 50 mM NH_4HCO_3 and 50 ml of absolute methanol in a glass bottle and then mix thoroughly.
10. 5 % H_2O_2 solution. Pipette 5 ml of stock 30 % H_2O_2 into a 50 ml tube and then add 25 ml sterile deionized water and mix well by vortexing. Prepare freshly before use and keep in the dark.
11. 10 mM dithiothreitol (DTT) in 10 mM NH_4HCO_3. Weigh 15 mg of DTT powder into a falcon tube and dissolve in 10 ml of 10 mM NH_4HCO_3 (*see* **Note 5**).
12. 100 mM iodoacetamide (IAA) in 10 mM NH_4HCO_3. Weigh 180 mg of IAA powder into a falcon tube and dissolve in 10 ml of 10 mM NH_4HCO_3 (*see* **Note 6**).

13. 10 ng/μl trypsin in 10 mM NH_4HCO_3. Add 2 ml of 10 mM NH_4HCO_3 into a bottle containing 20 μg sequencing grade modified trypsin and then vigorously shake.
14. 50 % acetonitrile. Measure 50 ml of acetonitrile (analytical grade) into a glass bottle and add 50 ml of deionized water and then mix thoroughly.
15. 0.1 % formic acid in 50 % acetonitrile. Measure 49.95 ml of 50 % acetonitrile into a glass bottle and add 50 μl of 100 % formic acid, then mix thoroughly.

2.6 Protein Analysis

1. 0.1 % formic acid in LC/MS water. Measure 999 ml of water (LC-MS reagent grade) into a glass bottle and add 1 ml of 100 % formic acid. Then degas using a sonicator for 15 min.
2. 0.1 % formic acid in LC-MS acetonitrile. Measure 999 ml of acetonitrile (LC-MS reagent grade) into a glass bottle and add 1 ml of 100 % formic acid. Then degas using a sonicator for 15 min.
3. 1.7 ml low binding microcentrifuge tubes.
4. LCGC certified clear glass 12 × 32 mm screw neck total recovery vials, with caps and preslit PTFE/silicone septa, 1 ml volume.
5. 30 % acetonitrile. Measure 30 ml of acetonitrile (LC-MS reagent grade) into a glass bottle and add 70 ml of water (LC-MS reagent grade) and then mix thoroughly.
6. Liquid chromatography tandem mass spectrometry (LC-MS) system.

3 Methods

Conduct all procedures at room temperature unless otherwise stated.

3.1 Sample Preparation (For Cells Grown in a Tissue Culture Plate or Flask, See Note 7)

1. Remove the cell culture media and place in a 50 ml conical tube and wash cells with 5 ml 1× PBS.
2. Remove the 1× PBS and combine with the previously removed culture media and add 1 ml of 0.25 % trypsin–EDTA solution to the cell culture plate and incubate at 37 °C for 5 min (*see* **Note 8**).
3. Add 1 ml of culture media to the tissue culture plate (*see* **Note 9**) and swirl the plate to distribute the medium across the plate.
4. Using a pipette, pipette the solution up and down between 10 and 30 times to make a single cell suspension.
5. Remove the single cell suspension and combine with the previously removed solutions (original growth medium and PBS

wash solution) and wash the plate with a further 5 ml of 1× PBS. Remove the PBS and combine with other previously removed solutions.

6. Centrifuge the combined solutions at 1000 × *g* for 5 min (*see* **Note 10**), remove the supernatant and resuspend the cell pellet in 1.2 ml of 1× PBS (*see* **Note 11**).
7. Transfer 1 ml of the resuspended cells pellet solution to a 1.5 ml microcentrifuge tube and centrifuge at 1500 × *g* for 5 min. Wash the cell pellet twice with 1 ml of 1× PBS each time.
8. After the final PBS wash, resuspend the cell pellet in 300 μl of sterile deionized water and add 600 μl of 100 % ice cold acetone. Mix the tube by inverting and store the tube at −80 °C for at least 1 h.
9. Centrifuge the solution at 9200 × *g* for 30 min and remove the supernatant.
10. Dry the protein pellet in a fume hood for 30–60 min after which time resuspend the pellet in 100–500 μl of 0.5 % SDS by pipetting several times to mix.

3.2 Sample Preparation (Clinical Material: White Blood Cells)

1. Store white blood cell pellets (or other suitable clinical material) at −80 °C until required.
2. Add 50 μl of deionized water to the cell pellet and sonicate the sample until all the cells have lysed (approximately four cycles of 5 min each time until the solution is homogenous).
3. Add 200 μl of acetone and keep at −20 °C for overnight or until required.
4. Centrifuge the solution at 9200 × *g* for 30 min and discard the supernatant.
5. Dry the protein pellet in a fume hood for 30–60 min after which time resuspend the pellet in 25 μl of 0.5 % SDS by pipetting several times to mix.

3.3 Protein Concentration Determination by the Lowry Method (See Note 12)

1. Serially dilute protein samples with 0.5 % SDS. If the concentration of the samples cannot be approximately estimated, prepare 2–3 dilutions of samples spanning an order of magnitude.
2. Add 5 μl of diluted samples and BSA standard dilutions in triplicate to the wells of a 96-well plate and subsequently add 200 μl of Lowry reagent solution A and incubate for 30 min at a temperature above 25 °C to avoid precipitation of SDS.
3. Add 50 μl of Lowry reagent solution B and incubate at room temperature for 30 min.

4. Measure the absorbance of the wells in a microplate reader at 750 nm within 10 min of completion of incubation time (*see* **Note 13**).

3.4 Sodium Dodecyl Sulfate Polyacrylamide Gel Electrophoresis

Acrylamide is an irritant, a potent neurotoxin, a reproductive toxin and a probable human carcinogen. Avoid exposure by the use of protective clothing and appropriate disposable gloves. While toxicity is significantly reduced after polymerization, the possibility of unpolymerized acrylamide remains a potential hazard. The following protocol is for preparing one gel.

1. Assemble the gel casting components. Insert the gel comb in the glass plate sandwich to the desired depth and make a small mark on the glass plate with a marker pen some 5–6 mm lower than the ends of the comb teeth and then remove the comb.
2. Mix 4196.7 μl of deionized water, 3125 μl of 40 % acrylamide–bis solution (29:1), 2500 μl of 1.5 M Tris buffer pH 8.8, 125 μl of 10 % SDS and 50 μl of 10 % ammonium persulfate solution in a 25 ml Erlenmeyer flask and mix by swirling.
3. Add 3.3 μl of TEMED into the solution then mix by swirling again and introduce this separating gel solution into the glass sandwich to the level of mark made previously without trapping any air bubbles. Overlay deionized water to a depth of 5 mm, and leave the gel to polymerize for at least 40 min.
4. Mix 1828.3 μl of deionized water, 375 μl of 40 % acrylamide–bis solution (29:1), 742 μl of 0.5 M Tris buffer pH 6.8, 30 μl of 10 % SDS, and 23 μl of 10 % ammonium persulfate solution in a 25 ml Erlenmeyer flask and mix by swirling. Add 1.7 μl of TEMED into the solution then mix by swirling again.
5. Discard the water at the top of the polymerized separating gel and pour in the stacking gel solution to 1–2 mm lower than the edge of glass plate and insert the comb into the glass plate sandwich without trapping any air bubbles. Allow the gel sandwich to polymerize for at least 30 min.
6. After the gel has polymerized, attach the glass plate sandwich to the core electrophoresis apparatus and add running buffer to the top and bottom reservoirs. Ensure the bottom of the gel is submerged in the buffer in the bottom reservoir. Gently remove the comb and wash the wells with a small amount of running buffer from the top chamber using a 21 gauge needle.
7. To prepare samples for loading, mix 4 volumes containing 15–50 μg of sample protein suspension is mixed with 1 volume of 5× loading buffer and boil samples for 5 min before loading into individual wells of the polyacrylamide gel. In one lane of the gel, add 2 μg of low molecular weight protein marker (*see* **Note 14**).

8. Cover the lid and connect to a power supply and run the gel at a constant 20 mA per gel until the marker dye front (bromophenol blue) reaches the end of the separating gel (approximately 90 min).

3.5 Gel Staining and Destaining (Coomassie Blue)

1. After electrophoresis, remove the gel sandwich from the electrophoresis apparatus and carefully disassemble the glass sandwich and immerse the gel in Coomassie blue stain solution and gently rock for 1 min. Incubate the submerged gel at 70 °C for 5 min followed by a further 5 min at room temperature with gentle shaking.
2. Remove the Coomassie blue staining solution and replace the liquid with destaining solution. Incubate the gel with gentle shaking or rocking with regular replacement of the destaining solution for 1–2 h or overnight until the gel background has become transparent (*see* **Note 15**).
3. When the gel background is clear, wash the gel one time with deionized water and scan the gel for a documentary record (*see* **Note 16**).

3.6 Gel Staining and Destaining (Silver Staining)

If the protein lanes after Coomassie blue staining are not clear, particularly towards the bottom of the gel, the gel can be re-stained by silver staining. Carry out all steps at room temperature.

1. Soak the previously Coomassie blue stained gel in silver staining fixing solution for 30 min and then wash the gel in washing solution twice for 5 min each time.
2. Discard the final washing solution and submerge the gel in sensitizing solution for 1 min, wash the gel twice for 5 min each time with deionized water.
3. Immerse the gel in silver nitrate solution for 30 min and subsequently wash the gel with deionized water for 1 min.
4. Immerse the gel in developing solution until a dark brown color appears, then discard the solution.
5. Immerse the gel in stopping solution for 20 min, then rinse the gel once with deionized water.
6. Scan the gel for a documentary record (*see* **Note 16**) and if the gel is to be stored before further processing, immerse it in a solution of 0.1 % acetic acid in deionized water and keep at room temperature or 4 °C for longer storage.

3.7 Reduction/Alkylation and In-Gel Digestion

1. Clean a clear glass plate with distilled water and then rinse with 95 % ethanol. Allow to air-dry, then place the gel to be processed on top of the glass plate.
2. Using a clean scalpel, divide the gel horizontally into 10–13 sections (Fig. 1), and subsequently remove the protein slice for

each lane of the gel from the surrounding gel matrix. Then dice each gel slice into 1 mm^3 small gel cubes, again using a clean scalpel. Divide the cubes equally into three individual wells of a 96-well low binding plate (*see* **Note 17**). Thus each lane slice will generate three individual wells (Fig. 2).

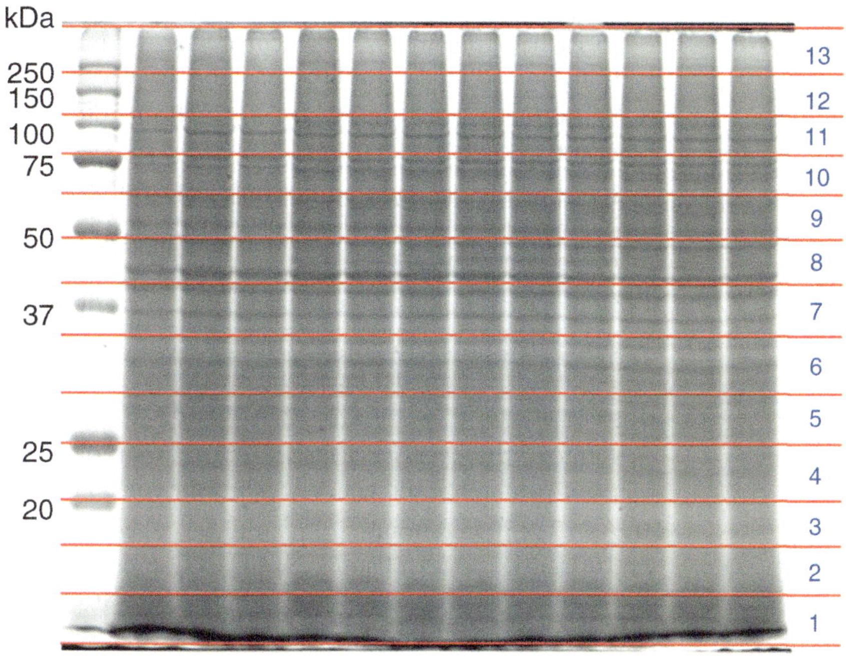

Fig. 1 A standard Coomassie blue stained SDS-polyacrylamide gel. The gel was cut into 13 sections as shown (*red lines*) and the protein bands were subsequently removed from the surrounding gel matrix. The bands were then diced into 1 mm^3 gel cubes

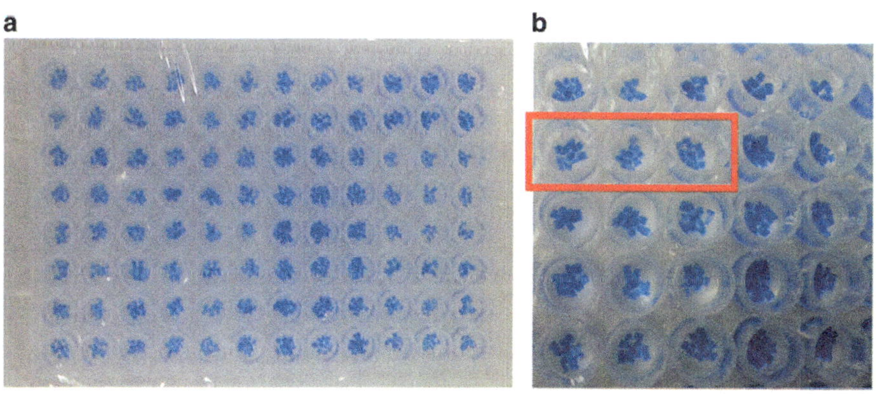

Fig. 2 (**a**) Example of a 96-well plate containing gel cubes prepared from Coomassie blue stained SDS-polyacrylamide gel. (**b**) Enlargement of a portion of the plate shown in (**a**). The three wells *boxed* in *red* are from a single gel slice

3. For gel cubes that have been stained only with Coomassie blue, add 200 μl of 25 mM ammonium bicarbonate (NH_4HCO_3)/50% methanol to each well and incubate at room temperature with occasional agitation for 15 min. Remove the solution and repeat until the gel cubes become light blue in color.
4. For gel cubes that have been stained by double staining with Coomassie blue and silver staining, add fresh 5 % H_2O_2 solution to each of the wells, and incubate for 15 min. Remove the solution and repeat until the gel cubes have a pale yellow color.
5. Wash the gel cubes once with sterile distilled water with agitation for 5 min.
6. Remove the water and add 200 μl of 100 % acetonitrile and incubate the gel cubes for 10 min with occasional agitation at room temperature.
7. Remove the liquid and allow gel cubes to air-dry for 5–10 min.
8. Add 20–25 μl (enough to cover the cubes) of 10 mM dithiothreitol (DTT) in 10 mM NH_4HCO_3 to each of the wells and incubate at 56 °C for 1 h (*see* **Note 18**).
9. Remove the excess DTT/NH_4HCO_3 solution and allow the plate containing the gel cubes to come to room temperature.
10. Add 20–50 μl (enough so that the gel cubes are submerged) of 100 mM IAA in 10 mM NH_4HCO_3 and incubate the plate containing the gel cubes in the dark (*see* **Note 6**) for 1 h at room temperature.
11. Remove the solution and add 200 μl of 100 % acetonitrile to each well and incubate samples for 10 min with occasional agitation.
12. Remove the solution and allow the samples to air-dry for 5–10 min at room temperature.
13. Add 20–50 μl (sufficient to cover the dried gel cubes) of 10 ng/μl trypsin in 10 mM NH_4HCO_3 solution and incubate plates at 4 °C for 10–15 min to rehydrate the gel cubes.
14. Add 5–10 μl of 10 mM ammonium bicarbonate and allow digestion to occur at 37 °C for 3 h to overnight.
15. After trypsin digestion, spin the plates briefly to bring the contents to the bottom of the well (*see* **Note 19**). Remove the liquid to a well in a new 96-well plate (*see* **Note 20**).
16. Add 50 μl of 0.1 % formic acid/50 % acetonitrile to the gel residue and incubate with agitation at room temperature for 15 min.
17. Remove the liquid and combine with the liquid from the previous step in the new 96-well plate. Repeat the extraction with 50 μl of 0.1 % formic acid/50 % acetonitrile twice more, and combine with the previously extracted liquid (*see* **Note 20**).

18. Evaporate the pooled liquid at 40 °C until the samples have dried. Store plates at –20 °C until LC-MS/MS analysis, at which point bring the plates to room temperature and resuspend the extracted peptides in 15–20 μl of 0.1 % formic acid by pipetting the solution around all areas of the wells. Transfer the solution into a 1.7 ml low-binding microcentrifuge tube and centrifuge for 10 min at 9200 × *g*. Transfer the supernatant into a glass vial for HPLC Q-TOF (avoiding any small pieces of gel plug or air bubbles) before injecting 4–5 μl into the NanoLC column (*see* **Note 21**). Use the data generated for bioinformatic analysis (*see* **Note 22**).

4 Notes

1. The disodium salt of EDTA will not dissolve completely until the pH of the solution is adjusted to pH 8.0 by the addition of NaOH. This can be achieved with either 10 N NaOH or NaOH pellets added directly to the solution.
2. Initial pH adjustment of the solution can be made with concentrated (12 N) HCl, but lower ionic strength HCl (6 N, 1 N) should be used for final adjustments to provide better control.
3. Inhalation of SDS powder can cause respiratory tract irritation and possibly a severe allergic reaction. Powdered SDS should be weighed and handled inside a functioning fume hood, and personnel should wear full protective clothing including gloves, eye protection, and a face mask.
4. Ammonium persulfate decays slowly in solution and so for best results it should be made freshly each time. The solution can be stored at 4 °C and used for 2–3 weeks, or at –20 °C and kept for up to a year.
5. Only freshly prepared dithiothreitol (DTT) solution should be used.
6. Only freshly prepared iodoacetamide (IAA) solution should be used. IAA is unstable and light-sensitive. Prepare solutions immediately before use and perform alkylation in the dark.
7. Protocol is based on ten million cells grown in a 100 mm^2 tissue culture plate. Equal numbers of cells between samples is critical for subsequent accurate quantitative analysis. Cells should be counted before harvesting.
8. Trypsinization of cells will degrade cell surface proteins. If single cell suspensions are not required for additional analysis, cells can be scraped directly in PBS omitting trypsinization.
9. This inhibits the activity of trypsin.

10. High-speed centrifugation should be avoided to prevent breaking the cells.
11. In our previous publication [6] 100 μl of cell suspension was used to monitor cell death by annexin V and PI staining, while another 100 μl of cell suspension was taken to monitor infection percentage by flow cytometry. If additional analysis is not required, cells can be resuspended in 1 ml of PBS.
12. There are several methods to quantitate total proteins such as the Bradford assay, the bicinchoninic acid (BCA) protein assay, and the Lowry method. We commonly use the Lowry method to quantitate proteins because of its broad range in sensitivity. However, no protein quantitation method is ideal for all samples, and experimenters should determine for themselves which method is most suitable for their samples.
13. Variation in the third decimal value and an R^2 of 0.99 are acceptable.
14. Marker resolution can be improved by boiling the solution prior to loading, as with the samples.
15. Small material or dust can clog NanoLC columns, so users should not use tissue paper or other materials to blot away the excess stain solution.
16. Prior to scanning, the overall amount of protein should be approximately equal between the lanes, and that there is no significant "necking" or other distortion of the protein lanes. If there is significant "necking" or unequal protein abundance between the lanes, re-run samples on a fresh gel before proceeding.
17. This protocol uses 96-well low binding plates, but the same protocol can be followed using microcentrifuge tubes. If using microcentrifuge tubes, it is important that to spin the tubes briefly between each stage to bring the liquid contents to the bottom of the tube. During steps that require a specific temperature, ensure that the temperature is the same at the bottom and at the top of the tube.
18. This step reduces protein disulfide bonds.
19. This step can be omitted if the equipment to spin the 96-well plates is not available.
20. It is very important to keep the orientation of the samples when transferring from an old plate to a new plate.
21. There are many LC-MS/MS systems available. Each system has its own specific requirements which users should adhere to when using.
22. Peptide mass data can be analyzed in a number of ways, the authors of this protocol use the MASCOT (Matrix Science)

database search program. This on line resource will allow free searches of up to 1200 spectra with some restrictions on usage. However, automated and large search submissions (more than 1200 spectra) require the user to license an in-house copy of the Mascot server.

Acknowledgements

D.R.S. is supported by grants from Mahidol University and the Office of the Higher Education Commission under the National Research Universities Initiative, Mahidol University and the Thailand Research Fund.

References

1. Thiberville SD, Moyen N, Dupuis-Maguiraga L, Nougairede A, Gould EA, Roques P, de Lamballerie X (2013) Chikungunya fever: epidemiology, clinical syndrome, pathogenesis and therapy. Antiviral Res 99:345–370
2. Powers AM (2011) Genomic evolution and phenotypic distinctions of Chikungunya viruses causing the Indian Ocean outbreak. Exp Biol Med 236:909–914
3. Smith DR (2015) Global protein profiling studies of chikungunya virus infection identify different proteins but common biological processes. Rev Med Virol 25:3–18
4. Schirle M, Heurtier MA, Kuster B (2003) Profiling core proteomes of human cell lines by one-dimensional PAGE and liquid chromatography-tandem mass spectrometry. Mol Cell Proteomics 2:1297–1305
5. Wikan N, Khongwichit S, Phuklia W, Ubol S, Thonsakulprasert T, Thannagith M, Tanramluk D, Paemanee A, Kittisenachai S, Roytrakul S, Smith DR (2014) Comprehensive proteomic analysis of white blood cells from chikungunya fever patients of different severities. J Transl Med 12:96
6. Abere B, Wikan N, Ubol S, Auewarakul P, Paemanee A, Kittisenachai S, Roytrakul S, Smith DR (2012) Proteomic analysis of chikungunya virus infected microgial cells. PLoS One 7:e34800

Chapter 17

Bioinformatics Based Approaches to Study Virus–Host Interactions During Chikungunya Virus Infection

Sreejith Rajasekharan and Sanjay Gupta

Abstract

The limitations of high-throughput genomic methods used for studying virus–host interactions make it difficult to directly obtain insights on virus pathogenesis. In this chapter, the central steps of a protein structure similarity based computational approach used to predict the host interactors of Chikungunya virus are explained by highlighting the important aspects that need to be considered. Identification of such conserved set of putative interactions that allow the virus to take control of the host has the potential to deepen our understanding of the virus-specific remodeling processes of the host cell and illuminate new arenas of disease intervention.

Key words Chikungunya virus, Protein interactions, Protein structure similarity, Structural bioinformatics, Virus–host interactions

1 Introduction

A complex network of specific protein–protein interactions is involved at the interface of viral infection and host response; with the host attempting to eradicate the invading viral pathogen, and the pathogen whilst evading the host immune surveillance continues to proliferate. The experimental challenges of identifying virus–host protein interactions are numerous and therefore efforts to elucidate interaction maps between viruses and their host are far from complete. Although large-scale screens of virus–host interactions using either yeast two-hybrid (Y2H) provide substantial resources for the computational analysis and modeling of processes involved in virus infection and proliferation, the large size and varying quality of the high-throughput screens make it difficult to directly obtain insights on virus infection processes from the screening results [1]. Computational approaches can accelerate these experimental efforts to identify protein interactions by enabling the integration of results from different screens as well as to identify

Justin Jang Hann Chu and Swee Kim Ang (eds.), *Chikungunya Virus: Methods and Protocols*, Methods in Molecular Biology, vol. 1426, DOI 10.1007/978-1-4939-3618-2_17,

general trends and connections among the targeted proteins such as common pathways and biological processes involved.

Computational methods have been proven to be very helpful in identifying protein interactions within a single organism (intraspecies; 2) or between two (interspecies). Although there are ample studies on predicting the intraspecies protein interactions, the research on interspecies interaction prediction has been limited primarily because of the scarcity of data sources. However, over the past decade, with the availability of genomic and proteomic data, studies using computational approaches for predicting virus–host interactions have increased considerably [3–5]. As HIV is among the most studied virus, several different computational methods along with experimental studies had also been employed for predicting the interactions of HIV and its human host [6–8]. Some of the recent studies involve both sequence and structural homology to identify protein interactions among the virus and its host [9, 10]. Besides viral infections, a few studies have been done for nonviral pathogens also based on the structural similarity of pathogen and host proteins [4, 11].

This chapter explains the implementation of a protein structure similarity based computational approach to predict the interactions between Chikungunya virus (CHIKV) and its hosts (human and mosquito). This approach has been used earlier for HIV–human [8], Dengue virus (DENV)–human [12], and Chandipura virus (CHPV)–human interaction prediction [13] as well as for nonviral pathogens causing tropical diseases [5] and involves the mapping of the host proteins with defined structure and known interactions with structurally similar viral proteins. Together with the knowledge of intraviral protein interactions of CHIKV previously reported by the authors [14, 15], this approach puts forth a platform for the better understanding of CHIKV biology.

2 Methods

2.1 Viral Protein Structures

1. Obtain the experimentally determined three dimensional structures of CHIKV envelope proteins [E1 (PDB ID: 3N42-F), E2 (PDB ID: 3N42-B), and E3 (PDB ID: 3N42-A)] from PDB (Protein Data Bank) archives. Since the structures of capsid, 6K, and all four non-structural proteins of the CHIKV are not available in PDB, use I-TASSER to model the structures (16, 17 *see* **Note 1**).
2. Generate the computational models of proteins using CHIKV genes sequences [18] [GenBank Accession No., JF272473 (nsP1), JF272474 (nsP2), JF272475 (nsP3), JF272476 (nsP4), JF272477 (capsid), JF272481 (6K), 2011].

2.2 Identification of Structurally Similar Proteins Among Chikungunya Virus and Its Hosts

1. Obtain the structurally similar human and Drosophila (*see* **Note 2**) counterparts of CHIKV proteins by submitting the viral protein structures (either known or I-TASSER generated) to DaliLite v.3 Web server (*see* **Note 3**).
2. Consider all the *H. sapiens* and *D. melanogaster* proteins with *Z*-score ≥2 as structurally similar proteins of the respective CHIKV protein. Refer these proteins as hCHIKV and dCHIKV proteins, respectively.

2.3 Prediction of CHIKV-Host Protein Interactions

1. Identify the putative human host interactors of CHIKV (interspecies or exogenous interactions) by obtaining the cellular protein partners of hCHIKV (intraspecies or endogenous interactions) from HPRD (19, Human Protein Reference Database), BIOGRID (20, Biological General Repository for Interaction Datasets) and STRING [21] databases. All the literature curated interactions established through in vitro and/or in vivo methods among human proteins are enlisted in these datasets.
2. Obtain the interactions of dCHIKV with other *D. melanogaster* proteins (endogenous interactions) using database DroID with a cutoff confidence value of 0.4 to identify the possible interactions among CHIKV and Drosophila proteins (exogenous interactions, 22).
3. Use the FlyBase database to obtain the *Ae. aegypti* orthologs of *D. melanogaster* [23] and consider these proteins (aCHIKV—*Ae. aegypti* orthologs of *D. melanogaster*) as putative mosquito interactors of CHIKV.

2.4 Gene Ontology and Interaction Validation

1. Annotate the putative host protein interactors of CHIKV proteome on the basis of their Gene Ontology (GO) analysis (functionality and cellular compartmentalization).
2. Primarily short-list the interaction datasets on the basis of protein localization, since two proteins theoretically must share at least one cellular compartment for direct interaction among them. Obtain the cellular compartment of the putative host protein interactors from GOanna Web server, provided by AgBase v. 2.00. This Web server generates the GO annotations based on sequence homology [24].
3. Convert its output into annotation summary file using GOanna2ga tool. Obtain the results by interpretation of summary file using GOslim viewer. The assignment of GO annotation terms is based on previously annotated data similar to the input data identified by BLAST.
4. Further screen the cellular compartments assigned to individual protein on the basis of reported literature that suggest the localization of viral proteins in the host cell.

5. Submit short-listed proteins to Database for Annotation, Visualization and Integrated Discovery (DAVID, 25–27) and Reactome databases [28] to obtain the list of terms enriched among these proteins and the biological pathways involving them.
6. Organize the DAVID annotation charts as tree structures, and as the distance from root increases the terms are considered to be more specific. Use GO level 4 terms to make the study more specific and informative.
7. Use Bonferroni procedure to obtain corrected *p*-values. Transform these values into –log 10 terms for graphical representation of data.
8. Make the interaction networks using Cytoscape [29].
9. A single protein is represented by multiple PDB accession number. Similarly, in Dali database each protein might be represented by multiple PDB structures. As all of these structures correspond to the same protein, the interaction dataset generated will also be the same.
10. Eliminate the redundancy obtained in the data by representing each protein with a single PDB structure. In addition to this, multiple hCHIKV (human proteins structurally similar to CHIKV proteins) proteins can have common cellular partners and among these only unique pairs of interactions between human uniprot accession and CHIKV proteins are considered.
11. Compare the putative interactions obtained with the experimental studies carried out on CHIKV and related alphaviruses Sindbis virus (SINV) and Semliki Forest virus (SFV) to validate these results.

2.5 Results Interpretation

Using the protein structures described in Subheading 2.1 as an example of data input, the computational approach described in this chapter will predict 2028 human and 86 mosquito proteins that interact with those of CHIKV through 3918 and 112 unique interactions, respectively. The functional processes involving target proteins will be broadly categorized among structural (involved in intracellular signaling; JAK-STAT pathway and enzyme linked receptor mediated signaling pathways) and nonstructural proteins (involved in translation–transcription regulation, programmed cell death, and stress responses) of CHIKV. The predicted interactions will be highlighted on the basis of their functional relevance during CHIKV infection and prioritized based on the interaction data of other related alphaviruses [30]. This bioinformatics based protocol can also be used to predict protein–protein interactions between a host and other pathogens during infections.

3 Notes

1. The viral protein structures can also be generated by other bioinformatic tools such as SWISS-MODEL Workspace, Modeller v9.11, and FG-MD (fragment guided-molecular dynamics).
2. In the dearth of ample knowledge about the mosquito proteome, perform the interaction study using drosophila proteins and then identify the mosquito orthologous proteins.
3. DaliLite v.3 Web server uses a weighted sum-of-similarity (sum-of-pairs method) for comparing the intra-molecular distances between amino acids. The comparison correlates with the expert classification as the structures of homologues proteins get higher scores for similarity relative to the structures of evolutionary unrelated proteins. DALI or distance alignment matrix method server strategically screens all the structures available in PDB to ensure that no significant similarity is missed. This server divides the input structures into hexapeptide fragments and calculates a distance matrix by evaluating the contact patterns between successive fragments. While determining the similarity between two proteins, the 3D structural coordinates of these proteins can be compared by alignment of alpha carbon distance matrices. When two proteins distance matrices share the same or similar features in approximately the same positions, they can be said to have similar folds with similar-length loops connecting their secondary structure elements.

References

1. Friedel CC (2013) Computational analysis of viral-host interactome. In: Bailer SM, Leiber D (eds) Virus-host interactions: methods and protocols, methods in molecular biology, 1st edn. Springer, New York, pp 115–130
2. Shoemaker BA, Panchenko AR (2007) Deciphering protein-protein interactions. Part ii. Computational methods to predict protein and domain interaction partners. PLoS Comput Biol 3(4):e43
3. Aloy P, Russell RB (2003) InterPreTS: protein interaction prediction through tertiary structure. Bioinformatics 19(1):161–162
4. Lu L, Lu H, Skolnick J (2002) MULTIPROSPECTOR: an algorithm for the prediction of protein-protein interactions by multimeric threading. Proteins 49:350–364
5. Davis FP, Braberg H, Shen MY et al (2006) Protein complex compositions predicted by structural similarity. Nucleic Acids Res 34:2943–2952
6. Tastan O, Qi Y, Carbonell JG, Klein-Seetharaman J (2009) Prediction of interactions between HIV-1 and human proteins by information integration. Pac Symp Biocomput 14:516–527
7. Evans P, Dampier W, Ungar L et al (2009) Prediction of HIV-1 virus host protein interactions using virus and host sequence motif. BMC Med Genomics 2:27–39
8. Doolittle JM, Gomez SM (2010) Structural similarity-based predictions of protein interactions between HIV-1 and Homo sapiens. Virol J 7:82–96
9. Davis FP, Barkan DT, Eswar N et al (2007) Host-pathogen protein interactions predicted by comparative modeling. Protein Sci 16(12):2585–2596
10. Franzosa EA, Xia Y (2011) Structural principles within the human-virus protein-protein interaction network. Proc Natl Acad Sci U S A 108(26):10538–10543

11. Zhang QC, Petrey D, Deng L et al (2012) Structure based prediction of protein-protein interactions on a genome wide scale. Nature 490:556–561
12. Doolittle JM, Gomez SM (2011) Mapping protein interactions between Dengue virus and its human and insect hosts. PLoS Negl Trop Dis 5(2):e954
13. Rajasekharan S, Rana J, Gulati S et al (2013) Predicting the host protein interactors of Chandipura virus using a structural similarity-based approach. FEMS Pathog Dis 69(1):29–35
14. Sreejith R, Rana J, Dudha N et al (2012) Mapping of interactions among Chikungunya virus nonstructural proteins. Virus Res 169(1):231–236
15. Dudha N, Rana J, Rajasekharan S et al (2014) Host-pathogen interactome analysis of Chikungunya virus envelope proteins E1 and E2. Virus Genes 50(2):200–209. doi:10.1007/s11262-014-1161-x
16. Berman HM, Westbrook J, Feng Z et al (2000) The Protein Data Bank. Nucleic Acids Res 28:235–242
17. Zhang Y (2008) I-TASSER server for protein 3D structure prediction. BMC Bioinformatics 9:40–47
18. Dudha N, Appaiahgari MB, Bharati K et al (2012) Molecular cloning and characterization of Chikungunya virus genes from Indian isolate of 2006 outbreak. J Pharm Res 5(7):3860–3863
19. Mishra GR, Suresh M, Kumaran K et al (2006) Human protein reference database—2006 update. Nucleic Acids Res 34:D411–D414
20. Stark C, Breitkreutz BJ, Reguly T et al (2006) BioGRID: a general repository for interaction datasets. Nucleic Acids Res 34:D535–D539
21. Szklarczyk D, Franceschini A, Kuhn M et al (2011) The STRING database in 2011: functional interaction networks of proteins globally integrated and scored. Nucleic Acids Res 39:D561–D568
22. Aranda B, Achuthan P, Alam-Faruque Y et al (2009) The IntAct molecular interaction database in 2010. Nucleic Acids Res 38:D525–D531
23. Crosby MA, Goodman JL, Strelets VB et al (2006) FlyBase: genomes by the dozen. Nucleic Acids Res 35:D486–D491
24. McCarthy F, Wang N, Magee GB et al (2006) AgBase: a functional genomics resource for agriculture. BMC Genomics 7:229–241
25. Ashburner M, Ball CA, Blake JA et al (2000) Gene ontology: tool for the unification of biology. Nat Genet 25:25–29
26. Dennis G, Sherman B, Hosack D et al (2003) DAVID: database for annotation, visualization, and integrated discovery. Genome Biol 4:P3
27. Huang DW, Sherman BT, Lempicki RA (2009) Systematic and integrative analysis of large gene lists using DAVID bioinformatics resources. Nat Protoc 4:44–57
28. Matthews L, Gopinath G, Gillespie M et al (2009) Reactome knowledgebase of human biological pathways and processes. Nucleic Acids Res 37:D619–D622
29. Shannon P, Markiel A, Ozier O et al (2003) Cytoscape: a software environment for integrated models of biomolecular interaction networks. Genome Res 13:2498–2504
30. Rana J, Sreejith R, Gulati S et al (2013) Deciphering the host-pathogen interface in Chikungunya virus-mediated sickness. Arch Virol 158(6):1159–1172

Chapter 18

T-Cell Epitope Prediction of Chikungunya Virus

Christine Loan Ping Eng, Tin Wee Tan, and Joo Chuan Tong

Abstract

There has been a growing demand for vaccines against Chikungunya virus (CHIKV), and epitope-based vaccine is a promising solution. Identification of CHIKV T-cell epitopes is critical to ensure successful trigger of immune response for epitope-based vaccine design. Bioinformatics tools are able to significantly reduce time and effort in this process by systematically scanning for immunogenic peptides in CHIKV proteins. This chapter provides the steps in utilizing machine learning algorithms to train on major histocompatibility complex (MHC) class I peptide binding data and build prediction models for the classification of binders and non-binders. The models could then be used in the identification and prediction of CHIKV T-cell epitopes for future vaccine design.

Key words Chikungunya/CHIKV, MHC, Antigens, Peptides, Epitopes, Prediction, Machine learning

1 Introduction

There has been growing concerns of public health threat over the reemergence of Chikungunya virus (CHIKV). Despite the intensified research activities in response to this, there are still no antiviral treatments or vaccines available for the infection [1–3]. Epitope-based vaccines are a promising alternative to traditional vaccines, where it is hoped that accurate selection of immunogenic epitopes would elicit the desired immune responses. It has been demonstrated that CHIKV infection evokes strong innate immune responses with $CD8^+$ T cell activation [4]. The selection of T cell epitopes is therefore crucial in the design of CHIKV epitope-based vaccines.

There are various bioinformatics approaches designed for T cell epitope prediction. This could greatly expedite the process of epitope selection by screening all possible epitopes from CHIKV proteins. Many of these software are publicly available, with various methods and algorithms, including simple sequence-based predictions as well as complex structure-based predictions [5, 6]. Machine learning approaches can be utilized to construct computational models to predict the binding of T cell epitopes to major

Justin Jang Hann Chu and Swee Kim Ang (eds.), *Chikungunya Virus: Methods and Protocols*, Methods in Molecular Biology, vol. 1426, DOI 10.1007/978-1-4939-3618-2_18,

histocompatibility complex (MHC) class I molecules. These prediction models can then be used to identify possible immunogenic proteins from CHIKV. This chapter outlines the process of constructing MHC class I T cell epitope prediction models using machine learning approaches to aid in the identification of possible CHIKV T cell epitopes for vaccine design.

2 Materials

2.1 Data

Sufficient MHC class I peptide binding and non-binding data are necessary for the training of machine learning algorithms. There are public databases with large amount of MHC class I peptide data, and these include Immune Epitope Database (IEDB) [7] and MHCBN [8]. The data required are the peptide sequences with corresponding peptide binding affinity to MHC class I molecules (*see* **Note 1**).

2.2 Software

WEKA is an open source platform with a suite of machine learning algorithms (http://www.cs.waikato.ac.nz/ml/weka/) [9]. The software can be used to experiment with various machine learning algorithms and select for the best method for the dataset.

3 Methods

3.1 Definition of Binders and Non-binders

Classify peptide sequences in the MHC class I peptide dataset into two groups of binders and non-binders based on experimentally determined binding affinities. Peptides with binding affinities below 500 nM can be classified as binders with the capacity to stimulate $CD8^+$ T cell responses, while non-binders are those with binding affinities exceeding the threshold [10].

3.2 Transformation of Peptide Sequence to Feature Vectors

The next step involves transforming the peptide sequences into feature vectors as input for machine learning (*see* **Note 2**). Amino acid physicochemical properties can be extracted from each peptide sequence based on amino acid property groups such as hydrophobicity, charge, and polarity (11–13; *see* **Note 3**). Table 1 lists the amino acid physicochemical property groups based on amino acid indices by Tomii and Kanehisa [14].

Next, the amino acid physicochemical properties for each peptide sequence can be globally described using the composition-transition-distribution (CTD) method. The CTD method developed by Dubchak et al. [15] extracts information on each amino acid physicochemical property using three descriptors, which are composition (C), transition (T), and distribution (D). Calculations for each descriptor are as described below:

1. Composition describes the frequency in percentage of amino acid property groups within the sequence. This is calculated by

Table 1
Amino acid physicochemical properties group

Amino acid property	Groups 1	2	3
Hydrophobicity	*Polar* KEDQNR	*Neutral* ASTPHYG	*Hydrophobic* VLIMFWC
Charge	*Positive* KR	*Neutral* NCQGHILMFPSTWYVA	*Negative* DE
Polarity	*Polarity value 4.9–6.2* IFWCMVYL	*Polarity value 8.0–9.2* ATGSP	*Polarity value 10.4–13.0* QRKNEDH

Amino acids are grouped into three divisions for each amino acid physicochemical property based on amino acid indices by Tomii and Kanehisa [11–14]

the percentage of the total number of amino acids in each property group in relation to the entire length of the peptide sequence:

$$C = \left(\frac{n_1 \times 100}{N}, \frac{n_2 \times 100}{N}, \frac{n_3 \times 100}{N} \right),$$

where, $N = \sum_{i=1}^{m} n_i$, $m = 3$ is the number of groups for each amino acid property, and ni represents the amino acid count of ith group [16, 17].

2. Transition measures the percentage frequency of changes between amino acid property groups in the peptide sequence. T can thus be calculated as the percent frequency of group i transitioning to group j, or j followed by i, for $i, j \in \{n_1, n_2, n_3\}$:

$$T = \left(\frac{T_{G_1G_2} \times 100}{N-1}, \frac{T_{G_1G_3} \times 100}{N-1}, \frac{T_{G_2G_3} \times 100}{N-1} \right),$$

where, $T_{G_iG_j}$ counts the number of transitions of amino acids in group i followed by group j, or the transitions of group j to i, and $N-1$ is the total number of transitions in the peptide sequence [16–18].

3. Distribution represents the locations of the first, 25, 50, 75, and 100 % amino acids in the property group within the peptide sequence. The position of the amino acids at each fraction is noted, and calculated in relation to the peptide length.

$$D = (D_1, D_2, D_3),$$

$$D_i = \left(\frac{P_{i0} \times 100}{N}, \frac{P_{i25} \times 100}{N}, \frac{P_{i50} \times 100}{N}, \frac{P_{i75} \times 100}{N}, \frac{P_{i100} \times 100}{N} \right),$$

where, *Pij* ($j = 0, 25, 50, 75, 100$) is the length at which *j*% of the amino acids in group *i* is found [16, 17].

Therefore, for each of the three amino acid property group, 21 descriptors of *C*, *T* and *D* can be calculated, resulting in a total of 63 feature vectors describing the amino acid physicochemical properties of each peptide sequence. WEKA accepts data in ARFF file format, with an example shown in Fig. 1. The dataset consisting of peptide sequences transformed into feature vectors would have to be formatted accordingly.

3.3 Construction of Prediction Models

There are numerous machine learning algorithms which can be used in the classification of binders and non-binders. Several common machine learning algorithms, available from the Classify tab in WEKA Explorer are listed below (*see* **Note 4**):

1. Artificial neural network—classifiers\functions\MultilayerPerceptron.

```
% 1. Title: MHC Class I Peptide Binding Data
%
% 2. Allele: HLA-A_0201
%
% 3. Source: Peters B, Bui HH, Frankild S, Nielsen M, Lundegaard C,
% Kostem E, Basch D, Lamberth K, Harndahl M, Fleri W, Wilson SS,
% Sidney J, Lund O, Buus S, Sette A. A community resource
% benchmarking predictions of peptide binding to MHC-I molecules.
% PLoS Comput Biol. 2006 Jun 9;2(6):e65.
%
@relation HLA-A_0201

@attribute 'hydroc1' real
@attribute 'hydroc2' real
@attribute 'hydroc3' real
 .
 .
 .
@attribute 'polard15' real
@attribute 'class' {binder, nonbinder}

@data
26.67,33.33,40.0,21.43,28.57,21.43,13.33,13.33,33.33,80.0,100.0,26.6
7,26.67,66.67,86.67,93.33,6.667,20.0,40.0,53.33,73.33,40.0,40.0,20.0
,35.71,14.29,21.43,6.667,26.67,60.0,86.67,93.33,13.33,20.0,33.33,73.
33,80.0,40.0,40.0,53.33,53.33,100.0,40.0,33.33,26.67,21.43,28.57,21.
43,6.667,20.0,40.0,53.33,73.33,26.67,26.67,66.67,86.67,93.33,13.33,1
3.33,33.33,80.0,100.0,binder
26.67,33.33,40.0,21.43,21.43,28.57,20.0,20.0,26.67,66.67,86.67,6.667
,6.667,40.0,80.0,93.33,13.33,46.67,53.33,73.33,100.0,40.0,26.67,33.3
3,35.71,28.57,21.43,6.667,13.33,26.67,73.33,93.33,20.0,20.0,46.67,60
.0,86.67,33.33,33.33,66.67,80.0,100.0,46.67,13.33,40.0,14.29,35.71,7
.143,13.33,40.0,53.33,60.0,100.0,6.667,6.667,6.667,93.33,93.33,20.0,
26.67,33.33,80.0,86.67,nonbinder
```

Fig. 1 ARFF file format for WEKA input

2. Support vector machine—classifiers\functions\LibSVM.
3. Random forest—classifiers\trees\RandomForest.
4. Naïve Bayes—classifiers\bayes\NaiveBayes.

The default test option for training in WEKA is cross-validation with ten folds. In tenfold cross-validation, ten subsets are randomly partitioned from the dataset with one subset reserved as testing data for model evaluation and the rest used for training. The cycle repeats ten times, where each subset would be tested exactly once. With sufficient data, cross-validation may overcome the problem of over-fitting where the prediction model is over-fitted to the training data and is not able to generalize beyond training data to future data (*see* **Note 5**).

Parameter optimization is an important step in the construction of prediction models. Performance of the prediction models largely depends on the parameters for the training process, for example, the complexity parameter and type of kernel for support vector machine. This can be tuned by clicking on the classifier in Weka Explorer and changing the parameter options accordingly.

3.4 Evaluation of Prediction Models

Upon completion of the training process, the prediction models can be evaluated with several performance measures. These include accuracy (ACC), sensitivity (SN), specificity (SP), area under the curve (AUC), and Matthews correlation coefficient (MCC) described as follows:

1. Accuracy represents the freedom from error in the classification, which in this case the number of binders and non-binders are classified correctly by the model.
2. $\text{ACC} = \left(\dfrac{\text{TP} + \text{TN}}{\text{TP} + \text{FP} + \text{TN} + \text{FN}} \right).$
3. Sensitivity and specificity measure the proportion of correctly identified positive and negative samples. SN measures the ability of the model to correctly identify binders among all binders while SP gives the ratio of correctly predicted non-binders from all non-binders in the dataset.
4. $\text{SN} = \left(\dfrac{\text{TP}}{\text{TP} + \text{FN}} \right).$
5. $\text{SP} = \left(\dfrac{\text{TN}}{\text{TN} + \text{FP}} \right).$
6. Area under the curve (AUC) is a common measure for binary classification which describes the classifier performance over all possible thresholds. This gives the probability of a positive sample chosen randomly ranking higher than a randomly chosen negative sample [19, 20].

7. Matthews correlation coefficient assesses the quality of binary classifications, as it calculates the correlation coefficient between the observed and predicted samples in the dataset [21].

8. $$CC = \left(\frac{(TP \times TN) - (FP \times FN)}{\sqrt{(TN + FN)(TN + FP)(TP + FN)(TP + FP)}} \right).$$

3.5 Prediction of CHIKV T Cell Epitopes

Upon successful construction of MHC class I epitope binding prediction models, T cell epitopes can be identified and predicted from CHIKV proteins. The steps are listed below:

1. Generate a list of all possible peptide sequence of window length 9 or 10 from CHIKV proteins.
2. Extract feature vectors from peptide sequences with CTD method.
3. Format test sequences into ARFF format, assigning unknown ("?") class for each sequence.
4. Load prediction model onto WEKA Explorer.
5. Under Test options, select "Supplied test set" for test sequences.
6. Check the "Output predictions" under "More options".
7. Predict classifications of new test sequences by right-clicking the prediction model and selecting "Re-evaluate model on current test set".
8. Analyze prediction results from the classifier output in "Predictions on test set".

4 Notes

1. In general, the larger the number of samples in the dataset the better. It would also be advisable to ensure that there is an approximately equal number of positive and negative samples in the dataset. An unbalanced dataset may result in bias in the training process as well as during performance evaluation.
2. There are many ways of feature transformation. Most machine learning algorithms accept numbers as an input. Additional features that can be extracted from the protein sequence are standard amino acid composition and secondary structure information among others.
3. There are also several other amino acid property groups which can be included in the feature transformation process. These include normalized van der Waals volume, polarizability and solvent accessibility [16].

4. Different machine learning algorithms are well-suited to different types of dataset. It is possible to compare and contrast the performance of various prediction models constructed using different machine learning algorithms to select the best model for the dataset.
5. The cross-validation process can be optimized by selecting the appropriate number of folds which depends on the number of samples in the dataset. For smaller datasets, less number of folds such as fivefold would be more advisable.

References

1. Weaver SC (2014) Arrival of chikungunya virus in the new world: prospects for spread and impact on public health. PLoS Negl Trop Dis 8(6):e2921. doi:10.1371/journal.pntd.0002921
2. Weaver SC, Osorio JE, Livengood JA et al (2012) Chikungunya virus and prospects for a vaccine. Expert Rev Vaccines 11(9):1087–1101. doi:10.1586/erv.12.84
3. Kaur P, Chu JJ (2013) Chikungunya virus: an update on antiviral development and challenges. Drug Discov Today 18(19-20):969–983. doi:10.1016/j.drudis.2013.05.002
4. Wauquier N, Becquart P, Nkoghe D et al (2011) The acute phase of Chikungunya virus infection in humans is associated with strong innate immunity and T CD8 cell activation. J Infect Dis 204(1):115–123. doi:10.1093/infdis/jiq006
5. Tong JC, Tan TW, Ranganathan S (2007) Methods and protocols for prediction of immunogenic epitopes. Brief Bioinform 8(2):96–108. doi:10.1093/bib/bbl038
6. Lafuente EM, Reche PA (2009) Prediction of MHC-peptide binding: a systematic and comprehensive overview. Curr Pharm Des 15(28): 3209–3220
7. Vita R, Overton JA, Greenbaum JA et al (2015) The immune epitope database (IEDB) 3.0. Nucleic Acids Res 43(Database issue):D405–D412. doi:10.1093/nar/gku938
8. Lata S, Bhasin M, Raghava GP (2009) MHCBN 4.0: a database of MHC/TAP binding peptides and T-cell epitopes. BMC Res Notes 2:61. doi:10.1186/1756-0500-2-61
9. Hall M, Frank E, Holmes G et al (2009) The WEKA Data Mining Software: an update. SIGKDD Explor 11(1):10–18. doi:10.1145/1656274.1656278
10. Sette A, Vitiello A, Reherman B et al (1994) The relationship between class I binding affinity and immunogenicity of potential cytotoxic T cell epitopes. J Immunol 153(12):5586–5592
11. Grantham R (1974) Amino acid difference formula to help explain protein evolution. Science 185(4154):862–864
12. Engelman DM, Steitz TA, Goldman A (1986) Identifying nonpolar transbilayer helices in amino acid sequences of membrane proteins. Annu Rev Biophys Biomol Struct 15:321–353. doi:10.1146/annurev.bb.15.060186.001541
13. Klein P, Kanehisa M, DeLisi C (1984) Prediction of protein function from sequence properties. Discriminant analysis of a data base. Biochim Biophys Acta 787(3):221–226
14. Tomii K, Kanehisa M (1996) Analysis of amino acid indices and mutation matrices for sequence comparison and structure prediction of proteins. Protein Eng 9(1):27–36
15. Dubchak I, Muchnik I, Mayor C et al (1999) Recognition of a protein fold in the context of the Structural Classification of Proteins (SCOP) classification. Proteins 35(4):401–407
16. Cui J, Han LY, Lin HH et al (2007) Prediction of MHC-binding peptides of flexible lengths from sequence-derived structural and physicochemical properties. Mol Immunol 44(5):866–877. doi:10.1016/j.molimm.2006.04.001
17. Li ZR, Lin HH, Han LY et al (2006) PROFEAT: a web server for computing structural and physicochemical features of proteins and peptides from amino acid sequence. Nucleic Acids Res 34(Web Server Issue):W32–W37. doi:10.1093/nar/gkl305
18. El-Manzalawy Y, Dobbs D, Honavar V (2008) On evaluating MHC-II binding peptide prediction methods. PLoS One 3(9):e3268. doi:10.1371/journal.pone.0003268
19. Fawcett T (2006) An introduction to ROC analysis. Pattern Recogn Lett 27(8):861–874. doi:10.1016/j.patrec.2005.10.010
20. Linden A (2006) Measuring diagnostic and predictive accuracy in disease management: an introduction to receiver operating characteristic (ROC) analysis. J Eval Clin Pract 12(2):132–139. doi:10.1111/1365-2753.2005.00598.X
21. Matthews BW (1975) Comparison of the predicted and observed secondary structure of T4 phage lysozyme. Biochim Biophys Acta 405(2): 442–451

Part III

Immunology and Animal Model Studies

Chapter 19

Mouse Models of Chikungunya Virus

Lara J. Herrero, Penny A. Rudd, Xiang Liu, Stefan Wolf, and Suresh Mahalingam

Abstract

The majority of medical advances have been made using animals. Studies using mouse models of chikungunya-induced disease have proven invaluable for dissecting the intricate nature of the immune response to this viral infection and identifying potential targets for the development of treatment strategies. Herein we describe the common mouse models used to research the pathobiology of chikungunya virus infection to date.

Key words Chikungunya, *Mus musculus*, In vivo infections, Intracerebral, Intranasal, Subcutaneous

1 Introduction

Chikungunya virus (CHIKV) is an old world alphavirus transmitted by the *Aedes* mosquitoes. Like most old world alphaviruses, it is an arthrogenic virus causing debilitating pain in those afflicted. CHIKV was first isolated in Tanzania in the early 1950s and its name means "that which bends up" and reflects the stoop posture seen in patients [1]. Other signs of illness include fever, rash, vomiting and myalgia. In the last 10–15 years, CHIKV has reemerged causing numerous important outbreaks in new areas where the virus had not been previously found. One such region is La Réunion , where in 2005–2006 over 255,000 cases were reported in a total population of about 770,000. Furthermore, as of April 2015, the WHO estimated there to be over 1,379,780 suspected cases of chikungunya in the Caribbean Islands, Latin America, and the USA [2]. Over the past years, the distribution of CHIKV has changed primarily due to its ability to be transmitted by *Aedes albopictus* or the Asian tiger mosquito. The worldwide distribution of this species of mosquito can explain why CHIKV is now endemic to new areas such as the south of France and Florida state in the USA [3]. Currently, there are neither specific treatments nor vaccines against CHIKV. Patients receive supportive care and are

Justin Jang Hann Chu and Swee Kim Ang (eds.), *Chikungunya Virus: Methods and Protocols*, Methods in Molecular Biology, vol. 1426,DOI 10.1007/978-1-4939-3618-2_19, © Springer Science+Business Media New York 2016

Table 1
Immunocompetent mouse models of chikungunya virus infection/disease

Mouse strain	Age	Mode of infection	Measureable outcomes	Protocol	References
Swiss	Neonatal	Intracerebral (i.c)	Mortality	3.1	[2]
BALB/c C57BL/6 NIH Swiss	3 weeks to adult	Intranasal (i.n.)	Histopathology/clinical signs/viremia	3.2	[7, 8, 9, 10]
ICR/ CD-1 BALB/c	Newborn	Subcutaneous (s.c) in the scruff of the neck	Mortality/histopathology	3.3	[8, 9]
C57BL/6	14 days to adult	Subcutaneous (s.c.) in the footpad	Histopathology/viremia/ joint and foot swelling and inflammation	3.4	[13, 16,19]

most often administered simple analgesics (acetaminophen, paracetamol) and/or nonsteroidal anti-inflammatory drugs (ibuprofen, naproxen), which help to relieve pain and fever.

The use of animal models is imperative for the development of novel therapeutics or vaccines. They allow us to better understand interactions between virus and host. By using animal models, we can gain insights into immune responses, pathogenesis as well as examine potential drug efficacy. Overall, they help us learn about new aspects of disease biology in a complex system, which cannot be replicated *in vitro*. In this chapter, we discuss the methodology of the more prominent mouse models used in CHIKV research. Among these, we describe an intracranial neonatal model (Albino Swiss mice), an intranasal 3 -week-old to adult mouse model (BALB/c, Swiss, and C57BL/6 mice), a subcutaneous newborn model (nape of the neck, ICR and CD-1 mice), and a subcutaneous 2-week-old to adult model (footpad, C57BL/6 mice).

Please refer to Table 1 for an overview of protocols outlined in this chapter.

2 Materials

2.1 Intracerebral Infection of Neonatal Mice

1. 0.22 μm filters.
2. Endotoxin free phosphate buffer saline (PBS).
3. 27 G or smaller surgical needles.
4. 1 ml syringes.
5. Sterile 1.5 ml tubes.
6. P200 pipette and 200 μl pipette tips.
7. 80 % ethanol to sterilize work area.

8. Class 2 biological safety cabinet.
9. Anesthesia machine/chamber.
10. Anesthetics: ketamine, xylazine, and avertin.
11. CHIKV stock. Consider the cell type used to generate virus stock (mammalian origin or mosquito origin).
12. Neonatal albino Swiss mice (*see* **Note 1**) [4–6].

2.2 Intranasal Infection of CHIKV in Mice Aged 3 Weeks to Adult

1. Consumable **items 1–10** as in Subheading 2.1.
2. Mice. Several strains of mice from 3-week-old to adult can be used:
   (a) 5-week-old or older NIH Swiss [7].
   (b) 6- to 8-week-old female BALB/c [8, 9].
   (c) 3- to 5-week-old C57BL/6 mice [7, 10].

2.3 Subcutaneous Inoculation of Newborn Mice

1. Consumable **items 1–8** as in Subheading 2.1.
2. Mice. This protocol can be used with several strains of newborn mice between 2 and 3 days old:
   (a) BALB/c [11].
   (b) CD-1 [12].
   (c) ICR [12].

2.4 Subcutaneous Feet Infections of 14-Day-old to Adult C57BL/6 Mice

1. Consumable **items 1–8** as in Subheading 2.1.
2. Endotoxin free RPMI 1640 media containing 2 % fetal calf serum (FCS).
3. Mouse restrainer.
4. 0.5 ml BD ultra fine U-100 insulin syringe with 29 G × 12.7 mm needle or a 30 G needle.
5. Disposable scalpel blades No. 11.
6. Greiner Bio-One Minicollect® serum tubes heparin or EDTA treated (0.5–1 ml).
7. Facial tissues.
8. Digital vernier calipers.
9. Mice. 2-week- old to adult C57BL/6 mice or any knockout mice on a C57BL/6 background can be used.

3 Methods

Before commencing experiments please consider the following:

1. All the mice for experimental use must be bred under barrier conditions with a standard 12-h light/12-h dark cycle.
2. The experiments must be carried out according to the relevant national regulations.

3. Protocols and the experimental conditions must be approved by the animal care and use committee of the university/organization where the experiments will be conducted.
4. All experiments are to be performed within an appropriate biological containment facility according to the standards/levels of the country in which the experiments are being performed.
5. Please note that the physical/biological containment level for CHIKV may change according to the country's regulations.

3.1 Intracerebral Infection of Neonatal Mice

1. Prior to virus stock preparation, prepare the work area and disinfect all biological safety cabinet surfaces with 80 % ethanol (*see* **Note 2**).
2. Filter virus stock through a 0.22 μm filter to ensure no microbial contamination.
3. Aliquot stocks into 1.5 ml sterile tubes and store at −80 °C until use.
4. Prior to inoculation, thaw the CHIKV stock on ice and keep on ice until use.
5. Meanwhile, maintain mice in the appropriate biological safety level 2 (BSL-2) facility with 40–60 % humidity, 22 °C and 12 h light–dark cycle [10].
6. Arrange mice into appropriate groups making sure that there are enough animals across all groups (control and experimental) to ensure statistical validity of the experimental results.
7. Prior to infection, prepare the work area and disinfect all biological safety cabinet surfaces with 80 % ethanol.
8. Anesthetize each mouse prior to infection (*see* **Note 3**).
9. Verify that the mice are adequately anesthetized by gently pinching the toes. If the animal withdraws their paw, this indicates that a deep plane of sleep has not been reached.
10. Mice need to be restrained manually on a solid surface by firmly placing your thumb and index finger on either side of the head just at the nape of the neck (Fig. 1).
11. Insert the needle (intradermal needle) into the cranium approximately half way between the eye and the ear just off center from the midline.
12. Inoculate suckling mice (1–3-day-old) with 1×10^5 LD_{50} of CHIKV in a volume equal or inferior to 10 μl [5, 13] (Fig. 1). For weanling or older mice, 30 μl may be injected via the i.c. route (*see* **Notes 4–5**).
13. Put mice on a heat mat to avoid drop of body temperature during anesthesia and monitor mice until recovery (approximately 30 min).

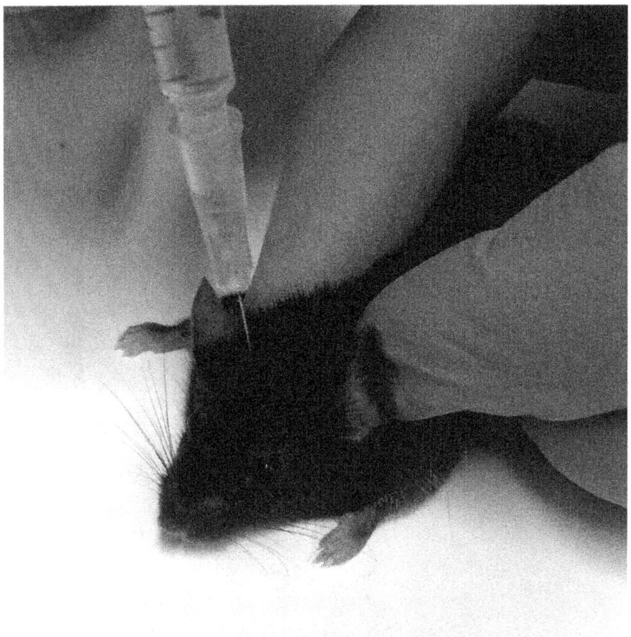

Fig. 1 Intracerebral inoculation (i.c.). Mice must be anesthetized via the intraperitoneal or subcutaneous routes prior to i.c. infections. Mice are restrained firmly to ensure maximum stability during inoculation. The needle is inserted into the cranium approximately half way between the eye and the ear just off center from the midline. Volumes of 0.01–0.03 ml may be administered via the i.c. route

14. After injections, place the infected mice back in a clean cage and observe for another 15–30 min to ensure recovery from both the needle puncture and anesthetic.
15. Monitor the infected mice (including mock-infected) every 12 h for:

(a) Weight gain/loss (calculated using starting weights prior to infection).

(b) Survival/moribund state. If any animal shows signs of being moribund or reaches experimental end points, euthanize immediately according to a method which has been approved by the ethics governing body.

3.2 Intranasal Infection of CHIKV in Mice Aged 3 Weeks to Adult

1. Prior to virus stock preparation, prepare the work area and disinfect all biological safety cabinet surfaces with 80 % ethanol.
2. Filter virus stock through a 0.22 μm filter to ensure no microbial contamination and prepare the work area and disinfect all biological safety cabinet surfaces with 80 % ethanol.
3. Aliquot stocks into 1.5 ml sterile tubes and store at −80 °C until use.

4. Prior to inoculation, thaw the CHIKV stock on ice and keep on ice until use.
5. Dilute virus stock in endotoxin free PBS to 1.26×10^7 pfu/ml (equivalent to $10^{6.5}$ pfu in 25 μl) for NIH and C57BL/6 mice infections [7] or 4×10^8 pfu/ml (equivalent to 1×10^7 pfu in 25 μl) for BALB/c infections [9].
6. Meanwhile, maintain and prepare mice as outlined in Subheading 3.1.
7. Prior to infection, prepare the work area and disinfect all biological safety cabinet surfaces with 80 % ethanol.
8. Anesthetize mice (*see* **Note 6**).
9. Hold the animal in a supine position on the work surface area with its head slightly elevated.
10. Intranasally (i.n.) inoculate the mice with no more than 0.05 ml of CHIKV by administering the virus slowly dropwise into the nares with a micropipette alternating sides (Fig. 2). Precaution must be taken to not exceed the recommended volumes and to avoid rapid inoculation in order to prevent suffocation and possible death.
11. Put mice on a heat mat to avoid drop of body temperature during anesthesia and monitor mice until recovery (approximately 30 min).

Fig. 2 Intranasal inoculation (i.n.). Mice need to be lightly anesthetized to achieve i.n. infections. Animals are held in a supine position and a maximum of 0.05 ml of inoculum is slowly delivered dropwise into the nares of the animals alternating sides

12. Monitor mice every 24 h for:
    (a) Weight loss (calculated using starting weights prior to infection).
    (b) Clinical disease signs including: ruffled fur, hunched posture, and lethargy (*see* **Notes 7–9** for details on potential clinical presentations).
13. Please strictly adhere to the humane endpoints applied by the ethics governing body. If mice show clinical signs of being moribund or excessive weight loss as defined by the animal ethics committee, mice should then be euthanized by a method approved by the ethics governing body.

3.3 Subcutaneous Inoculation of Newborn Mice

1. Prior to virus stock preparation, prepare the work area and disinfect all biological safety cabinet surfaces with 80 % ethanol.
2. Filter virus stock through a 0.22 μm filter to ensure no microbial contamination.
3. Aliquot stocks into 1.5 ml sterile tubes and store at −80 °C until use.
4. Prior to inoculation, thaw the CHIKV stock on ice and keep on ice until use.
5. Dilute virus stock in endotoxin free PBS to 2×10^7 pfu/ml (equivalent to 10^6 pfu in 50 μl) for BALB/c infection [11] or 1×10^6 pfu/ml (equivalent to 5×10^4 pfu in 50 μl) for ICR and CD-1 infection [12]
6. Meanwhile, maintain and prepare mice as outlined in Subheading 3.1. Allow the mothers ad libitum access to food and water while having the pups suckle freely [11].
7. Prior to infection, prepare the work area and disinfect all biological safety cabinet surfaces with 80 % ethanol.
8. Inoculate the mice with CHIKV subcutaneously in the loose skin on their back (Fig. 3) ensuring to pinch the skin away from the body and avoiding the hand (*see* **Note 10**). Mice DO NOT require anesthetic for this route of inoculation unless specifically requested by the ethics governing body.
9. Monitor mice every 12–24 h for:
    (a) Weight gain (calculated using starting weights prior to infection).
    (b) Clinical disease signs including: lethargy, loss of balance and difficulty walking, dragging of the hind limbs, and hair loss around the inoculation site on the back (*see* **Notes 11–14**).
10. Strictly adhere to the humane endpoints applied by the ethics governing body. If mice show clinical signs of being moribund, mice should be euthanized by a method approved by the ethics governing body.

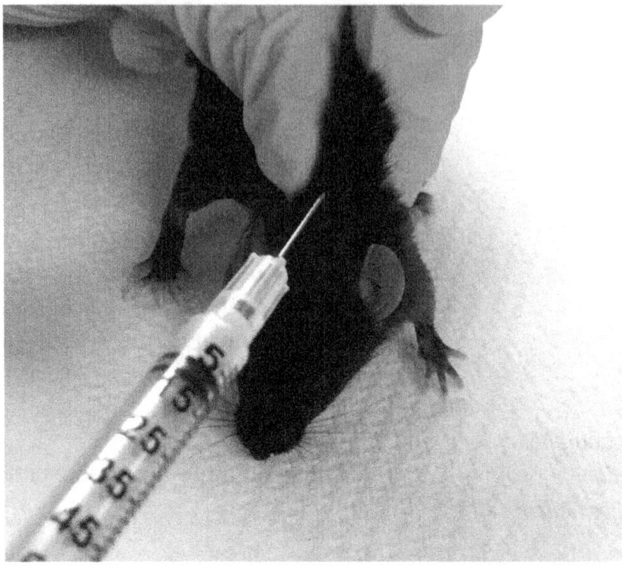

Fig. 3 Subcutaneous inoculation (s.c.) of newborn mice. For this technique, scruff the animals by the nape of the neck and gather the loose skin of the neck by pinching firmly. No anesthetic is required. Insert the needle into this area carefully by avoiding your hand and moving away from the body and slowly deliver 50 μl of inoculum

3.4 Subcutaneous Feet Infections of 2-Week-Old to Adult C57BL/6 Mice

1. Prior to virus stock preparation, prepare the work area and disinfect all biological safety cabinet surfaces with 80 % ethanol.
2. To ensure clean viral stocks free of contamination, filter supernatant using a 0.22 μm filter.
3. Aliquot stocks into 1.5 ml sterile tubes and store at −80 °C until use.
4. Prior to inoculation, thaw the CHIKV stock on ice and keep on ice until use.
5. For adult mice, each injection consists of 6.9×10^3 pfu (equivalent to 1×10^4 $CCID_{50}$) of CHIKV virus prepared in a final volume of 40 μl of RPMI 1640 media containing 2 % fetal calf serum (*see* **Note 15**). Confirm the inoculation dose by assaying residual preparation in the same manner as the virus stock, i.e., by $CCID_{50}$ assay using C6/36 and Vero cells *(16)* or by plaque assay. Alternatively, the virus can also be prepared by diluting in PBS. Younger animals are usually given a range of doses varying from 1×10^3 pfu to 1×10^5 pfu diluted in PBS [14, 15] in a final volume of 20 μl (*see* **Note 16**).
6. Meanwhile, group house adult female C57BL/6 mice (Adult animals are aged 6-weeks and older *see* **Notes 17-18**) together with up to 5–6 animals per cage using standard mouse cages on a ventilated rack under controlled conditions. Supply food and water ad libitum. Also, group house younger mice under the

same controlled conditions as adults. However, wean the animals prior to experimentation (can be as early as 17 days after birth). For younger mice, male and female mice can be used equally without any disparity in disease onset, severity or duration.

7. Arrange mice into appropriate groups. Make sure you have the appropriate replicate number of RPMI 1640 supplemented with 2 % FCS or PBS (virus diluent) control mice and the infected mice so that experimental results will provide statistical validity.
8. Prior to infection prepare the work area and disinfect all biological safety cabinet surfaces with 80 % ethanol.
9. Put conscious animals into a mouse restrainer and gently pull their hind leg through the central opening and then hold firmly (Fig. 4).

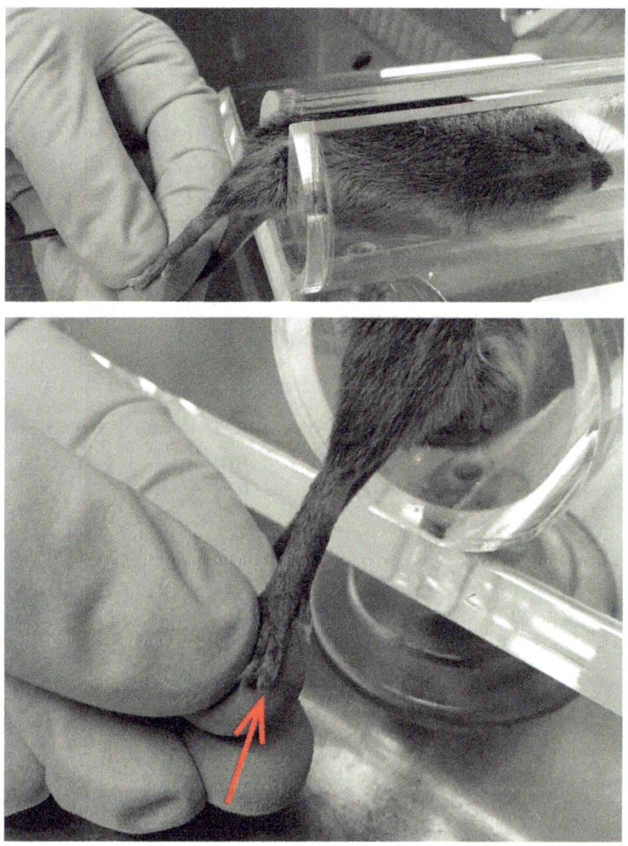

Fig. 4 Subcutaneous (s.c.) feet infections. Animals are placed into a mouse restraint and their foot is gently pulled through the central opening. 10–40 μl of virus may be administered in each foot depending on the age of the mice. Older mice are injected on the ventral side of the foot going towards the ankle while younger mice are injected into the footpads. Care should be taken as to not penetrate too far into the skin, thereby delivering the inoculum intradermally rather than subcutaneously

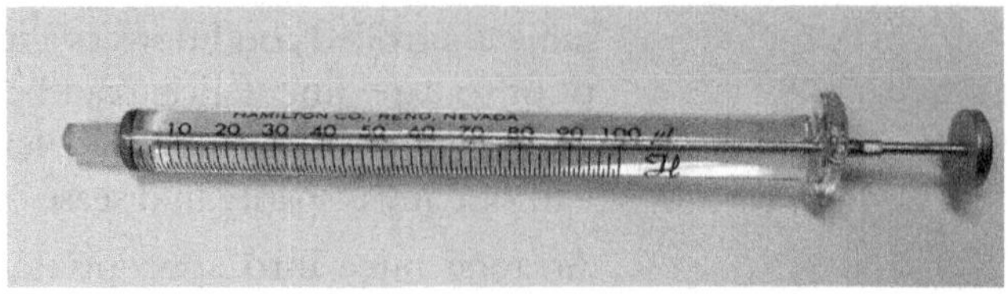

Fig. 5 Choice of needles and syringes. For optimal results, it is important to choose adequately sized syringes and needles. Always dispose of used needles in a safe manner. Inactivation of CHIKV prior to disposal may be required by using bleach or Biodyne. Always abide by the regulations and recommendations of the local/institutional biosafety and animal ethics committees

10. Administer virus to the adults using a 0.5 ml BD ultra fine inserted subcutaneously on the ventral side of each hind foot towards the ankle [16]. Care should be taken as not to insert the needle deep into the tissue and deliver the inoculum intradermally. Subcutaneous injections are easy to execute and rarely painful. Therefore, anesthetic is not required for this procedure.
11. For younger animals, use a 30 G needle to inoculate the mice subcutaneously in either one or both hind footpads with a total volume of 10–20 μl of inoculum per foot (*see* Fig. 5 for syringe options).
12. Advance the needle several millimeters through the subcutaneous tissue and then remove gently, in order to minimize leakage of the inoculum.
13. Monitor mice every 24 h for:
    (a) Viremia: following inoculation, bleed adult animals once a day from the tail vein to monitor viremia (*see* **Note 19**). Animals usually clear the virus by day 5 or 6 p.i., after which sampling is no longer required. Younger animals show a peak at day 2–3 p.i. and clear the virus by day 4–5 p.i. These animals can also be tail bled, however smaller amounts are to be collected in order not to exceed 1 % of total body weight or the amount that is approved by the local animal ethics committee. If a substantial amount is needed, sacrifice animals at the desired time points.
    (b) Foot swelling: Measure inflammation caused by CHIKV by multiplying the height by the width of the perimetatarsal area of each hind foot using digital vernier calipers. The overall inflammation seen in the foot is expressed as the percent increase over the basal measurements taken one day prior to inoculation. Foot measurements should be taken daily. A small increase of swelling is normally seen on day 2 p.i. and then resolves followed by an additional and stronger peak seen at day 6–7 p.i. Inflammation usually

subsides around day 12 p.i. For younger animals, perform monitoring in the same way. However, peak foot swelling is normally seen on day 3 p.i and subsides by day 5 p.i., and resolves by day 10 p.i.

(c) Overall well-being: inflammation of the hind feet is the only observable sign of disease in this model. Animals otherwise appear normal. Animals do not succumb to the infection.

14. End of experimentation—Euthanasia. Following the completion of the study or to terminate the experiment at specific time points where pathology or immunological responses need to be assessed, adult animals can be euthanized by inhalation of CO_2 (*see* **Notes 20** and **21**). Younger animals are generally sedated with isoflurane (*see* **Note 22**) and euthanized by thoracotomy and exsanguinations [15].

4 Notes

1. Traditionally i.c. infection of neonatal mice is performed as a means to isolate virus directly from the sera from human CHIKV infected patients.
2. All the in vitro and in vivo experiments must be undertaken in a class 2 biological safety cabinet.
3. Mice must be anesthetized prior to i.c. infection. To anesthetize neonatal mice younger than 7 days, inject the mice with ketamine + xylazine (50–120 mg/kg + 5–10 mg/kg; max injection volume is 0.5 ml per mouse) by intraperitoneal injection route. Ensure to weigh the mice prior to injection and calculate the dose accordingly. For 7–12 day-old neonatal mice, inject the mice with ketamine + xylazine (50–150 mg/kg + 5–10 mg/kg; max injection volume is 0.5 ml per mouse) via the intraperitoneal injection route.
4. This mouse model was used in the late 1960s when LD_{50} was still an adequate way of measuring virus load. The authors realize that LD_{50} are widely considered no longer ethically acceptable and this mouse model is not commonly used in the CHIKV research field. If this model is to be reproduced, please adhere to the ethical standards of the research institution where the study is being conducted and use a humane dose 50 % (HD_{50}).
5. In order to deliver smaller volumes with accuracy, use an appropriate sized syringe (Fig. 5).
6. Anesthetize mice by administering 240 mg/kg avertin or 80–100 mg/kg ketamine intraperitoneally [17]. To assure sterility, it is recommended to wipe the peritoneal area with 80 % ethanol before injection.

7. 10-week-old NIH Swiss mice only show viremia on day 1 and 2 post infection (p.i.) with La Réunion and Ross strains of CHIKV. There will be no detectable viremia thereafter in any of the cohorts. There will also be no febrile response, nor weight loss or other signs of disease [7].
8. For 5-week-old C57BL/6 mice, viremia lasts 2–3 days with a peak titre of about 10^3 pfu/ml when infected with the La Réunion and Ross strains [7]. The highest titre of Ross-infected C57BL/6 mice may be found in the brain with titers of 10^7 pfu/g. Mice may also show detectable virus in the heart, kidneys and skeletal muscle of the leg [7]. The Ross strain may produce clinical signs of disease and ongoing reduction in body weight gain beginning by day 5 p.i. Disease signs include ruffled fur, a hunched posture, and lethargy by day 6 p.i. Mice will reach humane endpoint by days 7 or 8 p.i. [7].
9. 8-week-old BALB/c mice may show severe weight loss from 1 to 3 days p.i. which continues during the course of infection. Mice may also show clinical signs such as lethargy and hind limb weakness by day 6 p.i. Viremia can be detected on day 5 p.i. with a mean titer of $10^{1.5}$ pfu/ml [9]. BALB/c mice inoculated with Ross CHIKV may also show infection of neurons, which can cause neuronal necrosis by day 5 p.i. [18].
10. Due to the nature of this injection and the potential risk to the handler, it is advisable to wear protective gloves (such as HexArmor® glove) to prevent needle stick injury.
11. Clinical scores can be measured according to the following scale: 0, no disease signs; 1, ruffled fur; 2, mild hind limb weakness; 3, moderate hind limb weakness; 4, severe hind limb weakness and dragging; and 5, moribund. The clinical score of 5 should be set as the humane endpoint.
12. Clinical scores should peak around 7–10 days post infection.
13. The clinical signs in the CD-1 mice are usually more severe; however, the range of clinical manifestations are generally similar for both groups. In CD-1 mice, the alopecia on the back is often more severe and sometimes includes skin vesicles. In the ICR mice, alopecia can be observed, but it is often less diffuse and without blistering of the skin.
14. Mortality from CHIKV infection should be low, and the surviving mice eventually recover hair growth and use of their hind limbs within 6 weeks of infection.
15. Prepare total amount of inoculum in one tube and mix thoroughly to ensure homogeneity of the virus. Keep inoculum on ice to prevent a decrease in titer.
16. Caution should be taken when using either IFNAR$^{-/-}$ or IRF 3/7$^{-/-}$ mice as CHIKV is lethal in these strains. Extra monitoring will also be required.

17. In our laboratory, animals are either bred *in house* or acquired from the animal resource center (Murdoch, WA).
18. There are no significant differences in disease onset, duration, or severity in animals up to 8 months of age. However, ideally animals aged 6–8-weeks should be used routinely.
19. Tail vein collection is a quick and simple way to collect blood to monitor viremia. The animal should be placed comfortably into a restraint. Then, blood may be collected by generating a small nick into the lateral tail vein using a scalpel. 50–200 μl of blood can be collected into a heparinized or EDTA treated Minicollect® tube using this method. Blood flow should be stopped by applying light pressure, with tissues, to the area for approximately 30 s before the animal is returned to its cage. Collect blood from areas closest possible to the tip of the tail, and move to areas away from the tip of the tail on each subsequent day of sampling.
20. Perfusions may be performed on animals using 4 % paraformaldehyde diluted in PBS if only histology samples are to be collected.
21. The method of euthanasia must adhere to local animal ethics committee's regulations and approvals. Several different methods can be used. The choice may be dependent on samples being collected and downstream applications.
22. When using isoflurane, it should be administered as a vapor as a mixture of 4–5 % isoflurane and 95 % oxygen for induction and lowered to 1–3 % isoflurane for maintenance.

References

1. Weaver SC, Lecuit M (2015) Chikungunya virus and the global spread of a mosquito-borne disease. N Engl J Med 372(13):1231–1239. doi:10.1056/NEJMra1406035
2. Lumsden WH (1955) An epidemic of virus disease in Southern Province, Tanganyika Territory, in 1952–53. II. General description and epidemiology. Trans R Soc Trop Med Hyg 49(1):33–57
3. World Health Organization (2015) Chikungunya, Fact sheet N°327 http://www.who.int/mediacentre/factsheets/fs327/en/. Accessed 22 October 2015
4. Mavale M, Parashar D, Sudeep A, Gokhale M, Ghodke Y, Geevarghese G, Arankalle V, Mishra AC (2010) Venereal transmission of Chikungunya virus by Aedes aegypti mosquitoes (Diptera: Culicidae). Am J Trop Med Hyg 83(6):1242–1244. doi:10.4269/ajtmh.2010.09-0577
5. Ross RW (1956) The Newala epidemic. III. The virus: isolation, pathogenic properties and relationship to the epidemic. J Hyg 54(2):177–191
6. Singharaj P, Simasathien P, Halstead SB (1966) Recovery of dengue and Chikungunya viruses from Thai haemorrhagic fever patients by passage in sucking mice. Bull World Health Organ 35(1):66
7. Wang E, Volkova E, Adams AP, Forrester N, Xiao SY, Frolov I, Weaver SC (2008) Chimeric alphavirus vaccine candidates for Chikungunya. Vaccine 26(39):5030–5039. doi:10.1016/j.vaccine.2008.07.054
8. Kumar M, Sudeep AB, Arankalle VA (2012) Evaluation of recombinant E2 protein-based and whole-virus inactivated candidate vaccines against Chikungunya virus. Vaccine 30(43):6142–6149. doi:10.1016/j.vaccine.2012.07.072

9. Mallilankaraman K, Shedlock DJ, Bao H, Kawalekar OU, Fagone P, Ramanathan AA, Ferraro B, Stabenow J, Vijayachari P, Sundaram SG, Muruganandam N, Sarangan G, Srikanth P, Khan AS, Lewis MG, Kim JJ, Sardesai NY, Muthumani K, Weiner DB (2011) A DNA vaccine against Chikungunya virus is protective in mice and induces neutralizing antibodies in mice and nonhuman primates. PLoS Negl Trop Dis 5(1), e928. doi:10.1371/journal.pntd.0000928
10. Parashar D, Paingankar MS, Kumar S, Gokhale MD, Sudeep AB, Shinde SB, Arankalle VA (2013) Administration of E2 and NS1 siRNAs inhibit Chikungunya virus replication *in vitro* and protects mice infected with the virus. PLoS Negl Trop Dis 7(9), e2405. doi:10.1371/journal.pntd.0002405
11. Dhanwani R, Khan M, Lomash V, Rao PV, Ly H, Parida M (2014) Characterization of Chikungunya virus induced host response in a mouse model of viral myositis. PLoS One 9(3), e92813. doi:10.1371/journal.pone.0092813
12. Ziegler SA, Lu L, da Rosa AP, Xiao SY, Tesh RB (2008) An animal model for studying the pathogenesis of Chikungunya virus infection. Am J Trop Med Hyg 79(1):133–139
13. Shimizu S (2004) Routes of administration. In: Hedrich H (ed) The laboratory mouse. Elsevier, Berlin, pp 527–542
14. Chen W, Foo SS, Taylor A, Lulla A, Merits A, Hueston L, Forwood MR, Walsh NC, Sims NA, Herrero LJ, Mahalingam S (2015) Bindarit, an inhibitor of monocyte chemotactic protein synthesis, protects against bone loss induced by Chikungunya virus infection. J Virol 89(1):581–593. doi:10.1128/JVI.02034-14
15. Pal P, Fox JM, Hawman DW, Huang YJ, Messaoudi I, Kreklywich C, Denton M, Legasse AW, Smith PP, Johnson S, Axthelm MK, Vanlandingham DL, Streblow DN, Higgs S, Morrison TE, Diamond MS (2014) Chikungunya viruses that escape monoclonal antibody therapy are clinically attenuated, stable, and not purified in mosquitoes. J Virol 88(15):8213–8226. doi:10.1128/JVI.01032-14
16. Gardner J, Anraku I, Le TT, Larcher T, Major L, Roques P, Schroder WA, Higgs S, Suhrbier A (2010) Chikungunya virus arthritis in adult wild-type mice. J Virol 84(16):8021–8032. doi:10.1128/JVI.02603-09, JVI.02603-09 (pii)
17. Gargiulo S, Greco A, Gramanzini M, Esposito S, Affuso A, Brunetti A, Vesce G (2012) Mice anesthesia, analgesia, and care, Part I: anesthetic considerations in preclinical research. ILAR J 53(1):E55–E69. doi:10.1093/ilar.53.1.55
18. Powers AM, Logue CH (2007) Changing patterns of Chikungunya virus: re-emergence of a zoonotic arbovirus. J Gen Virol 88(Pt 9):2363–2377. doi:10.1099/vir.0.82858-0
19. Morrison TE, Oko L, Montgomery SA, Whitmore AC, Lotstein AR, Gunn BM, Elmore SA, Heise MT (2011) A mouse model of Chikungunya virus-induced musculoskeletal inflammatory disease: evidence of arthritis, tenosynovitis, myositis, and persistence. Am J Pathol 178(1):32–40. doi:10.1016/j.ajpath.2010.11.018

Chapter 20

Generation of Mouse Monoclonal Antibodies Specific to Chikungunya Virus Using ClonaCell-HY Hybridoma Cloning Kit

Chow Wenn Yew and Yee Joo Tan

Abstract

Monoclonal antibodies offer high specificity and this makes it an important tool for molecular biology, biochemistry and medicine. Typically, monoclonal antibodies are generated by fusing mouse spleen cells that have been immunized with the desired antigen with myeloma cells to create immortalized hybridomas. Here, we describe the generation of monoclonal antibodies that are specific to Chikungunya virus using ClonaCell-HY system.

Key words Monoclonal antibody, Chikungunya virus, Fusion, Screening, Immunofluorescence

1 Introduction

It has been almost 40 years since monoclonal antibodies were first generated by Georges Köhler and César Milstein in 1975 by fusing mouse myeloma cells and mouse splenocytes [1]. This discovery was awarded the Nobel Prize in Physiology and Medicine in 1984. Since then, monoclonal antibody generation process has undergone numerous modifications and optimizations to allow creation of antibodies in a shorter time and with higher efficiency [2–9]. Limitations of the fusion method include the low successful rate of fusion and the risk of some hybridomas overgrowing, with the faster growing hybridomas having the tendency to be non-antibody-secreting cells [10].

ClonaCell-HY is a monoclonal antibody generating system that is specially optimized for the selection and cloning of hybridoma cells as soon as after fusion. This is achieved by using a methylcellulose-based HAT selection semisolid media that reduces the possibility of fast-growing clones overgrowing, and minimizes the number of clones to be screened. Thus, we use the ClonaCell-HY system to generate monoclonal antibodies that are specific against

Justin Jang Hann Chu and Swee Kim Ang (eds.), *Chikungunya Virus: Methods and Protocols*, Methods in Molecular Biology, vol. 1426,DOI 10.1007/978-1-4939-3618-2_20, © Springer Science+Business Media New York 2016

the Chikungunya virus. Subsequently, we employ immunofluorescence analysis (IFA) to screen for good hybridoma clones that produce the desired antibodies.

Mouse monoclonal antibodies that are specific for Chikungunya virus can be used as research or diagnostic reagents for detecting infected cells. Besides that, these antibodies can be tested further for neutralizing activity against the Chikungunya virus. Neutralizing antibodies can be used to identify essential epitopes for Chikungunya virus infection [11] and could be modified to human–mouse chimeric form which could be a potential tool to fight Chikungunya virus infection through passive immunotherapy [12–14].

In this protocol, we describe steps for generating mouse monoclonal antibodies specific to Chikungunya virus. We provide the protocols for immunization of mice, fusion of mouse splenocytes and screening by immunofluorescence to identify the successful hybridoma clones.

2 Materials

Pre-thaw the cell medium and reagents to room temperature before use. Waste materials are to be treated as biohazardous and disposed of according to institutional regulations.

2.1 Biological Materials, Cell Culture Medium, and Reagents

1. Chikungunya virus SGEHICHD122508 (GenBank; FJ445502): a virus isolate from the serum of an infected patient in Singapore provided by Environmental Health Institute, National Environmental Agency of Singapore.
2. 4-week-old Balb/c mice.
3. SP2/0 myeloma cells (ATCC CRL-1581).
4. Baby hamster kidney (BHK) fibroblasts.
5. ClonaCell-HY hybridoma kit (*see* **Note 1**): Medium A (pre-fusion medium and hybridoma expansion medium), Medium B (fusion medium), Medium C (hybridoma recovery medium), Medium D (hybridoma selection medium), Medium E (hybridoma growth medium), polyethylene glycol (PEG).
6. Dulbecco's Modified Eagle's Medium (DMEM) supplied with 10 % (v/v) fetal bovine serum (FBS).
7. Roswell Park Memorial Institute medium (RPMI) supplied with 10 % (v/v) FBS.
8. 0.4 % trypan blue.
9. Sterile phosphate-buffered saline (PBS):137 mM NaCl, 2.7 mM KCl, 10 mM Na_2HPO_4, 1.8 mM KH_2PO_4.
10. PBST: sterile PBS with 0.1 % (v/v) Triton X.

11. Blocking buffer: PBST with 3 % FBS.
12. FITC goat anti-mouse IgG/IgM.
13. DAPI solution: 1 mg/ml in PBS.
14. Freezing medium: 90 %, 10 % dimethylsulfoxide (DMSO).

2.2 Additional Equipment and Supplies

1. Single and multichannel pipettes.
2. Pipette-aid with sterile serological pipettes: 1, 2, 5, 10 ml.
3. 5 and 10 ml syringes.
4. 18 G needles.
5. 50 ml sterile conical tubes.
6. 100 mm sterile petri dishes.
7. 24, 96-well sterile culture plates.
8. T75 sterile tissue culture flask.
9. Bench centrifuge.
10. Biosafety cabinet.
11. 37 °C incubator with >95 % humidity and 5 % CO_2.
12. Countess® Automated Cell Counter.
13. ImageXpress Micro XLS.
14. 40 μm nylon cell strainer.
15. Plastic container with lid to hold at least a dozen 100 mm petri dishes.

3 Methods

Carry out all procedures in a certified biosafety cabinet to maintain sterility. Centrifugations are done at room temperature.

3.1 Immunization of Mouse by Chikungunya Virus

1. Inject Chikungunya virus through the intraperitoneal route at a dose of 10^6 plague-forming unit/100 μl into 4-week-old Balb/c mice.
2. Repeat the immunization procedure twice over a period of 14 days and carry out a final immunization out 7 days before sacrificing the mice for harvesting of the spleen.

3.2 Preparation of SP2/0 Myeloma Cells

1. Thaw SP2/0 cells in complete DMEM and culture in Medium A for at least 1 week prior to fusion.
2. On the day before fusion, split the cells so that at least 2×10^7 cells are available next day (*see* **Note 2**).
3. Harvest myeloma cells on the day of fusion in a 50 ml conical tube by flushing the cells off using serological pipettes, collect the cell suspension and centrifuge at $300 \times g$ for 10 min.

4. Wash the cell pellet with 30 ml Medium B by resuspending in 1 ml Medium B first before topping up to 30 ml. Repeat centrifugation to pellet the cells down again and repeat wash for another two times (*see* **Note 3**).
5. After three washes, resuspend the cells in 25 ml of Medium B. Count the cells viability and number by using trypan blue stain with a cell counter. Viability of the cells should be higher than 95 %.
6. Calculate the volume of the cell suspension so that it contains 2×10^7 cells.

3.3 Preparation of Mouse Splenocytes

1. Sacrifice immunized mouse and remove spleen using sterile instruments and aseptic technique.
2. Keep spleen in 5 ml of ice-cold Medium A (*see* **Note 4**).
3. Pre-wet a cell strainer with 2 ml Medium B and place it on top of a 50 ml conical tube. Place the spleen into the cell strainer and using the plunger of a 5 ml syringe, mash the cells out of the spleen. Rinse the strainer with Medium B to make sure all cells are washed into the conical tube. Disrupt any cell clumps by pipetting up and down.
4. Top up the cell suspension with Medium B to 30 ml to wash the cells before pelleting the cells by centrifuging at $400 \times g$ for 10 min.
5. Remove the supernatant and wash the cells for another two times with 30 ml Medium B before resuspending the cells in 25 ml Medium B (*see* **Note 3**).
6. Count the cells using a cell counter; it is not necessary to add trypan blue for cell counting (*see* **Note 5**).
7. Calculate the volume of the cell suspension so that it contains 1×10^8 splenocytes (*see* **Note 6**).

3.4 Fusion

1. Add 2×10^7 SP2/0 cells and 1×10^8 splenocytes (as calculated) into a 50 ml conical tube. Centrifuge at $400 \times g$ for 10 min. Remove as much supernatant as possible to prevent dilution of PEG during fusion later.
2. To the cell pellet (which is the mixture of SP2/0 and splenocytes), slowly add 1 ml of PEG over a period of 1 min by using a 1 ml pipette. Stir the cells continuously with the pipette tip over the next 1 min.
3. While continuously stirring, add 4 ml of Medium B into the mixture slowly over a period of 4 min (*see* **Note 7**).
4. Add 10 ml of Medium B slowly into the mixture and incubate in a 37 °C water bath for 15 min.
5. Add 30 ml of Medium A slowly into the mixture after 15 min of incubation. Centrifuge the mixture at $400 \times g$ for 7 min.

Discard the supernatant and wash the cell pellet in 40 ml Medium A.

6. After second wash, resuspend the cell pellet in 10 ml of Medium C slowly. Transfer the cell suspension to a T75 tissue culture flask containing 40 ml of Medium C. Incubate at 37 °C for 18 h.

3.5 Plating of Fused Cells

1. Thaw Medium D in a fridge overnight on the day fusion is performed.
2. Shake vigorously to mix contents in Medium D bottle and set it aside for 1 h to let it warm up to room temperature.
3. Transfer the fused cell suspension into a 50 ml conical tube. Flush the flask to dislodge any cells that might stick to the flask (*see* **Note 8**). Centrifuge at 400 × *g* for 10 min.
4. Remove as much supernatant as possible and resuspend the cell pellet in 10 ml of Medium C (*see* **Note 9**).
5. Transfer the resuspended cells into the 90 ml of Medium D (in its original bottle). Mix thoroughly by gently inverting the Medium D bottle for 5 times and incubate in 37 °C water bath for 15 min (*see* **Note 10**).
6. Use a 10 ml syringe with 18 G needle to slowly plate 9.5 ml of cell suspension into a 100 mm petri dish (ten dishes in total). Tilt the dishes to level the cell suspension and try not to introduce air bubbles. Plate as much cell suspension as possible.
7. Place the dishes into a plastic container with a lid. Fill a 100 mm dish (without lid) with sterile water and place it in the container as well. Cap the lid of the container but ensure that it is not air tight. Place the whole container into a 37 °C incubator and incubate for 10 days without any disturbance (*see* **Note 11**).

3.6 Colonies Picking

1. After 10 days of undisturbed incubation, examine the plates for presence of colonies that are visible to naked eyes.
2. Prepare 96-well plates with 150 μl of Medium E in each of the well.
3. Pick an isolated colony with a pipette set to 10 μl and resuspend the colony in Medium E into 1 well of the 96-well plate.
4. Continue picking until all the clones are picked (*see* **Note 12**). Typically, 1 fusion will produce around 1000–1500 clones.
5. Incubate the plates with clones for up to 4 days. Clones are ready to be screened when the culture media has turned yellow.

3.7 Screening of Clones and Subcloning

1. Prepare 96-well plates of Baby Hamster Kidney (BHK) fibroblasts, with alternating columns of mock-infected and Chikungunya virus-infected treatment (screening plates). Seed BHK cells in RPMI media on 96-well plates at 1.35×10^4 density.

2. On the next day, infect the confluent BHK monolayer with Chikungunya virus at multiplicity of infection of 10. Incubate the plate at 37 °C for 1.5 h with shaking every 15 min.
3. Wash cells with PBS twice (120 μl for each wash) and maintain in RPMI media with 2 % FCS.
4. At 24 h post-infection, fix cells by ice-cold methanol for 10 min.
5. Block cells using blocking buffer.
6. Remove the blocking buffer in the screening plate by decanting just prior to use.
7. Aspirate 50 μl of the cell supernatant from each column of the 96-well plate by using a multichannel pipette. Dispense the cell supernatant into a column on a screening plate with mock-infected BHK cells. Aspirate from the same column and dispense into the next column with infected BHK cells on the screening plate (*see* **Note 13**).
8. Incubate the plate in a 37 °C incubator for 1 h and wash the plate twice with 100 μl PBST.
9. Prepare FITC goat anti-mouse secondary antibody by diluting it 200× in PBST. Aspirate 50 μl of the diluted secondary antibody into each well of the screening plate. Incubate the plate in a 37 °C incubator for 1 h. Wash the plate twice with 100 μl PBST.
10. Stain with DAPI solution (1000× dilution in PBS) for 10 min at room temperature. Wash the plate twice with 100 μl PBST.
11. Use the ImageXpress Micro system to take fluorescent images (DAPI and FITC) of the screening plates. Analyze fluorescent images with MetaXpress software to identify clones that show specificity towards Chikungunya infected cells but not the mock infected cells (Fig. 1).
12. Subculture clones that are positive in the first screen into a new 96-well plate for a second screening using the same method as the first screening (*see* **Note 14**).
13. Subculture clones that are positive in both screens into a 24-well plate. Make frozen stock of the clones by resuspending two million cells in 1 ml of ice-cold freezing media when cells have grown to a suitable density.
14. Successful clones are further subcloned via the limiting dilution method (*see* **Note 15**). Resuspend 150 cells into 10 ml of Medium E before plating 100 μl of the cell suspension into each well of a 96-well plate. Incubate the plate in a 37 °C incubator for 1 week. Check each well, mark and resuspend wells with a single colony. When the supernatant of the marked wells turn yellow, perform screening again to pick out the best subclones for each of the successful clones.

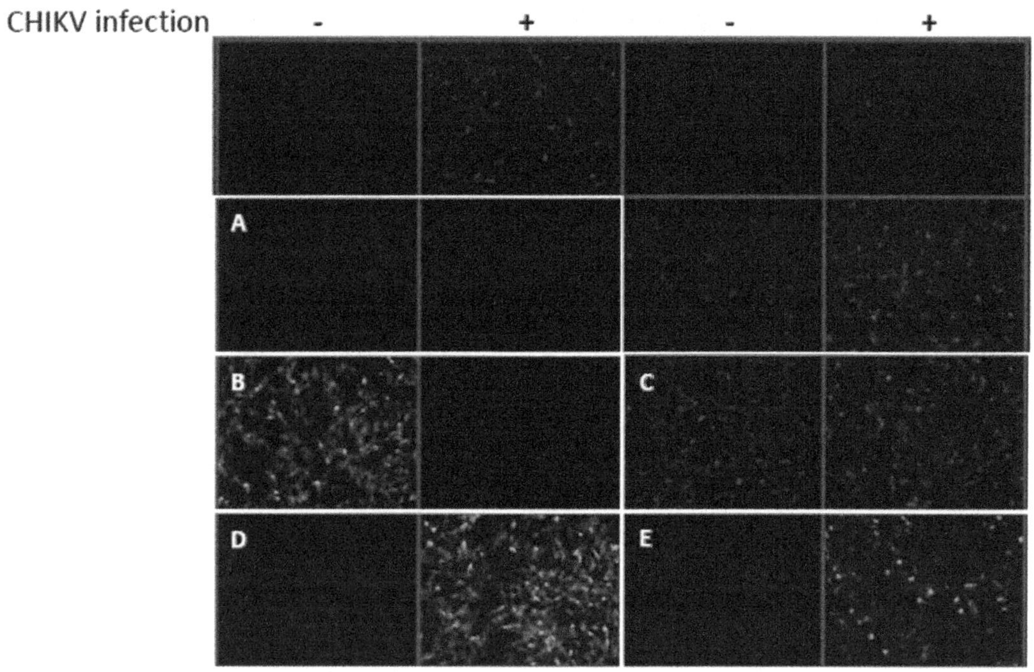

Fig. 1 An example of fluorescence images generated by the ImageXpress Micro system. Screening plates with alternating columns of mock-infected and CHIKV-infected are prepared to identify successful clones. (**a**) No fluorescence was observed for both mock and CHIKV-infected cells, suggesting that the clone does not produce antibodies. (**b**) Only the mock-infected cells showed fluorescence; or (**c**) both mock and CHIKV-infected cells showed fluorescence, suggesting that the antibody is targeting antigen that is unrelated to CHIKV. (**d**, **e**) Clones that produce antibody that targets CHIKV-infected but not mock-infected cells. These clones can be subcultured for further screening.

4 Notes

1. Keep medium frozen in a −20 °C freezer and thaw in a 4 °C fridge 1 day before use.
2. It is recommended to prepare more than the 2×10^7 cells needed in the case of contamination or any unforeseen circumstances.
3. It is important to remove the serum in the cell culture media as serum will interfere with PEG during fusion, thereby leading to low fusion frequency.
4. It is recommended to process the spleen as quickly as possible after removal from the mice.
5. Trypan blue is not necessary in this cell counting since splenocytes consist of different cell types of different sizes and the smaller cells tend to look stained under trypan blue even if they are alive. Hence only calculate total cell number with the assumption that all splenocytes are still viable.

6. In the event where more or less than 1×10^8 splenocytes are harvested, adjust the number of SP2/0 cells accordingly so that the ratio of SP2/0 to splenocytes remains at 1:5.
7. Set the pipette-aid to dispense at the lowest setting and dispense carefully into the tube.
8. It is important to collect all the cells in the flask after fusion as each and every single cell might be a good antibody producing clone.
9. Do not exceed 10 ml of final volume as this will dilute the methylcellulose concentration, leading to failure to form semisolid matrix.
10. Bubbles formed during the mixing step are normal and they will rise to the top during the incubation.
11. Beware when opening and closing the incubator as shaking during incubation will lead to streaking of the clones.
12. Picking all the clones will eliminate the risk of losing potential clones that grow slower.
13. Each 96-well plate of clones will have two screening plates.
14. The purpose of the second screening is to eliminate unstable clones that may have lost their antibody-producing property over time.
15. Subcloning of the successful clones will further ensure that the clones are monoclonal. Besides that, subcloning can enhance the stability of the clones by eliminating the weaker cells.

Acknowledgments

This work was supported by intramural funding in IMCB and a grant from Exploit Technologies Pte Ltd (ETPL) (grant number: ETPL/11-R15COT-0005). We would like to thank Dr Jang-Hann Chu Justin for his contribution in the immunization of mice and the preparation of screening plates.

References

1. Kohler G, Milstein C (1975) Continuous cultures of fused cells secreting antibody of predefined specificity. Nature 256(5517): 495–497
2. McMahon MJ, O'Kennedy R (2001) The use of in vitro immunisation, as an adjunct to monoclonal antibody production, may result in the production of hybridomas secreting polyreactive antibodies. J Immunol Methods 258(1-2):27–36
3. Takahashi M, Fuller SA, Winston S (1991) Design and production of bispecific monoclonal antibodies by hybrid hybridomas for use in immunoassay. Methods Enzymol 203: 312–327
4. Gupta CK, Sokhey J, Gupta RK, Singh H (1991) Development of allogenic hybridomas for production of monoclonal antibodies against oral polio vaccine strains. Vaccine 9(11):853–854

5. Clark SA, Griffiths JB, Morris CB (1990) Large-scale production and storage of monoclonal antibodies and hybridomas. Methods Mol Biol 5:631–645
6. Posner MR, Elboim H, Santos D (1987) The construction and use of a human-mouse myeloma analogue suitable for the routine production of hybridomas secreting human monoclonal antibodies. Hybridoma 6(6):611–625
7. Cianfriglia M, Mariani M, Armellini D, Massone A, Lafata M, Presentini R, Antoni G (1986) Methods for high frequency production of soluble antigen-specific hybridomas; specificities and affinities of the monoclonal antibodies obtained. Methods Enzymol 121:193–210
8. Bankert RB (1983) Rapid screening and replica plating of hybridomas for the production and characterization of monoclonal antibodies. Methods Enzymol 92:182–195
9. Davis JM, Pennington JE, Kubler AM, Conscience JF (1982) A simple, single-step technique for selecting and cloning hybridomas for the production of monoclonal antibodies. J Immunol Methods 50(2):161–171
10. Kennett RH, McKearn TJ, Bechtol KB (1980) Monoclonal antibodies. Plenum Press, New York
11. Kam YW, Lum FM, Teo TH, Lee WW, Simarmata D, Harjanto S, Chua CL, Chan YF, Wee JK, Chow A, Lin RT, Leo YS, Le Grand R, Sam IC, Tong JC, Roques P, Wiesmuller KH, Renia L, Rotzschke O, Ng LF (2012) Early neutralizing IgG response to Chikungunya virus in infected patients targets a dominant linear epitope on the E2 glycoprotein. EMBO Mol Med 4(4): 330–343
12. ter Meulen J (2007) Monoclonal antibodies for prophylaxis and therapy of infectious diseases. Expert Opin Emerg Drugs 12(4): 525–540
13. Marasco WA, Sui J (2007) The growth and potential of human antiviral monoclonal antibody therapeutics. Nat Biotechnol 25(12):1421–1434
14. Casadevall A, Dadachova E, Pirofski LA (2004) Passive antibody therapy for infectious diseases. Nat Rev Microbiol 2(9):695–703

Chapter 21

Immunohistochemical Detection of Chikungunya Virus Antigens in Formalin-Fixed and Paraffin-Embedded Tissues

Chun Wei Chiam, I-Ching Sam, Yoke Fun Chan, Kum Thong Wong, and Kien Chai Ong

Abstract

Immunohistochemistry is a histological technique that allows detection of one or more proteins of interest within a cell using specific antibody binding, followed by microscopic visualization of a chromogenic substrate catalyzed by peroxidase and/or alkaline phosphatase. Here, we describe a method to localize Chikungunya virus (CHIKV) antigens in formalin-fixed and paraffin-embedded infected mouse brain.

Key words Chikungunya virus, Immunohistochemistry, Antibody, Formalin-fixed, Paraffin-embedded

1 Introduction

CHIKV pathogenesis has been studied using suckling mice and nonhuman primates, including infection of the central nervous system (CNS). Immunostaining of CHIKV has utilized human antisera against CHIKV, CHIKV-infected mouse hyperimmune ascitic fluid, and monoclonal antibodies. In mice, CHIKV mainly infects muscles, joints, skin, liver, spleen, and the CNS [1–3]. While CHIKV infection was not seen in the brains of subcutaneously infected ICR mice [2], others have demonstrated CHIKV infection of the choroid plexus, ependymal wall, lepto-meningeal cells, and astrocytes following intradermal or intracerebral inoculation of outbred OF1 mice, and inbred C57BL/6 and 129 s/v mice [1, 4].

In this chapter, we describe a method for immunohistochemistry in formalin-fixed and paraffin-embedded tissue, using rabbit anti-CHIKV capsid antibody to elucidate CHIKV infection and spread in suckling mice. After routine specimen and slide preparation, the tissue section will be deparaffinized and rehydrated. For immunohistochemistry, antigen retrieval will be first performed by pretreatment with heat retrieval, followed by a sequential application of endogenous enzyme blocking, serum blocking, primary antibody

Justin Jang Hann Chu and Swee Kim Ang (eds.), *Chikungunya Virus: Methods and Protocols*, Methods in Molecular Biology, vol. 1426,DOI 10.1007/978-1-4939-3618-2_21, © Springer Science+Business Media New York 2016

(specific antibody to the antigen of choice), and a labeled secondary antibody (against the primary antibody). The antigen will be revealed by an enzyme complex and chromogenic substrate with interposed washing steps before counterstaining. The antigen can then be viewed using a bright-field microscope to visualize the chromogen precipitates.

With minor modifications, immunofluorescence can be performed using a secondary antibody that is chemically conjugated with a fluorescent dye. The fluorescence can be visualized with fluorescence or confocal microscope [5]. Other methods include fluorescence *in situ* hybridization (FISH) and chromogenic *in situ* hybridization (CISH).

2 Materials

Prepare all solutions using reverse osmosis water. Prepare and store all reagents at room temperature (unless indicated otherwise). Use the methods presented below after obtaining suckling mice which have been infected with CHIKV according to appropriate animal ethics and biosafety protocols.

2.1 Paraffin Block Sectioning

1. Formalin-fixed and paraffin-embedded tissue blocks.
2. Silanized glass slide with 2 % (3-aminopropyl) triethoxysilane in acetone.
3. Rotary microtome.
4. Water bath, 45 °C.
5. Hot plate, 60 °C.

2.2 Immunohistochemistry

1. Tris–EDTA buffer: 10 mM Tris–HCl, pH 9, 1 mM EDTA, 0.05 % Tween 20.
2. Endogenous peroxidase blocking solution: 3 % H_2O_2 in methanol.
3. Tris-buffered saline (TBS) buffer: 50 mM Tris–HCl, pH 7.6, 150 mM NaCl.
4. Blocking solution: normal goat serum, 1:20 dilution in TBS.
5. Primary antibody: rabbit anti-CHIKV capsid antibody, 1:5000 dilution in TBS.
6. Secondary antibody system: EnVision/HRP Rabbit/Mouse (Dako).
7. DAB+ chromogen reagent, 1:50 dilution (Dako).
8. Hematoxylin.
9. Mounting media: DPX mountant for histology.
10. Rice cooker.
11. Upright bright-field microscope.

3 Methods

3.1 Paraffin Block Sectioning

1. Trim precooled newly embedded paraffin blocks at 10 μm thickness with a rotary microtome until the tissues are fully exposed.
2. Continue sectioning the blocks at 4 μm thickness (*see* **Note 1**).
3. Float the paraffin sections with the desired plane facing up in a 45 °C water bath.
4. Separate the paraffin sections individually with a splinter forceps.
5. Gently pick up each individual paraffin section onto a silanized microscope glass slide (*see* **Note 2**).
6. Either dry the tissue sections on a 60 °C hot plate for 20–30 min, or leave the tissue sections at room temperature, to allow the paraffin sections to attach to slides more strongly (*see* **Note 3**).

3.2 Immunohistochemistry

1. Preheat Tris–EDTA buffer, pH 9.0 for 30 min in a rice cooker (*see* **Note 4**).
2. Put the tissue section slides into a slide holder. Deparaffinize and rehydrate the sections following the steps listed in Table 1 (*see* **Note 5**).
3. Immerse the slides in preheated Tris–EDTA buffer, pH 9.0 for 30 min.
4. Cool down for 20 min at room temperature.
5. Wash in running tap water for 5 min.
6. Block endogenous peroxidase enzyme by incubating tissue sections in endogenous peroxidase blocking reagent for 20 min (*see* **Note 6**).
7. Wash in TBS, pH 7.6 for 5 min (*see* **Note 7**).

Table 1
Deparaffinization and rehydration of tissue sections

	Reagent	Time (min)
Deparaffinization	Xylene I	5
	Xylene II	5
	Xylene III	5
Rehydration	Absolute ethanol	2
	95 % ethanol I	2
	95 % ethanol II	1
	95 % ethanol III	1
	Running tap water	2–3

8. Block nonspecific binding by incubating tissue sections with 1:20 dilution of normal goat serum for 20 min at room temperature in a moist chamber, to avoid drying of the tissue sections (*see* **Notes 8** and **9**).
9. Wipe-dry the slides around the tissue but keep the tissue wet.
10. Incubate tissue sections overnight at 4 °C with primary antibody, rabbit anti-CHIKV capsid (1:5000 in TBS), in a moist chamber (*see* **Note 10**).
11. Incubate positive tissue control, negative tissue control, positive staining control, and negative staining control sections with either primary antibody or TBS only (*see* **Notes 11** and **12**).
12. Wash with TBS three times (5 min each).
13. Incubate tissue sections for 30 min with the secondary antibody, EnVision/HRP Rabbit/Mouse, at room temperature in a moist chamber.
14. Wash with TBS buffer two times (5 min each).
15. Develop color with 1:50 dilution of DAB reagent for 5–10 min.
16. Wash with running tap water for 5 min.
17. Counterstain tissue sections with hematoxylin for 1 min.
18. Wash with running tap water two times (5 min each).
19. Dehydrate tissue sections with a hair dryer, or using an oven at 60 °C.
20. Add DPX mounting media (*see* **Note 13**).
21. Place coverslips over specimen sections and gently press to expel excess mounting media.
22. Lie flat for 24 h to dry.
23. Examine with a bright-field microscope (Fig. 1).

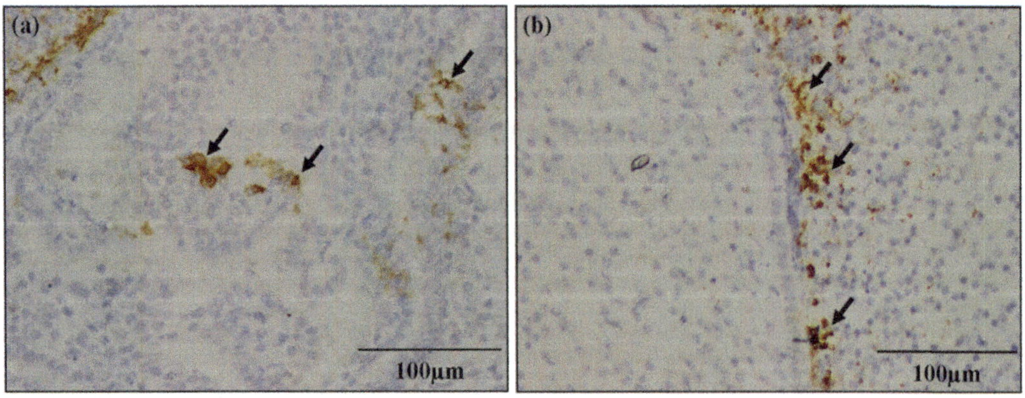

Fig. 1 Immunohistochemical staining of the brains of CHIKV-infected suckling mice. Viral antigens (*stained brown*, *arrows*) of various regions of the brain, including choroid plexus (**a**) and right lateral ventricle (**b**). Immunohistochemical staining was performed with DAB chromogen, polyclonal rabbit anti-CHIKV capsid antibody, and hematoxylin counterstain

4 Notes

1. The paraffin blocks should be sectioned at 3 or 4 μm in thickness, and no thicker than 5 μm. Thick sections will have multiple layers of cells and make interpretation of positive staining extremely difficult.
2. Using silanized coated slides will prevent or reduce the loss of tissue sections subjected to antigen retrieval treatments. Other coated slides such poly-L-lysine can also be used.
3. Antigens maybe destroyed if exposed to high temperatures.
4. Heat-induced antigen retrieval can be performed using a pressure cooker, a microwave, vegetable steamer, or a rice cooker. Optimal duration of antigen retrieval may vary from 10 to 60 min and usually depends on the length of formalin fixation. In general, 20 min appears to be satisfactory for most antigens and fixation protocols. Citrate buffer, pH 6 (10 mM citric acid, 0.05 % Tween 20) can be used as an alternative antigen retrieval buffer to Tris–EDTA buffer. It is important to optimize treatment with the antigen retrieval solution to obtain maximum and specific staining with minimal background.
5. Perform deparaffinizing and rehydrating in a fume hood. Do not let tissue sections dry after this step.
6. Incubation with 3 % hydrogen peroxide is to block endogenous peroxidase enzyme activity.
7. Use a Coplin or slide staining jar (for a small number of slides), or a staining dish (for a large number of slides) to wash tissue sections.
8. Incubation with normal goat serum is to prevent nonspecific primary and secondary antibodies binding.
9. Remove as much washing buffer as possible without drying the tissue sections. Drying at any stage will cause nonspecific binding and high background staining.
10. Different dilutions of primary antibody (e.g., 1:2500, 1:5000, 1:10,000) should be tested to optimally reduce background and amount of the antibody needed. The incubation time (e.g., 2 h or overnight) and temperature (e.g., room temperature or 4 °C) should also be tested with a known positive tissue control, negative tissue control, positive staining control, and negative staining control.
11. Prepare positive and negative tissue controls. Inoculate mice with CHIKV and serum-free media to produce CHIKV-infected and mock-infected mice, respectively.
12. Prepare positive and negative staining controls. Use CHIKV-infected Vero cells with 70 % cytopathic effect. Mix the cell supernatant with finely cut non-infected mouse lung tissue to

act as a positive staining control. Mix mock-infected Vero cells with finely cut non-infected mouse lung to act as a negative staining control. Centrifuge the positive and negative staining control at 10,000 × g for 5 min before fixing in 10 % neutral buffered formalin for a week and process as described above. Under microscopic observation, the CHIKV-infected Vero cells should be seen attached to the mouse lung.

13. To prevent air bubbles, ensure that the tissue section is completely dry. Soak in xylene before mounting. If air bubbles are still present, soak the tissue section in xylene for 24 h to remove coverslip. Remount the tissue section again.

Acknowledgments

We thank Andres Merits of Tartu University, Estonia for contributing rabbit anti-CHIKV capsid antibody. The authors' work was supported by European Union's Seventh Framework Program (Integrated Chikungunya Research (ICRES) grant agreement no. 261202), University of Malaya (HIR grant UM.C/625/1/HIR/016, PG030-2012B), and the Ministry of Higher Education, Malaysia (FRGS grant FP032-2010A).

References

1. Couderc T, Chrétien F, Schilte C et al (2008) A mouse model for Chikungunya: young age and inefficient Type-I interferon signaling are risk factors for severe disease. PLoS Pathog 4, e29
2. Ziegler SA, Lu L, da Rosa APAT et al (2008) An animal model for studying the pathogenesis of Chikungunya virus infection. Am J Trop Med Hyg 79:133–139
3. Labadie K, Larcher T, Joubert C et al (2010) Chikungunya disease in nonhuman primates involves long-term viral persistence in macrophages. J Clin Invest 120:894–906
4. Das T, Hoarau JJ, Bandjee MCJ et al (2015) Multifaceted innate immune responses engaged by astrocytes, microglia and resident dendritic cells against chikungunya neuroinfection. J Gen Virol 96:294–310
5. Robertson D, Savage K, Reis-Filho JS et al (2008) Multiple immunofluorescence labelling of formalin-fixed paraffin-embedded (FFPE) tissue. BMC Cell Biol 9:13

Part IV

Antivirals and Vaccines

Chapter 22

Antiviral Strategies Against Chikungunya Virus

Rana Abdelnabi, Johan Neyts, and Leen Delang

Abstract

In the last few decades the Chikungunya virus (CHIKV) has evolved from a geographically isolated pathogen to a virus that is widespread in many parts of Africa, Asia and recently also in Central- and South-America. Although CHIKV infections are rarely fatal, the disease can evolve into a chronic stage, which is characterized by persisting polyarthralgia and joint stiffness. This chronic CHIKV infection can severely incapacitate patients for weeks up to several years after the initial infection. Despite the burden of CHIKV infections, no vaccine or antivirals are available yet. The current therapy is therefore only symptomatic and consists of the administration of analgesics, antipyretics, and anti-inflammatory agents. Recently several molecules with various viral or host targets have been identified as CHIKV inhibitors. In this chapter, we summarize the current status of the development of antiviral strategies against CHIKV infections.

Key words Chikungunya virus, Antivirals, Chloroquine, Arbidol, Ribavirin, Favipiravir, Immune modulators

1 Introduction

Chikungunya virus (CHIKV), belonging to the *Alphavirus* genus of the *Togaviridae* family is an arthropod-borne virus transmitted by female mosquitoes of the *Aedes* species [1]. CHIKV infections cause Chikungunya fever which is characterized by abrupt fever, rash and bilateral symmetric arthralgia. In most of the CHIKV-infected patients the acute phase is followed by persistent disabling polyarthritis that can severely incapacitate the patient for weeks up to several months [1]. Despite the widespread emergence of CHIKV and the high morbidity rate associated with it, there is no approved vaccine or antiviral treatment available at the moment. The current therapy is therefore purely based on the relief of the patient's symptoms and consists of the administration of analgesics, antipyretics, anti-inflammatory agents such as paracetamol, and nonsteroidal anti-inflammatory drugs, and of bed rest and fluids intake [2]. The use of aspirin during CHIKV infection is to be avoided because of the risk of bleeding and the potential risk of

Justin Jang Hann Chu and Swee Kim Ang (eds.), *Chikungunya Virus: Methods and Protocols*, Methods in Molecular Biology, vol. 1426,DOI 10.1007/978-1-4939-3618-2_22, © Springer Science+Business Media New York 2016

developing Reye's syndrome [3]. In addition, the use of systemic corticosteroids is not recommended due to the strong rebound effect after cessation of treatment [1]. In severe cases, where patients have limited response to nonsteroidal anti-inflammatory drugs, disease-modifying antirheumatic drugs (DMARDs) such as methotrexate, hydroxychloroquine, or sulphasalazine can be administered to relieve the symptoms [3, 4]. The development of novel, potent antiviral drugs against CHIKV is thus urgently needed.

2 Antiviral Strategies for the Treatment of CHIKV Infection

2.1 Inhibitors of Viral Entry

Antiviral agents targeting the entry of enveloped viruses are of major interest since they inhibit an early step in the viral life cycle which minimizes the cell damage caused by intracellular viral replication. In addition, viral entry inhibitors may target extracellular components, which are more accessible; therefore, they could be effective in lower dosages with limited toxicity [5].

2.1.1 Chloroquine

Chloroquine is an antimalarial drug that has also been shown to inhibit the in vitro replication of several viruses, including HIV, severe acute respiratory syndrome (SARS) coronavirus, and alphaviruses [6]. Chloroquine was reported to inhibit CHIKV entry into cells, possibly by raising the endosomal pH and thus preventing the fusion of the CHIKV E1 protein with the endosomal membrane [7, 8]. The potential effect of chloroquine treatment was assessed in two clinical trials. In a first clinical trial improvement of symptoms in the chronic phase of CHIKV infection were reported following chloroquine treatment [7], whereas another study failed to prove the efficacy of chloroquine as a treatment for the acute phase of CHIKV infection [8]. Therefore, the use of chloroquine as anti-CHIKV antiviral requires further study to prove its effectiveness and to determine the appropriate dosage and length of treatment.

2.1.2 Arbidol and Its Derivatives

Arbidol is a broad-spectrum antiviral that has been licensed in Russia and China for the treatment and prophylaxis of influenza and other respiratory infections [9]. Arbidol was also reported as an inhibitor of CHIKV infection in MRC-5 cells [10]. The mechanism of its anti-CHIKV activity has not been totally elucidated. An arbidol-resistant CHIKV strain could be selected and was shown to have acquired a mutation at amino acid 407 (G407R) in the CHIKV E2 glycoprotein, which may be involved in binding to host receptors [10]. Recently, a series of arbidol analogues have been synthesized and evaluated for their anti-CHIKV activity [11]. Two analogues in this series (IIIe and IIIf) inhibited the CHIKV-induced cytopathic effect (CPE) with selectivity indices higher than that of the parent compound arbidol [11].

2.1.3 Phenothiazines

Phenothiazines are clinically approved antipsychotics. In a screening assay using Semliki Forest virus (SFV) as a bio-safe surrogate for CHIKV, six compounds containing a 10H-phenothiazine core, including chlorpromazine, perphenazine, ethopropazine, thiethylperazine, thioridazine, and methdilazine were identified as possible SFV entry inhibitors. The antiviral activity of these molecules was confirmed using a recombinant CHIKV strain carrying a luciferase reporter gene (CHIKV-Rluc). When compared to the SFV-based screening results, the molecules showed similar potencies against the CHIKV-Rluc; however, the EC_{50} values determined using the CHIKV-Rluc were higher. The antiviral target of these inhibitors still needs to be identified [12].

2.1.4 Epigallocatechin Gallate

Epigallocatechin gallate (EGCG) is the major constituent of green tea. Recently, EGCG has been reported to have a modest but significant antiviral activity against CHIKV [13]. The inhibition of CHIKV entry and attachment to the target cells by EGCG was confirmed using pseudo-particles carrying the CHIKV envelope proteins [13].

2.2 Inhibitors of Viral Protein Translation

2.2.1 RNA Interference

RNA interference is induced by small interfering RNAs (siRNA) that are homologous in sequence to the gene that needs to be silenced. Small interfering RNAs are 21-23 nucleotides long dsRNA molecules having 3′-overhangs of two nucleotides. Treatment of cells with exogenous siRNAs results in the assembly of a RNA-induced silencing complex (RISC) which degrades specific complementary mRNA molecules. Consequently, protein expression of the targeted gene is markedly reduced.

Small interfering RNA (siRNA) sequences targeting CHIKV nsP3 and E1 genes were reported to significantly reduce CHIKV titers at 24 h post-infection in transfected Vero cells [14]. However, the inhibitory effect of these siRNAs was transient and diminished after 3 days of infection. These siRNAs could thus be used in combination with other antivirals for more effective treatment. In a more recent study, nsP1 and E2 siRNAs were generated and their potential activity was evaluated in cell culture and in animal models [15]. SiRNAs directed against nsP1 and E2 as well as their combinations, reduced *in vitro* CHIKV replication in Vero cells with more than 90 %. Interestingly, when CHIKV-infected mice were injected 3 days post-infection with these siRNAs, CHIKV replication was completely inhibited at the highest dose of siRNA tested (1 mg/kg body weight; [15]).

Plasmid-based small hairpin RNAs (shRNAs) were also designed and evaluated as strategy to inhibit CHIKV replication. ShRNAs produced from the shRNA-plasmid construct resulted in their intracellular processing to siRNAs causing specific knockdown of viral RNA and subsequent inhibition of viral protein expression. Stable cell clones expressing shRNA against CHIKV E1 and nsP1 showed significant and sustained inhibition of CHIKV infection [16]. In addition, mice pretreated with E1 targeting shRNA were completely protected against CHIKV induced disease and their survival was observed up to 15 days post-infection (untreated animals died between 6 and 10 days post-infection; [16]).

2.2.2 Harringtonine and Homoharringtonine

Harringtonine, a cephalotaxine alkaloid derived from *Cephalotaxus harringtonia*, has been reported asa potent inhibitor of CHIKV with minimal cytotoxicity [17]. In addition, homoharringtonine, a more stable analogue of harringtonine, also showed anti-CHIKV activity. Homoharringtonine has been recently approved by the FDA for the treatment of chronic myeloid leukemia in 2012. Harringtonine and homoharringtonine were found to suppress the production of viral nsP3 and E2 proteins, most likely through the inhibition of the host cell protein translation machinery [17]. In addition, the decrease in nsP3 production resulted in a reduction of viral replicase complexes formation. Consequently, the level of negative-sense RNAs was decreased leading to the reduced synthesis of the viral positive-sense RNA genome [17].

2.3 Inhibitors of Viral Genome Replication

2.3.1 Ribavirin

Ribavirin, a structural analogue of guanosine, is a broad-spectrum antiviral drug that has been approved by the FDA for the treatment of respiratory syncytial virus in infants [18], and in combination with pegylated interferon alpha (IFN-α) for treatment of chronic hepatitis C virus infection [19]. Ribavirin was shown to inhibit the replication of CHIKV *in vitro* [20]. In addition, the combination of ribavirin and IFN-α2b was reported to result in asynergistic inhibitory effect against CHIKV replication in Vero cells [20]. The mechanism of the antiviral action of ribavirin is likely different for different viruses. The 5′-monophosphate metabolite of ribavirin acts a competitive inhibitor of inosine monophosphatedehydrogenase (IMPDH) resulting in a depletion of intracellular GTP (and dGTP) pools [21]. The predominant mechanism by which ribavirin inhibits the replication of other RNA viruses such as flaviviruses and paramyxoviruses has been shown to be mediated by depletion of GTP pools [22]. Other suggested mechanisms by which ribavirin inhibits RNA virus replication include the inhibition of viral RNA capping, inhibition of the viral polymerase, and lethal mutagenesis of the RNA genome [23].

2.3.2 6-Azauridine

6-Azauridine is a broad-spectrum anti-metabolite that inhibits the replication of both DNA and RNA viruses [24]. It is a uridine analogue that competitively inhibits the orotidine monophosphate decarboxylase enzyme, which is involved in the de novo synthesis of pyrimidines [24]. 6-Azauridine showed strong inhibitory effect against CHIKV replication in Vero cells with an EC_{50} value of 0.82 μM [20].

2.3.3 Mycophenolic Acid

Mycophenolic acid (MPA) is a non-competitive inhibitor of IMPDH which has been used clinically as an immunosuppressant to prevent the rejection of transplant organs. MPA inhibits in vitro CHIKV replication [21]. The inhibition of CHIKV replication appears to be due to the depletion of intracellular GTP pools [21].

2.3.4 Favipravir (T-705)

Favipiravir (T-705) is a broad-spectrum antiviral agent that was originally discovered as an inhibitor of influenza A virus replication. In the cell, T-705 is metabolized to its ribofuranosyl 5′-triphosphate form, which was shown to be a competitive inhibitor for the incorporation of ATP and GTP by the viral RNA-dependent RNA polymerase (RdRp; [25]). However, the exact mechanism of action of T-705 has not been totally clarified yet. Recently, it has been reported that T-705 and its defluorinated analogue, T-1105 inhibit the in vitro replication of CHIKV [26]. In addition, the oral treatment of CHIKV-infected AG129 mice with T-705 protected these mice from severe neurological disease and reduced the mortality rate by more than 50 %. Low-level T-705-resistant CHIKV variants have been selected. These variants carried a K291R mutation in the F1 motif of the RdRp that was shown to be responsible for the observed resistance to T-705. This position is highly conserved in the polymerase of+ssRNA viruses [26].

2.4 Inhibitors of CHIKV nsP2

The CHIKV nsP2 protein exhibits RNA triphosphatase/nucleoside triphosphatase activity, as well as helicase activity within the N-terminal half while the C-terminal half encodes the viral cysteine protease required for processing of the non-structural viral polyprotein [27, 28]. In addition, nsP2 plays an important role in shutting down host cell mRNA transcription via degradation of a subunit of the DNA-directed RNA polymerase II. It also inhibits the host antiviral response by suppressing transcription and type I/II interferon-stimulated JAK/STAT signaling [28]. In a high-throughput screening for CHIKV nsP2 inhibitors that target the nsP2-mediated transcriptional shut off, a natural compound derivative (ID1452-2) was shown to partially block the nsP2 activity resulting in inhibition of CHIKV replication in cell culture [29]. In another study, a number of nsP2 inhibitors were identified using a computer-aided screening procedure of which one lead compound (compound 1) showed a significant antiviral activity against CHIKV [30]. This compound was predicted to bind to the central portion of the nsP2 protease active site.

Recently, a number of arylalkylidene derivatives of 1,3-thiazolidin-4-one have been shown to inhibit the in vitro replication of CHIKV. The inhibition of the CHIKV protease was suggested to be the mechanism of action of these compounds [31].

2.5 Inhibitors of Host Targets

2.5.1 Furin Inhibitors

Cellular furins and furin-like proteases are involved in the cleavage of viral pE2 into mature E2 and E3 proteins. The inhibition of cellular furin may therefore be expected to inhibit the formation of mature viral particles. Decanoyl-RVKR-chloromethyl ketone (dec-RVKR-cmk) is an irreversible furin inhibitor that was shown to inhibit CHIKV infection in vitro via inhibition of viral glycoprotein maturation [32]. The combination of dec-RVKR-cmk and chloroquine resulted in an additive inhibitory effect on CHIKV replication. Surprisingly, pretreatment of cells with dec-RVKR-cmk revealed a significant inhibition of viral entry, indicating that dec-RVKR-cmk treatment could alter the cleavage of proteins involved in CHIKV endocytosis or early replication steps or that this molecule could even inhibit CHIKV receptor maturation [32].

2.5.2 Modulators of Cellular Kinases

Protein Kinase C(PKC) Activators

Prostratin and 12-*O*-tetradecanoylphorbol 13-acetate (TPA) are well-known tigliane diterpenoids with a basic phorbol carbon skeleton esterified at position 13 [33]. Due to their chemical structure, they act as natural analogues of diacylglycerol that induce the activation of protein kinases C. Previously, prostratin and TPA were reported to have antiviral activity against HIV [34]. Prostratin and TPA were also identified as potent and selective CHIKV inhibitors in vitro [33]. Further studies are required to determine their mode of action against CHIKV.

Kinase Inhibitors

In a cell-based screening of a kinase inhibitor library, six kinase inhibitors were found to inhibit CHIKV-associated cell death in a dose-dependent manner [35]. Of these molecules, four compounds have a benzofuran core scaffold, one a pyrrolopyridine and one a thiazol-carboxamide scaffold. Using image analysis, it was shown that CHIKV-infected cells treated with these molecules had less prominent apoptotic blebs, which are typical for the CHIKV cytopathic effect. Moreover, these compounds reduced viral titers up to 100-fold. It was suggested that the inhibition of the virus-induced CPE by these compounds was the result of inhibition of kinases involved in apoptosis [35].

2.5.3 HSP-90 Inhibitors

HSP-90 is a family of highly conserved molecular chaperones which includes two cytoplasmic isoforms: stress-induced HSP-90α and constitutively expressed HSP-90β. In general, HSP-90 is involved in maturation, localization, and turnover of its client proteins in a cell. HSP-90 has been reported to play an important role in the replication of many DNA and RNA viruses such as hepatitis B virus, hepatitis C virus, human cytomegalovirus and influenza virus. Consequently, HSP-90 inhibitors may have a role as broad(er)-spectrum antiviral agents. Two HSP-90 inhibitors, HS-10 and SNX-2112, were reported as CHIKV replication inhibitors. The treatment of CHIKV-infected mice (SvA129) with HS-10 and SNX-2112 significantly reduced the serum viral load at

48 h post-infection and protected against the CHIKV-induced inflammation in the limb of infected mice [36]. In co-immunoprecipitation studies CHIKV nsP3 and nsP4 were shown to interact with HSP-90. Interestingly, the knockdown of the HSP-90α subunit resulted in a more pronounced inhibition of viral replication than targeting the HSP-90β subunit. HSP-90α is thought to be involved in the stabilization of CHIKV nsP4 and the formation of the CHIKV replication complex [36]. Further mechanistic studies are required to unravel the role of HSP-90 in the replication cycle of CHIKV.

2.5.4 Modulators of Host Immune Response

The innate immune system plays an important role in the acute phase of CHIKV infection. Detection of CHIKV RNA by Toll-like receptors (TLRs) 3, 7, and 8, as well as RIG-I like receptors during the acute phase of infection is believed to trigger the production of type I IFNs. Consequently, type I IFNs activate the transcription of interferon-stimulated genes (ISGs), which encode proteins involved in the host antiviral defense leading to clearance of the infection [37]. Therefore, activation of the innate immune response could be interesting for the treatment of CHIKV infections.

IFN-α

Treatment with IFN-α2a and IFN-α2b inhibited CHIKV replication in Vero cells in a dose-dependent manner [20]. The combination of IFN-α2b and ribavirin resulted in a synergistic antiviral effect on in vitro CHIKV replication. A CHIKV strain carrying the E1 A226V mutation was reported to be more sensitive to the antiviral activity of recombinant IFN-α than wild-type virus [38].

2′,5′-Oligoadenylate Synthetase 3 (OAS3)

The role of OAS3 in the innate immunity towards CHIKV was investigated using a stable HeLa cell line expressing OAS3 [39]. The expression of OAS3 by this cell line efficiently inhibited CHIKV infection by blocking the early stages of virus replication. A CHIKV variant with a glutamine-to-lysine mutation at position 166 of the envelope E2 glycoprotein proved resistance to the antiviral activity of OAS3 [40].

Polyinosinic Acid–Polycytidylic Acid

Polyinosinic acid–polycytidylic acid (poly (I:C)), a synthetic analogue of dsRNA, is a potent stimulant of IFN that interacts with TLR3. Treatment of human bronchial epithelial cells with poly (I:C) before CHIKV infection suppressed virus-induced CPE up to 72 h post-infection. Poly (I:C) resulted in a significant up regulation of IFN-α, IFN-β, OAS, and MxA in uninfected cells [41].

RIG-I Agonists

RIG-I (retinoic acid-inducible gene I) is a member of the RIG-I like receptor family which recognizes viral dsRNA leading to activation of multiple antiviral factors that block viral infection at different stages. Interestingly, chemically or enzymatically synthesized dsRNA molecules with an exposed 5′-triphosphate end (5′ ppp) were reported to induce RIG-I [42, 43]. It has been recently shown that pretreatment of MRC-5 cells with an optimized 5′ triphosphorylated RNA molecule triggered RIG-I stimulation resulting in protection against CHIKV infection [44]. Moreover, the protective response against CHIKV induced by this 5′ ppp RNA was largely independent of the type I IFN response. These results suggest the potential efficacy of RIG-I agonists as an antiviral treatment for CHIKV infection.

2.6 Inhibitors with an Unknown Target

2.6.1 Trigocherrierin A

Trigocherrierin A is a new daphnane diterpenoid orthoester isolated from the leaves of *Trigonostemon cherrieri* [45]. Trigocherrierin A inhibits CHIKV in cell culture but the mechanism of its action remains elusive.

2.6.2 Aplysiatoxin Derivatives

Debromoaplysiatoxin and 3-methoxydebromoaplysiatoxin are marine toxins isolated from the marine cyanobacterium, *Trichodesmium erythraeum* [46]. Both compounds had significant antiviral activity against CHIKV at non-toxic concentrations. The compound was reported to block a post-entry step in the CHIKV lifecycle.

2.6.3 5,7-Dihydroxyflavones

A number of 5,7-dihydroxyflavones (apigenin, chrysin, naringenin, and silybin) were identified as inhibitors of the CHIKV subgenomic replicon [12]. The molecular target of these compounds is still unknown.

3 Conclusion

The global re-emergence of CHIKV and the high morbidity rate associated with its infection emphasizes the need to develop potent antiviral agents against CHIKV. So far, a number of classes of compounds that inhibit viral replication by targeting either a viral or a host factor have been reported. Most of the compounds have relatively modest activity and for most of them, activity in infection models (in mice) was not assessed. Some of these classes may serve as a starting point for the design of more specific and selective inhibitors of CHIKV replication. Also, to the best of our knowledge, no information is available yet on the effect of antivirals on the chronic stage of CHIKV infection. Recently, mouse models for CHIKV-induced arthritis and chronic joint disease have been developed which will help the evaluation of CHIKV antiviral agents in different stages of CHIKV infection [47, 48]. Several other viruses belonging to the *Alphavirus* genus, in particular the equine encephalitis viruses are considered to be a potential bioterroristic threat. When designing/developing antivirals against the Chikungunya virus it may be important to develop classes of compounds that have pan-alphavirus activity and that could thus also be used for the treatment of alphaviruses other than CHIKV.

Among the reported CHIKV inhibitors, favipravir, ribavirin, arbidol, and IFN-α have been approved previously for the treatment of other viral infections. This could markedly facilitate their evaluation for clinical use in CHIKV-infected patients. Favipravir, a drug with a broad-spectrum antiviral activity, has been approved in Japan for the treatment of influenza virus infections. It is currently also being evaluated in Western Africa for the treatment of Ebola virus infection. If its activity is demonstrated against this infection, this compound may be considered for treatment of other infections such as those caused by CHIKV. However, given the growing number of patients suffering from CHIKV infections, it may be justified to develop specific CHIKV/alphavirus drugs. Highly potent drugs are today available for the treatment of infections with herpes viruses, HIV, the hepatitis B and C virus. Without a doubt, it should also be possible, when investing sufficient effort, to develop highly effective and safe drugs for the treatment (and perhaps even prophylaxis) of infections with alphaviruses.

Acknowledgments

The original work of the authors is supported by EU FP7 SILVER (contract no. HEALTH-F3-2010-260644) and the BELSPO IUAP consortium BELVIR (Belgium). Leen Delang is funded by the Scientific Fund for Research of Flanders (FWO).

References

1. Simon F, Javelle E, Oliver M et al (2011) Chikungunya virus infection. Curr Infect Dis Rep 13:218–228
2. Cavrini F, Gaibani P, Pierro AM et al (2009) Chikungunya: an emerging and spreading arthropod-borne viral disease. J Infect Dev Ctries 3:744–752
3. Kucharz EJ, Cebula-Byrska I (2012) Chikungunya fever. Eur J Intern Med 23:325–329
4. Ali Ou Alla S, Combe B (2011) Arthritis after infection with Chikungunya virus. Best Pract Res Clin Rheumatol 25:337–346
5. Teissier E, Zandomeneghi G, Loquet A et al (2011) Mechanism of inhibition of enveloped virus membrane fusion by the antiviral drug arbidol. PLoS One 6(1), e15874
6. Kaur P, Chu JJ (2013) Chikungunya virus: an update on antiviral development and challenges. Drug Discov Today 18:969–983
7. Brighton SW (1984) Chloroquine phosphate treatment of chronic Chikungunya arthritis. An open pilot study. S Afr Med J 66: 217–218
8. De Lamballerie X, Boisson V, Reynier J-C et al (2008) On Chikungunya acute infection and chloroquine treatment. Vector Borne Zoonotic Dis 8:837–839
9. Blaising J, Polyak SJ, Pécheur EI (2014) Arbidol as a broad-spectrum antiviral: an update. Antiviral Res 107:84–94
10. Delogu I, Pastorino B, Baronti C et al (2011) In vitro antiviral activity of arbidol against Chikungunya virus and characteristics of a selected resistant mutant. Antiviral Res 90: 99–107
11. Di Mola A, Peduto A, La Gatta A et al (2014) Structure-activity relationship study of arbidol derivatives as inhibitors of Chikungunya virus replication. Bioorg Med Chem 22: 6014–6025
12. Pohjala L, Utt A, Varjak M et al (2011) Inhibitors of alphavirus entry and replication identified with a stable Chikungunya replicon cell line and virus-based assays. PLoS One 6, e28923
13. Weber C, Sliva K, von Rhein C et al (2015) The green tea catechin, epigallocatechin gallate

inhibits Chikungunya virus infection. Antiviral Res 113:1–3

14. Dash PK, Tiwari M, Santhosh SR et al (2008) RNA interference mediated inhibition of Chikungunya virus replication in mammalian cells. Biochem Biophys Res Commun 376: 718–722
15. Parashar D, Paingankar MS, Kumar S et al (2013) Administration of E2 and NS1 siRNAs inhibit Chikungunya virus replication in vitro and protects mice infected with the virus. PLoS Negl Trop Dis 7(9), e2405
16. Lam S, Chen KC, Ng MML, Chu JJ (2012) Expression of plasmid-based shRNA against the E1 and nsP1 genes effectively silenced Chikungunya virus replication. PLoS One 7(10), e46396
17. Kaur P, Thiruchelvan M, Lee RCH et al (2013) Inhibition of Chikungunya virus replication by harringtonine, a novel antiviral that suppresses viral protein expression. Antimicrob Agents Chemother 57:155–167
18. Turner TL, Kopp BT, Paul G et al (2014) Respiratory syncytial virus: current and emerging treatment options. Clin Outcomes Res 6:217–225
19. Pawlotsky JM (2014) New hepatitis C therapies: the toolbox, strategies, and challenges. Gastroenterology 146:1176–1192
20. Briolant S, Garin D, Scaramozzino N et al (2004) In vitro inhibition of Chikungunya and Semliki Forest viruses replication by antiviral compounds: synergistic effect of interferon-α and ribavirin combination. Antiviral Res 61:111–117
21. Khan M, Dhanwani R, Patro IK et al (2011) Cellular IMPDH enzyme activity is a potential target for the inhibition of Chikungunya virus replication and virus induced apoptosis in cultured mammalian cells. Antiviral Res 89:1–8
22. Leyssen P, De Clercq E, Neyts J (2006) The anti-yellow fever virus activity of ribavirin is independent of error-prone replication. Mol Pharmacol 69:1461–1467
23. Paeshuyse J, Dallmeier K, Neyts J (2011) Ribavirin for the treatment of chronic hepatitis C virus infection: a review of the proposed mechanisms of action. Curr Opin Virol 1:590–598
24. Rada B, Dragún M (1977) Antiviral action and selectivity of 6-azauridine. Ann N Y Acad Sci 284:410–417
25. Furuta Y, Gowen BB, Takahashi K et al (2013) Favipiravir (T-705), a novel viral RNA polymerase inhibitor. Antiviral Res 100:446–454
26. Delang L, Segura Guerrero N, Tas A et al (2014) Mutations in the Chikungunya virus non-structural proteins cause resistance to favipiravir (T-705), a broad-spectrum antiviral. J Antimicrob Chemother 69:2770–2784
27. Solignat M, Gay B, Higgs S et al (2009) Replication cycle of Chikungunya: a re-emerging arbovirus. Virology 393:183–197
28. Fros JJ, van der Maten E, Vlak JM, Pijlman GP (2013) The C-terminal domain of Chikungunya virus nsP2 independently governs viral RNA replication, cytopathicity, and inhibition of interferon signaling. J Virol 87:10394–10400
29. Lucas-Hourani M, Lupan A, Despres P et al (2012) A phenotypic assay to identify Chikungunya virus inhibitors targeting the nonstructural protein nsP2. J Biomol Screen 18(2):172–179
30. Bassetto M, De Burghgraeve T, Delang L et al (2013) Computer-aided identification, design and synthesis of a novel series of compounds with selective antiviral activity against Chikungunya virus. Antiviral Res 98:12–18
31. Jadav SS, Sinha BN, Hilgenfeld R et al (2015) Thiazolidone derivatives as inhibitors of Chikungunya virus. Eur J Med Chem 89: 172–178
32. Ozden S, Lucas-Hourani M, Ceccaldi P-E et al (2008) Inhibition of Chikungunya virus infection in cultured human muscle cells by furin inhibitors: impairment of the maturation of the E2 surface glycoprotein. J Biol Chem 283: 21899–21908
33. Bourjot M, Delang L, Nguyen VH et al (2012) Prostratin and 12-O-tetradecanoylphorbol 13-acetate are potent and selective inhibitors of Chikungunya virus replication. J Nat Prod 75:2183–2187
34. McKernan LN, Momjian D, Kulkosky J (2012) Protein kinase C: one pathway towards the eradication of latent HIV-1 reservoirs. Adv Virol 2012:805347
35. Cruz DJM, Bonotto RM, Gomes RGB et al (2013) Identification of novel compounds inhibiting Chikungunya virus-induced cell death by high throughput screening of a kinase inhibitor library. PLoS Negl Trop Dis 7(10), e2471
36. Rathore APS, Haystead T, Das PK et al (2014) Chikungunya virus nsP3 & nsP4 interacts with HSP-90 to promote virus replication: HSP-90 inhibitors reduce CHIKV infection and inflammation in vivo. Antiviral Res 103:7–16
37. Schwartz O, Albert ML (2010) Biology and pathogenesis of Chikungunya virus. Nat Rev Microbiol 8:491–500
38. Bordi L, Meschi S, Selleri M et al (2011) Chikungunya virus isolates with/without

A226V mutation show different sensitivity to IFN-alpha, but similar replication kinetics in non human primate cells. New Microbiol 34:87–91

39. Bréhin A-C, Casadémont I, Frenkiel M-P et al (2009) The large form of human 2′,5′-Oligoadenylate Synthetase (OAS3) exerts antiviral effect against Chikungunya virus. Virology 384:216–222
40. Gad HH, Paulous S, Belarbi E et al (2012) The E2-E166K substitution restores Chikungunya virus growth in OAS3 expressing cells by acting on viral entry. Virology 434:27–37
41. Li Y-G, Siripanyaphinyo U, Tumkosit U et al (2012) Poly (I:C), an agonist of toll-like receptor-3, inhibits replication of the Chikungunya virus in BEAS-2B cells. Virol J 9:114
42. Hornung V, Ellegast J, Kim S et al (2006) 5′-Triphosphate RNA is the ligand for RIG-I. Science 314:994–997
43. Pichlmair A, Schulz O, Tan CP et al (2006) RIG-I-mediated antiviral responses to single-stranded RNA bearing 5′-phosphates. Science 314:997–1001
44. Olagnier D, Scholte FEM, Chiang C et al (2014) Inhibition of dengue and Chikungunya virus infection by RIG-I-mediated type I IFN-independent stimulation of the innate antiviral response. J Virol 88:4180–4194
45. Bourjot M, Leyssen P, Neyts J et al (2014) Trigocherrierin a, a potent inhibitor of Chikungunya virus replication. Molecules 19: 3617–3627
46. Gupta DK, Kaur P, Leong ST et al (2014) Anti-Chikungunya viral activities of aplysiatoxin-related compounds from the marine cyanobacterium Trichodesmium erythraeum. Mar Drugs 12:115–127
47. Hawman DW, Stoermer KA, Montgomery SA et al (2013) Chronic joint disease caused by persistent Chikungunya virus infection is controlled by the adaptive immune response. J Virol 87:13878–13888
48. Morrison TE, Oko L, Montgomery SA et al (2011) A mouse model of Chikungunya virus-induced musculoskeletal inflammatory disease: evidence of arthritis, tenosynovitis, myositis, and persistence. Am J Pathol 178:32–40

Chapter 23

A Real-Time Cell Analyzing Assay for Identification of Novel Antiviral Compounds against Chikungunya Virus

Keivan Zandi

Abstract

Screening of viral inhibitors through induction of cytopathic effects (CPE) by conventional method has been applied for various viruses including Chikungunya virus (CHIKV), a significant arbovirus. However, it does not provide the information about cytopathic effect from the beginning and throughout the course of virus replication. Conventionally, most of the approaches are constructed on laborious end-point assays which are not capable for detecting minute and rapid changes in cellular morphology. Therefore, we developed a label-free and dynamical method for monitoring the cellular features that comprises cell attachment, proliferation, and viral cytopathogenicity, known as the xCELLigence real-time cell analysis (RTCA). In this chapter, we provide a RTCA protocol for quantitative analysis of CHIKV replication using an infected Vero cell line treated with ribavirin as an in vitro model.

Key words Cytopathic effects, Chikungunya virus, Arbovirus, Label-free, Proliferation, RTCA, Ribavirin

1 Introduction

Chikungunya virus (CHIKV) is an alphavirus from *Togaviridae* family with a positive sense single stranded RNA as genome [1] that particularly is transmitted to humans by the infected *Aedes* mosquito bites. The name of the virus is derived from a Makonde word meaning "that bends up," describing the contorted posture and rheumatic expressions by the infected patients [2]. Globally, from year 2004 till to date, millions of CHIKV infection cases were recorded from different parts of the world such as Americas, Africa, Asia, Europe, Indian and Ocean Islands [3]. Yet there are no vaccines and no definite treatments against CHIKV infection discovered [4]. Hence, the development of antiviral drugs against CHIKV is necessary. In previous studies, conventional cell-based techniques are generally enzymatic-based assays, in which both fluorescent and nonfluorescent probes are extensively used to

Justin Jang Hann Chu and Swee Kim Ang (eds.), *Chikungunya Virus: Methods and Protocols*, Methods in Molecular Biology, vol. 1426, DOI 10.1007/978-1-4939-3618-2_23, © Springer Science+Business Media New York 2016

detect virus replication [5]. The assays comprise comprehensive handling of the cells including cell fixation, labeling, and manual cell counting. Furthermore, these approaches are unable to deliver important information about the dynamics of virus replication in the cells.

The xCELLigence, real-time cell analyzer (RTCA) is a high-end technology approach that permits real-time cell growth recording using a label free cell-based assay which generates impedance variations in the culture media. The microelectrodes are fused at the bottom of special cell culture plates, named E-Plates [6]. In this system, microelectrodes regulate the electronic impedance prior to cellular activities through their sensors to record any minute changes. The RTCA system has been used on numerous research fields such as microbiological research [7], environmental toxicity [8] and cellular function [9]. The xCELLigence system can be applied in screening of antiviral drugs for CHIKV. This technology is useful, as CHIKV is a virus with fast replication cycle and such an accurate and real time assay is capable of detecting virus infection at an early stage.

In this chapter, we illustrate the application of RTCA system for quantitative analysis of CHIKV infection based on CPE induction using CHIKV-infected Vero cell line treated with ribavirin. This method is capable of screening potential antivirals in a shorter period of time, providing detailed information on cell activities throughout the experiment compared to conventional methods in CPE based screenings using colorimetric detection. Here, we provide the protocol of a RTCA system to monitor cell growth proliferation to determine the optimal seeding density of Vero cells, to determine the cytotoxicity of ribavirin with various concentration and as well as the antiviral activity of ribavirin.

2 Materials

Make all solutions by using radical-free water generated by refining deionized water at 18 MΩ cm sensitivity. Make and keep all reagents at room temperature unless specified otherwise.

2.1 Tissue Culture

1. 2× Minimal Essential Medium (MEM) with Earle's balanced salts and 2 mM l-glutamine, pH 7.4. Add 100 ml water into a 1 l graduated cylinder (*see* **Note 1**). Add 19 g of MEM powder and 4.4 g of sodium bicarbonate (*see* **Note 2**). Adjust pH to 7.4 using HCl or NaOH, and top up to 1 l using milliQ H_2O. Sterilize using a 0.2 μm pore-sized filter (*see* **Note 3**) and store at 4 °C.
2. MEM with Earle's balanced salts supplemented with 10 % fetal bovine serum (FBS). Add 250 ml of 2× MEM, 190 ml of

milliQ H_2O, 5 ml of l-glutamine (200 mM), 5 ml of nonessential amino acids (NEAA; 100× at 10 mM), and 50 ml of FBS into a 500 ml laboratory glass bottle. Store at 4 °C.

3. MEM with Earle's balanced salts supplemented with 2 % FBS. As above but with 2 % FBS.
4. Trypsin–Versene 1.2 % (*see* **Note 4**). Add 88.8 ml of 1× PBS, 10 ml of 10× Versene (0.53 mM), 1.2 ml of 10 % trypsin, and 1 ml of penicillin–streptomycin (10,000 U) into a 100 ml laboratory glass bottle. Store at 4 °C.
5. 1× phosphate buffer saline (PBS).

2.2 Cells

1. Vero cell line CCL81 (ATCC; *see* **Note 5**).

2.3 Virus

1. CHIKV strain MY/065/08/FN295485 cell supernatant at a titre of $10^{4.5}$ EID_{50}/ml. The virus strain belongs to the ECSA genotype and has the A226V mutation in E1 protein ([10]; *see* **Note 6**).

2.4 Antiviral Compound

1. 50 mg/ml ribavirin. Weigh 5 mg of the compound and dissolved in 100 μl of dimethylsulfoxide (DMSO) to make up 50 mg/ml stock solution. Vortex to dissolve and store aliquots at −20 °C until needed (*see* **Note 7**).

2.5 RTCA Components

1. RTCA SP System Bundle (ACEA Biosciences). The RTCA SP system consist of three main components: one unit of RTCA analyzer, model W380 (includes power and serial cables), one unit of RTCA SP station (includes a RTCA SP station, a RTCA protector shield 96, a RTCA frame 96, a RTCA contact pins, a RTCA contact pin extraction tool and a RTCA cleaning kit), and one unit of RTCA control (includes a notebook PC with RTCA software preinstalled).
2. E-Plate 96.
3. Incubator.

3 Methods

Perform all procedures at room temperatures except indicated.

3.1 Cell Proliferation Assay

1. Add 50 μl of MEM supplemented with 10 % FBS in each well of the E-Plate 96. Incorporate the E-Plate 96 with the RTCA SP station which is place in the incubator connected to the system as to get the background reading. Engage the E-Plate 96 from the system.

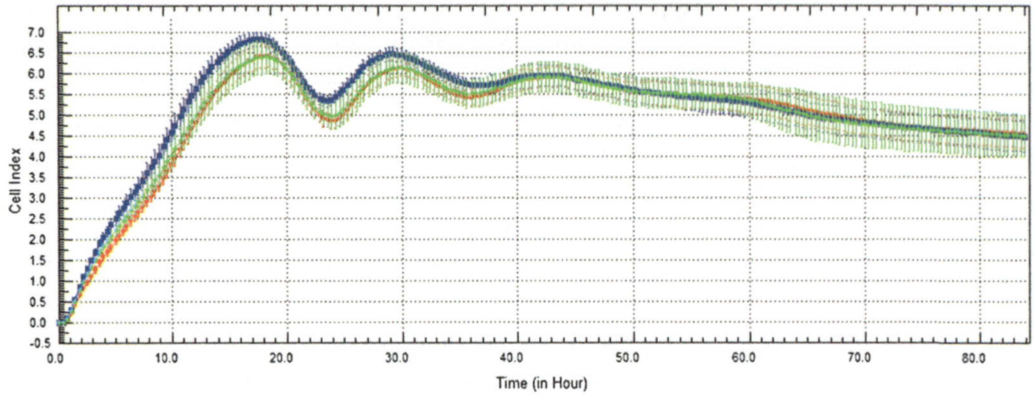

Fig. 1 Proliferation curve of Vero cells. The cells seeded in the E-Plate 96 were constantly observed by measuring CI values to optimize the cell seeding density and to determine the suitable time point for virus infection prior to the antiviral assay. The cell features such as adhesion, spreading, and proliferation were observed in the interval time of every 2 min. *Colored curves* represents the various numbers of cells seeded per well in the E-Plate 96: *red line*: 15,000 cells/well; *green line*: 18,000 cells/well; *blue line*: 20,000 cells/well. Each data point signifies the average ± standard deviation and was analyzed from triplicate record

2. Count and seed cells at three different densities of 1.5×10^4, 1.8×10^4, and 2.0×10^4 in MEM supplemented with 10 % FBS (*see* **Note 8**). Add 50 μl of the respective cell suspensions containing the amount of cells required into the 96-well E-Plate in four replicates. Incubate the E-Plate at room temperature for 30 min in the laminar flow (*see* **Note 9**).
3. Place the E-Plate 96 in the RTCA SP station at 37 °C for continuous impedance recording (Fig. 1; *see* **Note 10**).

3.2 Cytotoxicity Assay

1. Prepare and seed the cells in each well of E-Plate 96 at a density of 1.8×10^4 (*see* **Note 11**).
2. Take out 100 μl of MEM supplemented with 10 % FBS prior to each assay. Prepare five different concentrations of ribavirin in twofold dilutions (*see* **Note 12**). Prepare the compound in MEM supplemented with 2 % FBS and pass through a 0.2 mm filter. Add 100 μl of ribavirin and 100 μl of MEM supplemented with 2 % FBS in triplicates into each well, which will then give rise to final concentrations of ribavirin at 100 μg/ml, 50 μg/ml, 25 μg/ml, 12.5 μg/ml, and 6.25 μg/ml. Add triplicates of 200 μl of MEM supplemented with 2 % FBS as a negative control.
3. Place the E-Plate 96 in the RTCA SP station at 37 °C for continuous impedance recording (Fig. 2).

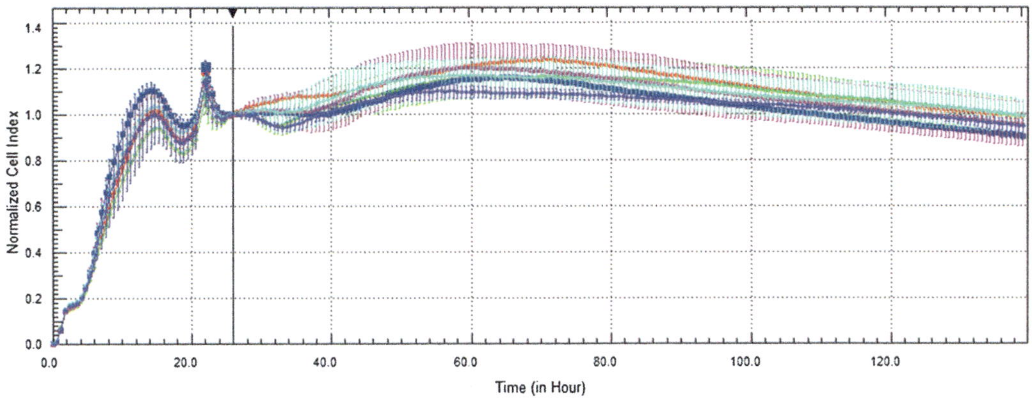

Fig. 2 The effect of the intensified compounds concentrations on Vero cells evaluated by RTCA. The cell proliferation after 120 h of incubation with increasing concentrations of ribavirin. The *black vertical line* signifies the ribavirin-treated Vero cells time point prior to seeding Vero cells for 18 h. *Colored curves* represent the various serial dilution of ribavirin. Each data point signifies the average ± standard deviation and was analyzed from triplicate record. *Red line*: 100 μg/ml; *blue line*: 50 μg/ml; *pink line*: 25 μg/ml; *turquoise line*: 12.5 μg/ml; *purple line*: 6.25 μg/ml and *green line*: 0 μg/ml. Each data point signifies the average ± standard deviation and was analyzed from triplicate records

3.3 Antiviral Assay

1. Use the same procedure for preparing and seeding the cells prior to the treatment assay.
2. Take out 100 μl of MEM supplemented with 10 % FBS in each well. Prepare three different concentrations of ribavirin in twofold dilutions (*see* **Note 14**). Prepare the compound in MEM supplemented with 2 % FBS and pass through a 0.2 mm filter.
3. Add 100 μl of MEM supplemented with 2 % FBS, followed by 50 μl of CHIKV. Incubate on the rocker for 30 min and 1 h later in the incubator at 5 % CO_2 atmosphere at 37 °C (*see* **Note 13**).
4. Take out the virus suspension and add 100 μl of compound for each concentration in triplicates. Add triplicates of 100 μl of MEM supplemented with 2 % FBS in each of the ribavirin-treated well.
5. Add 200 μl of 2 % of FBS media in triplicates for both CHIKV-infected cells and negative control.
6. Place the E-Plate 96 in the RTCA SP station at 37 °C for continuous impedance recording (Fig. 3).

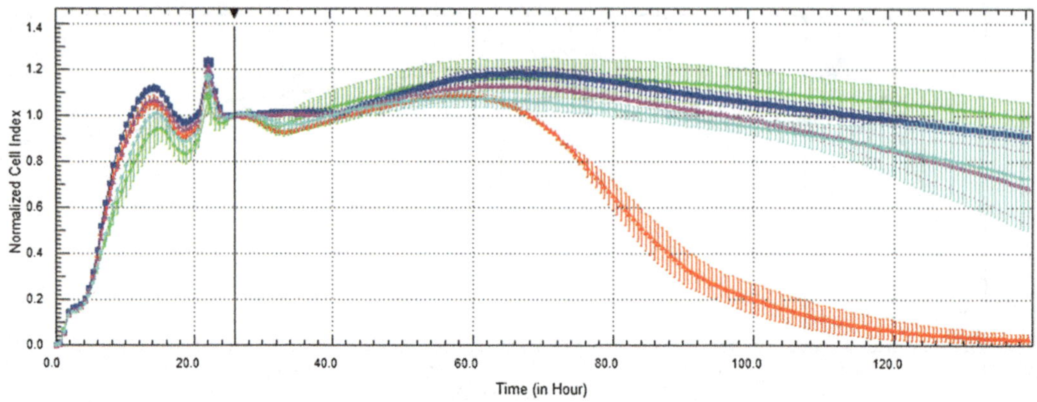

Fig. 3 The antiviral effect against CHIKV in ribavirin-treated Vero cells evaluated by RTCA. RTCA monitored antiviral effect at various concentrations of ribavirin on the proliferation, spreading, and adhesion of cells. The cell impedance was measured every 2 min. The *black vertical line* signifies treatment of infected Vero cells with ribavirin time point prior to seeding of Vero cells for 18 h. *Colored curves* represent the Vero cells, viral infected Vero cells, and ribavirin-treated viral infected Vero cells at different concentrations of ribavirin. *Red line*: positive control (CHIKV infection); *green line*: negative control (without CHIKV infection); *blue line*: 100 μg/ml; *pink line*: 50 μg/ml; *turquoise line*: 25 μg/ml. Each data point signified the average ± standard deviation and was analyzed from triplicate record

4 Notes

1. Autoclave the water for media preparation to avoid contamination of the culture media. Add some water in the cylinder before adding the other components to clumping of media powder and this allows the magnetic bar to stir easily and smoothly.
2. Cells in culture media will produce CO_2. High concentration of CO_2 will affect the pH of media and sodium bicarbonate acts as a buffer to stabilize the pH of media.
3. Use eye protection when handling vacuum-driven filters plastic bottles.
4. Versene acts as a boosting agent to remove the calcium and magnesium from the cell surface that permits trypsin to hydrolyze specific peptide bonds. The calcium and magnesium can be found in the extracellular matrix that supports in cell to cell adhesion and also disguises the peptide bonds that trypsin acts on.
5. Vero cell line is derived from an African green monkey kidney. Maintain this adherent cell line in EMEM supplemented with 10 % FBS in humidified air of 5 % CO_2 at 37 °C.
6. Propagate the virus on Vero cell line. Harvest by freeze–thaw method once it exhibits full CPE. To remove cell debris,

centrifuge at 1000 × *g* for 10 min. Titrate the virus stock from culture supernatant on Vero cell line by using tissue culture infectious dose 50 ($TCID_{50}$) methods according to Reed and Muench [11]. Aliquot and store at −80 °C until needed.

7. DMSO is a light sensitive reagent, so avoid direct exposure to light when preparing the solution.
8. Prepare the cells suspension in different concentrations before performing the background reading as it may take some time to do cell counting.
9. The incubation is important to allow the cells to dispense uniformly at the bottom of the wells.
10. Cell Index (CI) is a parameter that illustrates the changes in cell number, cell morphology, adhesion grade, and also the viability of the cells based on the measured electrical impedance. CI value at zero indicates the absence or non-adherence of cells onto the electrodes. In comparison, as more cells attach onto the electrodes it will lead to the increased of impedance measurement and thus the CI value increases [12].
11. Choose the best concentration based on Cell Index (CI) value. In the example given (Fig. 2), the best concentration for cells was at 1.8×10^4 for 18 h of cells incubation prior to treatment. Basically, all three concentrations of cells did not show much differences of CI value, and therefore, 1.8×10^4 cells number was chosen as it was the median number of the cell concentration.
12. The concentration of compound of each dilution should be prepared two times higher than needed as we have to mix with 100 μl of medium to give rise to the desired final concentration in a total volume of 200 μl.
13. This step is to ensure that the CHIKV penetrate the cell.
14. For antiviral assay, prepare twofold dilution of compound at three different concentrations. The concentration should be prepared two times higher than needed in order to give rise to desired final concentration following dilution in an equal volume of media containing 2 % FBS.

Acknowledgment

The author would like to thank University of Malaya for the UMRG flagship grant (Flagship: FL001-13HTM), High Impact Research (HIR) grant E000013-20001 and postgraduate (PPP) grant (Grant No. PG037-2013B) for funding the expenses of this work.

References

1. Rizzo F, Cerutti F, Ballardini M et al (2014) Molecular characterization of flaviviruses from field-collected mosquitoes in northwestern Italy, 2011–2012. Parasit Vectors 7:1–11
2. Ross RW (1956) The new ala epidemics III: the virus isolation, pathogenic properties and relationship to the epidemics. J Hyg 54: 177–191
3. Coffey LL, Failloux AB, Weaver SC (2014) Chikungunya virus-vector interactions. Viruses 6:4628–4663
4. Lundstrom K (2014) Alphavirus-based vaccines. Viruses 6:2392–2415
5. Ge Y, Deng T, Zheng X (2009) Dynamic monitoring of changes in endothelial cell-substrate adhesiveness during leukocyte adhesion by microelectrical impedance assay. Acta Biochim Biophys Sin 41:256–262
6. Solly K, Wang X, Xu X et al (2004) Application of real-time cell electronic sensing (RT-CES) technology to cell-based assays. Assay Drug Dev Technol 2:363–372
7. Slanina H, König A, Claus H et al (2011) Real-time impedance analysis of host cell response to meningococcal infection. J Microbiol Methods 84:101–108
8. Leme DM, Grummt T, Heinze R et al (2011) Cytotoxicity of water-soluble fraction from biodiesel and its diesel blends to human cell lines. Ecotoxicol Environ Saf 74:2148–2155
9. Keogh R (2010) New technology for investigating trophoblast function. Placenta 31: 347–350
10. Sam IC, Chan YF, Chan SY et al (2009) Chikungunya virus of Asian and Central/East African genotypes in Malaysia. J Clin Virol 46:180–183
11. Reed LJ, Muench HA (1938) Simple method of estimating fifty per cent end points. Am J Hyg 27:493–497
12. Xing JZ, Zhu L, Jackson JA et al (2005) Dynamic monitoring of cytotoxicity on microelectronic sensors. Chem Res Toxicol 18: 154–161

Chapter 24

Using Bicistronic Baculovirus Expression Vector System to Screen the Compounds That Interfere with the Infection of Chikungunya Virus

Szu-Cheng Kuo, Chao-Yi Teng, Yi-Jung Ho, Ying-Ju Chen, and Tzong-Yuan Wu

Abstract

Chikungunya virus (CHIKV) is the etiologic agent of Chikungunya fever and has emerged in many countries over the past decade. There are no effective drugs for controlling the disease. A bicistronic baculovirus expression system was utilized to co-express CHIKV structural proteins C (capsid), E2 and E1 and the enhanced green fluorescence protein (EGFP) in *Spodoptera frugiperda* insect cells (Sf21). The EGFP-positive Sf21 cells fused with each other and with uninfected cells to form a syncytium is mediated by the CHIKV E1 allowing it to identify chemicals that can prevent syncytium formation. The compounds characterized by this method could be anti-CHIKV drugs.

Key words Baculovirus, Chikungunya virus, Internal ribosome entry site, Syncytium

1 Introduction

Chikungunya virus (CHIKV) belongs to the *Alphavirus* of *Togaviridae* family [1]. CHIKV is an enveloped virus with a positive-stranded RNA genome about 11.8 kb, the genome size of the prototype CHIKV strain S27, AF369024 is 11,826 nucleotides. CHIKV genome contains two open reading frames (ORFs) embedded between the non-translated regions, 5′ UTR and 3′ UTR. Based on studies of other alphaviruses and sequence similarity, the first ORF located at the 5′-end encodes a polyprotein precursor of nonstructural proteins (nsP1, nsP2, nsP3, and nsP4), which function in viral replication [2]. The second ORF comprises the 26S subgenomic RNA of CHIKV and it encodes the polyprotein of the structural proteins (C, E2, and E1), and both E1 and E2 are glycoproteins [2].

CHIKV is transmitted by the same vector as dengue virus—the *Aedes* mosquitoes, such as *Aedes albopictus* and *Aedes aegypti*, and causes the disease known as Chikungunya fever [3]. Chikungunya

Justin Jang Hann Chu and Swee Kim Ang (eds.), *Chikungunya Virus: Methods and Protocols*, Methods in Molecular Biology, vol. 1426,DOI 10.1007/978-1-4939-3618-2_24,

fever usually gets into an acute phase with fever and skin rashes, followed by painful arthralgia that can last for months. Chikungunya fever was first identified in Tanzania and Uganda in 1953 [4]. Since then, outbreaks of Chikungunya infection have been reported in Africa, Southeast Asia, the Indian subcontinent, and the Indian Ocean. In August 2007, the first outbreak in European continent was documented in Italy with 217 laboratory-confirmed cases [5]. This was the first Chikungunya fever outbreak in a temperate climate country. Currently, Chikungunya infection has been identified in more than 40 countries. At present there is no effective vaccines or antiviral treatments for Chikungunya fever, and thus, it is an important issue to develop efficacy methods for development of anti-CHIKV drugs.

Insect cells based baculovirus expression vector system (BEVS) has been extensively used for the production of recombinant proteins [6]. In previous studies, we have demonstrated that the incorporation of an internal ribosome entry site (IRES) into the baculoviruses genome to generate bicistronic baculoviruses expression vectors [7, 8]. This IRES-based bicistronic baculoviruses expression vectors not only produces two genes of interest simultaneously in the same infected cells, and also facilitates the recombinant virus isolation and titer determination when the green fluorescent protein was co-expressed. Recently, Chikungunya virus glycoproteins E1 as well as E2 could be functionally expressed at high levels by BEVS in insect cells and could be properly glycosylated and cleaved by furin [9, 10]. In addition to producing E1 or E2 proteins of CHIKV to act as a vaccine candidate, expression of E1 proteins of CHIKV in baculovirus infected insect cells was essential and necessary to display membrane fusogenic activity and to induce syncytia formation [9, 11]. These findings demonstrated that the co-expression of EGFP and CHIKV structural proteins in Sf21 cells and the induction of syncytia could be easily monitored by the fluorescence microscope. Using this molecular "beacon" for the detection of membrane fusion, it has made it possible to analyze cell fusion events and to reveal CHIKV membrane fusion requiring low pH and cholesterol.

In this chapter, we introduce the application of this novel BEVS to act as a screening system for compounds that are potential use for treatment of Chikungunya fever. The materials for insect culture and recombinant baculovirus preparation by a bicistronic baculovirus transfer vector are presented. Then, we describe methods employing expressed structural proteins of CHIKV in insect cells by the recombinant baculovirus to screen for candidate compounds that prevent cell fusion event. Finally, we present the protocols of an in vitro anti-CHIKV activity assay to validate the compounds identified in the recombinant baculovirus-mediated methods.

2 Materials

2.1 Insect Cell Culture

1. Sf21 (IPLB-Sf21AE) cell line.
2. Grace's/TNM-FH Medium supplemented with 10 % fetal bovine serum (FBS; Caisson Laboratories).
3. 75-cm^2 tissue culture-treated flasks (T75 flasks).

2.2 Generation and Purification of Recombinant Baculoviruses

1. Bicistronic baculovirus transfer vector: pBac-CHIKV-26S-Rhir-E (*see* **Note 1**; Fig. 1).
2. BaculoGold linearized baculovirus DNA (BD; *see* **Note 2**).
3. Cellfectin II reagent (Invitrogen).
4. Serum-free TNM-FH medium.

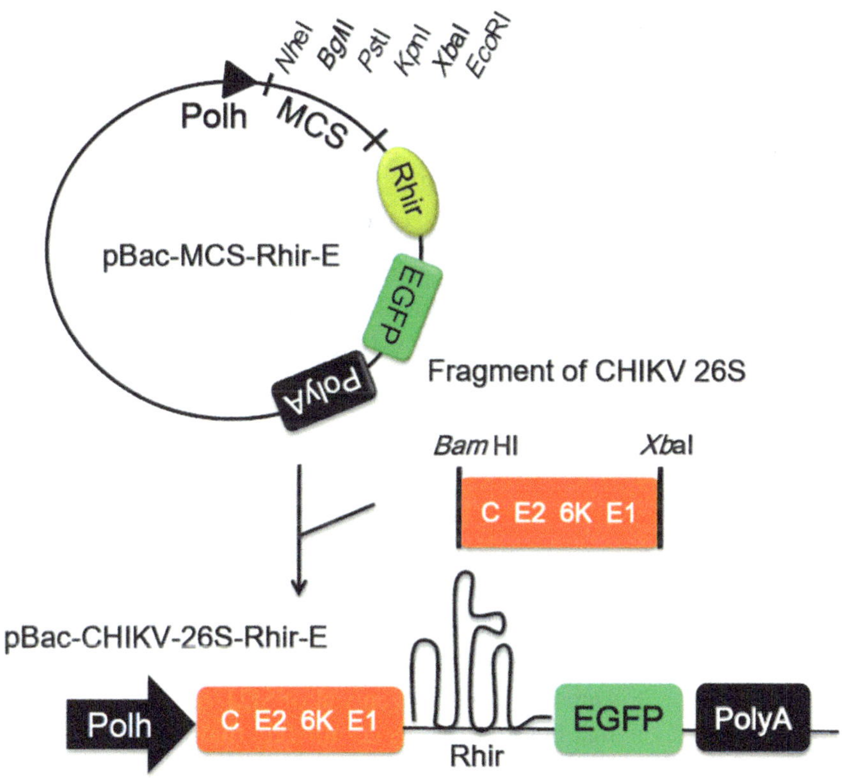

Fig. 1 Bicistronic baculovirus transfer vector, pBac-CHIKV-26S-Rhir-E. The bicistronic baculovirus transfer vector pBac-Rhir-E, in which the RhPV 5′-UTR IRES is located between the six MCS cloning sites (*Nhe*I, *Bgl*II, *Pst*I, *Kpn*I, *Xba*I, and *Eco*RI) and the EGFP genes. The cDNA for the 26S subgenome of the Chikungunya virus (CHIKV 26S RNA) was cloned into the multiple cloning site (MCS) between the polyhedrin promoter (Polh) and the RhPV-IRES-EGFP sequence. The cDNA fragment containing the full-length structural gene (7559–11293) of Chikungunya virus strain S27-African prototype (AF369024) flanked with *Bam*HI (5′-end) and *Xba*I (3′-end) was cloned into the *Bgl*II and *Xba*I sites of pBac-Rhir-E, and the resulting plasmid was named pBac-CHIKV-26S-Rhir-E

5. TNM-FH medium supplemented with 10 % FBS.
6. 24- and 96-well cell culture plate.
7. Inverted fluorescence microscope (Nikon).

2.3 Insect Cell–Cell Fusion Inhibition Assay

1. Sf21 (IPLB-Sf21AE) cell line.
2. Recombinant baculovirus vAc-CHIKV-26S-Rhir-E.
3. Inverted fluorescence microscope (Nikon).
4. FDA-approved drug library and selected compounds.
5. Sf-900 II SFM (pH 6.2, *see* **Note 3**) containing 5 % fetal calf serum (FCS) and 1× antibiotic–antimycotic (Gibco).
6. Sf-900 II SFM (pH 6.8) containing 2 % FCS and 1× antibiotic–antimycotic. Mix and adjust pH with 2 N NaOH.
7. Sf-900 II SFM (pH 5.8) containing 2 % FCS and 1× antibiotic–antimycotic and 100 μg/ml cholesterol. Mix and adjust pH with 2 N HCl.
8. 1× PBS with 3 % paraformaldehyde.

2.4 In Vitro Anti-CHIKV Activity Assay

1. BHK-21 cell line.
2. CHIKV 27S strain.
3. DMEM supplemented with 5 % FCS.
4. Alexa Fluor 594-conjugated goat anti-rabbit IgG (Invitrogen).
5. 1:1 methanol and acetone mixture. Store at −20 °C.
6. Rabbit anti-CHIKV E2 antibody as reported [9].
7. 1× PBS.

3 Methods

3.1 Culture of Insect Cells

1. Aseptically transfer 10 ml of TNM-FH to the 75-ml shake flask (T75 flask). Place the flask in an incubator at 26–28 °C. Loosen caps of flasks to allow oxygenation. Protect the medium from light exposure.
2. Quickly thaw a frozen vial of Sf21 cells in a 37 °C water bath. Just before the cells completely thaw, spray the vial with 70 % ethanol to decontaminate.
3. Gently transfer the entire content of the vial into the T75 flask containing 25 ml medium.
4. Replace with 10 ml fresh TNM-FH containing 10 % FBS after 4 h.
5. Leave the flask for 3–4 days until the culture density reaches $>2 \times 10^6$ viable cells/ml. Count total cell number using a manual cell counter.
6. Passage cells when cell density reaches 2×10^6.

7. Seed a new T75 flask at 3–5 × 10^5 cells/ml by diluting cells in 10 ml of TNM-FH.
8. Place the T75 flask in an incubator at 26–28 °C, non-humidified incubator. Gently swirl flasks to evenly distribute cells.
9. Once cells reach approximately 30 passages, discard the stock culture and thaw a fresh vial of cells (*see* **Note 4**).

3.2 Generation of Recombinant Baculoviruses in Sf21 Cells

1. Seed Sf21 cells (2 × 10^5) on a 24-well plate, allow cells to attach for 1–2 h before growing to 80 % confluency (*see* **Note 5**).
2. Take a 1.5 ml sterile tube and add 0.25 μg BaculoGold linearized baculovirus DNA and 1 μg plasmid DNA pBac-CHIKV-26S-Rhi-E (*see* **Note 6**).
3. Add 350 μl of serum free TNM-FH and 4 μl Cellfectin II reagent to the tube. Gently reverse 15–20 times and incubate at room temperature for 30 min.
4. Remove the culture medium from the 24-well and add total of transfection mixture (drop wise) to the insect cells on the well. After every two or three drops, gently swirl the plate to ensure even distribution.
5. Incubate the plate at 27 °C for 5 h.
6. After 5 h, remove the transfection solution from the wells and add fresh TNM-FH medium containing 10 % FBS and incubate the plate at 27 °C for 4–5 days.
7. After 4–5 days, monitor the cells by observation of green fluorescence under fluorescence microscopy. Collect the supernatant and infected cells for further virus purification (*see* **Note 7**).

3.3 Purification of Recombinant Baculoviruses

Sf21 cells have now been co-transfected with linearized baculovirus DNA and transfer vector (plasmid DNA) as described in Subheading 3.2, the culture medium will therefore contain a mixture of recombinant baculovirus (vAc CHIKV-26S-Rhir-E; Fig. 2) and wild-type baculovirus. To purify the recombinant baculovirus, use the expression of green fluorescent proteins mediated by the RhPV-IRES to guide the selection process (*see* **Note 8**) as described below:

1. Plate Sf21 cells on 96-well plate with 4 × 10^4 per well in 50 μl.
2. Make a serial dilution of virus sample from 10^{-1} to 10^{-9} by adding 100 μl of viral stock to 900 μl of TNM-FH with 10 % FBS. Mix thoroughly between dilutions.
3. Infect each row of cells with the same virus dilutions and incubate the plate at 27 °C for 3 days.
4. Monitor EGFP expressing cells under fluorescence microscope. Look for single plaque of EGFP expressing cells within each well in the 96-well plate.

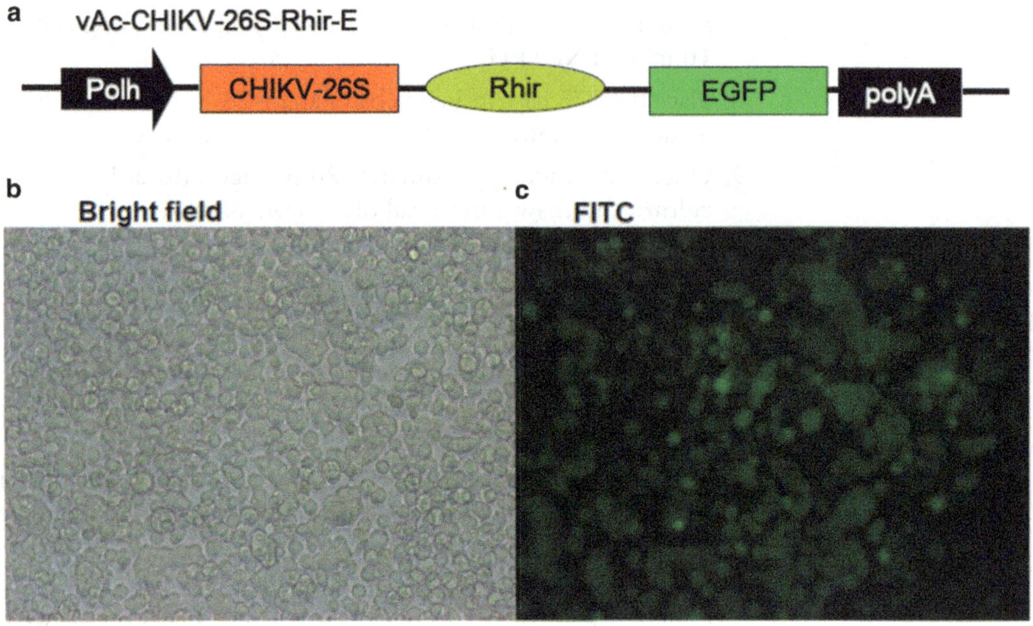

Fig. 2 Recombinant baculoviruses vAc-CHIKV-26S-Rhir-E induced syncytium formation in infected Sf21 cells. (**a**) The schematic presentation of the recombinant baculovirus vAc-CHIKV-26S-Rhir-E. (**b**) Sf21 cells infected by vAc-CHIKV-26S-Rhir-E were examined under a fluorescence microscope with a bright field (*left* panel) or a FITC channel (*right* panel)

5. Carefully aspirate the medium from the wells using 200 μl tips.
6. Add 30 μl aliquots of the medium to a new 24 well. Gently rock the plate to evenly distribute the virus. Incubate the plate at 27 °C and harvest medium until 80 % cells are EGFP expressing cells under microscope (*see* **Note 9**).

3.4 Insect Cell–Cell Fusion Inhibition Assay (Fig. 3)

1. Seed Sf21 cells (10^5 cells cultured in Sf-900 II SFM (pH 6.2) containing 5 % FCS at 27 °C) and infect cells with the recombinant baculovirus vAc-CHIKV-26S-Rhir-E (MOI of 1) in a 96-well tissue culture plate.
2. After 1-day post-infection (dpi), replace the culture medium with Sf-900 II SFM (pH 6.8) containing 2 % FCS and incubate the infected cells for another 24 h.
3. Pretreat infected Sf21 cells with Sf-900 II SFM (pH 6.8) containing 2 % FCS and the compounds to be tested (100 and 10 μM) at 27 °C for 1 h.
4. Trigger the insect cell–cell fusion by replacement of serial diluted hit compounds in the Sf-900 II SFM (pH 5.8) containing 2 % FCS and 100 μg/ml cholesterol, followed by an incubation for 2 h at 27 °C.
5. Add 50 μl PBS, 3 % paraformaldehyde per well for fixation.

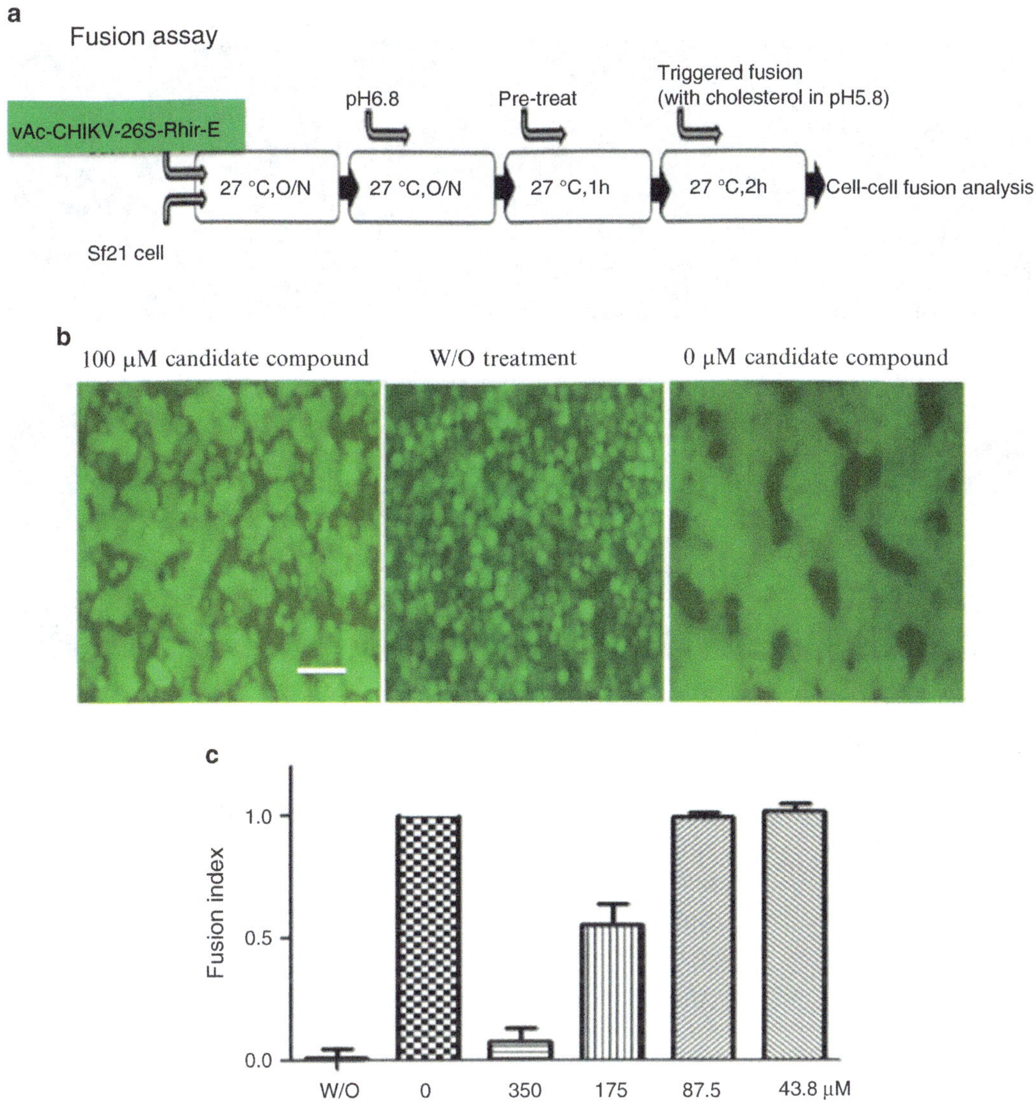

Fig. 3 Candidate compounds block CHIKV-induced Sf-21 cell–cell fusion. (**a**) Cell fusion inhibition assay. (**b**) vAc-CHIKV-26S-Rhir-E infected Sf21 cells were induced cell–cell fusion in the presence or absence of 100 μM candidate compound following the conditions outlined in (**a**). (**c**) vAc-CHIKV-26S-Rhir-E infected Sf21 cells induced cell–cell fusion in the presence of serial diluted candidate compound. Syncytial formations were quantitated by fusion index

6. Obtain images of insect cell–cell fusion using an inverted fluorescence microscope.
7. Calculate the fusion index: 1 – (number of EGFP positive cells/number of EGFP positive nuclei). In the comparisons of syncytial cell size, cell number and total area, count at least 100 EGFP positive single cells and measure as described in [11].

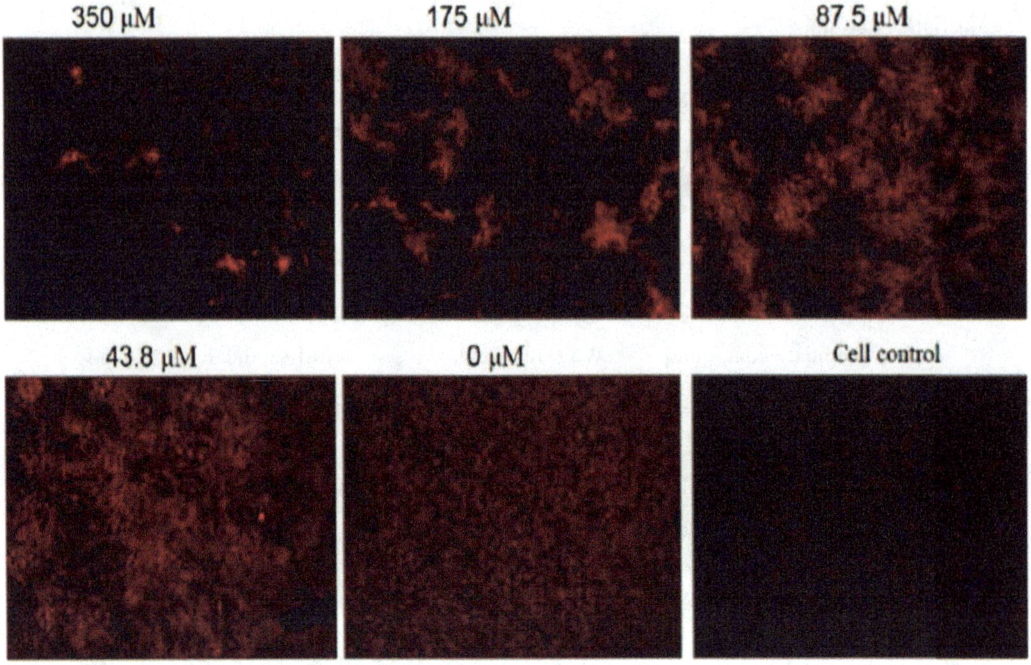

Fig. 4 Immunofluorescence assay of CHIKV E2 of CHIKV infected BHK-21 cells in the presence of serial diluted candidate compounds

3.5 In Vitro Anti-CHIKV Activity Assay (Fig. 4)

1. Seed BHK-21 cells (10^5 cells cultured in DMEM with 5 % FCS at 37 °C) in a well of 96-well tissue culture plate and incubate overnight.
2. Infect cells with CHIKV 27S strain (MOI of 1) in the presence of tested compound.
3. After 20 h incubation, fix the cells with ice-cold 1:1 acetone and methanol mixture for 5 min. Then, air-dry for 5 min.
4. Stain cells with rabbit anti-CHIKV E2 antibody (1:100) at room temperature for 1 h.
5. Wash three times with 100 μl PBS.
6. Stain cells with Alexa Fluor 594-conjugated goat anti-rabbit IgG at a dilution of 1:500 for 1 h at room temperature.
7. Wash three times with 100 μl PBS.
8. Capture and score signals of CHIKV E2 immunofluorescence using an inverted fluorescence microscope and a multimode microplate reader.

4 Notes

1. pBac-MCS-Rhir-E is a bicistronic baculovirus transfer vector in which the gene of interest (GOI, e.g., cDNA for the 26S subgenome of the Chikungunya virus (CHIKV 26S RNA))

can be cloned into the multiple cloning sites (MCS). This vector contains the internal ribosome entry site (IRES) derived from *Rhopalosiphum padi* virus [12, 13] to achieve bicistronic expression.

2. pBac-MCS-Rhir-E is based on the pBluebac 4.5 baculovirus transfer vector (Invitrogen). To generate recombinant baculoviruses, pBluebac 4.5 derived baculovirus transfer vector should be recombined with Bac-N blue linearized viral DNA (Invitrogen). However, the Bac-N-Blue linearized viral DNAs are not available now, and thus, BaculoGold linearized baculovirus DNA was used in this protocol.
3. The gp64 protein of baculovirus can induce the cell fusion when the value of pH was lower than 5.5 [14]. So, it is critical to control the value of pH above 5.5 during the Sf21 cells fusion assay induced by vAc-CHIKV-26S-Rhir-E infection.
4. The passage effect of insect cells on virus production is critical in recombinant baculovirus generation as well as in virus amplification. We found that the newly generated Sf21 cells are more accessible for baculoviruses infection. Besides, it was reported that higher passage may produce a lower virus titer [15].
5. The Sf9 cells can also be used to generate recombinant baculoviruses by the homologous recombination approach; however, we found the recombination rate was higher in Sf21 cells than in that of Sf9 cells.
6. The amount of viral DNA can be adjusted and changed accordingly. However, the total DNA (transfer vector plus viral DNA) should be kept lower than 1.5 μg. Based on our experience, if the amount of DNA is more than 1.5 μg, both the transfection and recombination rate will be reduced.
7. Homologous recombination between the transfer vector and baculovirus viral DNA is a rare event in insect cells (typically only about 0.1–1 %) and will always lead to a dominant wild type virus in the culture medium. And the recombinant viruses and wild type viruses are not easy to be distinguished in infected cells. The RhPV IRES along with the green fluorescent protein EGFP enables us to judge whether the recombinant baculoviruses have successfully been produced after co-transfection of transfer vector and viral DNA in Sf21 cells just by direct observation under a fluorescent microscope.
8. A conventional method to purify recombinant virus is by plaque assay. Based on the co-expression of GFP through RhPV IRES, the successful event of homologous recombination between baculovirus DNA with transfer vector in insect cells can easily be identified under a fluorescent microscope. Therefore, the cells that exhibit GFP correlate with recombinant virus-infected cells and thus can be used to purify the target viruses from wild type viruses.

9. Repeat **step 2–6** three or four times and select at least four different clones. To confirm whether the recombinant baculoviruses vAc-CHIKV-26S-Rhir-E can express the structural proteins of CHIKV, perform Western blot and probe with anti-E1 and anti-E2 antibodies of CHIKV as described previously [9].

Acknowledgement

This research was supported by the Ministry of Science and Technology (NSC103-2321-B-033-001,105-2321-B-033-001 & NSC 102-2632-M-033-001-MY3).

References

1. Griffin DE (2007) Alphaviruses. In: Knipe DM, Howley PM (eds) Fields virology, 5th edn. Lippincott-Williams & Wilkins, Philadelphia, pp 1023–1068
2. Solignat M, Gay B, Higgs S, Briant L, Devaux C (2009) Replication cycle of Chikungunya: a re-emerging arbovirus. Virology 393:183–197
3. Powers AM, Logue CH (2007) Changing patterns of chikungunya virus: re-emergence of a zoonotic arbovirus. J Gen Virol 88: 2363–2377
4. Ross RW (1956) A laboratory technique for studying the insect transmission of animal viruses, employing a bat-wing membrane, demonstrated with two African viruses. J Hyg (Lond) 54:192–200
5. Rezza G, Nicoletti L, Angelini R, Romi R, Finarelli AC, Panning M, Cordioli P, Fortuna C, Boros S, Magurano F (2007) Infection with Chikungunya virus in Italy: an outbreak in a temperate region. Lancet 370:1840–1846
6. Unger T, Peleg Y (2012) Recombinant protein expression in the baculovirus-infected insect cell system. Methods Mol Biol 800:187–199
7. Chen WS, Villaflores OB, Lu CF, Wu HI, Chen YJ, Teng CY, Chang YC, Chang SL, Wu TY (2012) Functional expression of rat neuroligin-1 extracellular fragment by a bi-cistronic baculovirus expression vector. Protein Expr Purif 81:18–24
8. Wu TY, Chen YJ, Teng CY, Chen WS, Villaflores O (2012) A bi-cistronic baculovirus expression vector for improved recombinant protein production. Bioeng Bugs 3:129–132
9. Kuo SC, Chen YJ, Wang YM, Kuo MD, Jinn TR, Chen WS, Chang YC, Tung KL, Wu TY, Lo SJ (2011) Cell-based analysis of Chikungunya virus membrane fusion using baculovirus-expression vectors. J Virol Methods 175:206–215
10. Metz SW, Geertsema C, Martina BE, Andrade P, Heldens JG, van Oers MM, Goldbach RW, Vlak JM, Pijlman GP (2011) Functional processing and secretion of Chikungunya virus E1 and E2 glycoproteins in insect cells. Virol J 8:353
11. Kuo SC, Chen YJ, Wang YM, Tsui PY, Kuo MD, Wu TY, Lo SJ (2012) Cell-based analysis of Chikungunya virus E1 protein in membrane fusion. J Biomed Sci 19:44
12. Woolaway KE, Lazaridis K, Belsham GJ, Carter MJ, Roberts LO (2001) The 5′ untranslated region of Rhopalosiphum padi virus contains an internal ribosome entry site which functions efficiently in mammalian, plant, and insect translation systems. J Virol 75:10244–10249
13. Chen YJ, Chen WS, Wu TY (2005) Development of a bi-cistronic baculovirus expression vector by the Rhopalosiphum padi virus 5′ internal ribosome entry site. Biochem Biophys Res Commun 335:616–623
14. Blissard GW, Wenz JR (1992) Baculovirus gp64 envelope glycoprotein is sufficient to mediate pH-dependent membrane fusion. J Virol 66:6829–35
15. Maruniak JE, Garcia-Canedo A, Rodrigues JJ (1994) Cell lines used for the selection of recombinant baculovirus. In Vitro Cell Dev Biol Anim 30:283–286

Chapter 25

Neutralization Assay for Chikungunya Virus Infection: Plaque Reduction Neutralization Test

Nor Azila Muhammad Azami, Meng Ling Moi, and Tomohiko Takasaki

Abstract

Neutralization assay is a technique that detects and quantifies neutralizing antibody in serum samples by calculating the percentage of reduction of virus activity, as the concentration of virus used is usually constant. Neutralizing antibody titer is conventionally determined by calculating the percentage reduction in total virus infectivity by counting and comparing number of plaques (localized area of infection due to cytopathic effect) with a standard amount of virus. Conventional neutralizing test uses plaque-reduction neutralization test (PRNT) to determine neutralizing antibody titers against Chikungunya virus (CHIKV). Here we describe the plaque reduction neutralization assay (PRNT) using Vero cell lines to obtain neutralizing antibody titers.

Key words Plaque reduction neutralization test, Antibody, Chikungunya, Neutralizing antibody, Neutralizing assay, Vero cell lines

1 Introduction

Chikungunya fever is an acute febrile illness characterized by presence of arthralgia caused by Chikungunya virus (CHIKV), a mosquito-borne alphavirus [1]. In CHIKV infection, it is speculated that antibodies play an important role in protection against the virus [2]. This virus-neutralizing antibody is central in protection against the disease. The interaction between virus and virus-neutralizing antibody is speculated to result in the reduction of virus capability to infect or replicate in cells [3]. Detection and quantification of antibodies reactive to viruses can be determined by using conventional and alternative methods including hemagglutination inhibition (HI) assay, focus reduction neutralization test (FRNT), indirect fluorescent antibody (IFA) assay, plaque reduction neutralization test (PRNT), and micro neutralization test (MnT). The PRNT assay developed by Henderson and Taylor in 1959 to detect arboviruses plaques and to measure the serum antibody neutralization titer is generally accepted as the most specific

Justin Jang Hann Chu and Swee Kim Ang (eds.), *Chikungunya Virus: Methods and Protocols*, Methods in Molecular Biology, vol. 1426,DOI 10.1007/978-1-4939-3618-2_25, © Springer Science+Business Media New York 2016

assay among conventional serological assays [4]. The technique detects and quantifies neutralizing antibody in samples including monoclonal antibody and polyclonal antibody by determining the percentage of reduction of virus activity, as the virus concentration is usually constant.

PRNT serves as an alternative way to determine the protection levels by measuring the levels of virus-neutralizing antibodies. These assays are based on experimental observations that neutralizing antibodies serve as a proxy to disease protection [3]. Plaque assay was developed by Dulbecco for studies of inactivation (or neutralization) of animal viruses and modified to measure serum antibody neutralization titer [4, 5]. Subsequently, other investigators developed an in vitro PRNT assay to determine the amount of antibody required to neutralize 50 % of the infectious virus particles ($PRNT_{50}$; [6]). In vaccine efficacy studies, PRNT assay is used to determine the protective capacity of immune response induced by a vaccine candidate by measuring the level of neutralizing antibody.

The basic design of the PRNT assay is to allow the formation of virus–antibody complexes to occur in vitro, and to measure the neutralizing effect by plaque formation on virus-susceptible cells [3]. Virus-antibody complex is formed by mixing CHIKV and test sample (monoclonal antibody, polyclonal antibody, serum, or plasma) at specific incubation conditions and time. The virus–antibody complexes are then added to virus-susceptible cells, which are subsequently overlaid with a semisolid medium and incubated for 3–5 days, or until cytopathic effect (CPE) and plaque formation are observed. The semisolid medium prevents the spread of virus progeny in the medium, resulting in a localized area of cytopathic effect (plaque). Plaques are revealed by direct staining of live cells, by adding dyes in overlay or adding dyes after the removal of the overlay medium.

However, in this method, dyes (stain or dye) are added after removal of overlay, which is different when compared to conventional PRNT assay method. In conventional method, dyes are usually added in the first or second overlay to monitor the development of plaques among live cells. Adding dyes after overlay removal has several advantages. Plaques do not have to be counted immediately and the chemically fixed cells can be stored stably for a long time [3].

Neutralizing antibody titer is determined by calculating the percentage of reduction in total virus infectivity, by counting and comparing number of plaques with standardized amount of virus. In samples with CHIKV-specific neutralizing antibody, the number of plaques will be less compared to that of control wells, because infection of host cells are prevented by neutralizing antibodies. In the absence of CHIKV-specific neutralizing antibodies, the number of plaques will be similar to those of control wells. Here we describe a conventional PRNT assay using patient's serum samples and Vero cell lines.

2 Materials

Prepare reagents on a clean bench and store them at 4 °C (unless indicate otherwise). Heat-inactivate fetal bovine serum (FBS) for 30 min at 56 °C in water bath before long-term storage at −20 °C.

1. Cell lines for PRNT test: Simian kidney cell lines (Vero and LLC-MK2) or mosquitoes cell lines (C6/36; *see* **Note 1**).
2. Complete medium: Eagles' Minimum Essential Medium (EMEM) supplemented with 10 % heat-inactivated fetal bovine serum (Hi-FBS). Optionally, add 20 U/ml penicillin and 20 μg/ml streptomycin into growth medium to prevent potential bacterial contamination.
3. 1× Dulbecco's phosphate buffer saline (DPBS).
4. 0.05 % (v/v) trypsin–EDTA solution.
5. CHIKV stock. In this assay, the optimum titer of viruses required is 2.5×10^3 pfu/ml. Dilute the viruses stock with EMEM supplemented with 10 % Hi-FBS if the virus titer is too high (*see* **Note 2**).
6. Serum or plasma samples. Heat-inactivate all the serum or plasma samples at 56 °C prior to performing this assay to inactivate complement factors. Allow the samples to cool to room temperature prior to use.
7. Maintenance medium: 1 % Methylcellulose-4000 supplemented with 0.94 % Eagle's MEM containing kanamycin, 2 % heat-inactivated FBS, 2 mM L-glutamine, and 7.5 % (w/v) sodium bicarbonate ($NaHCO_3$) solution. For 1 l of maintenance medium, prepare two 1 l media bottles and label the bottles with "Solution A" and "Solution B". Add 10 g of methylcellulose and put a magnetic stir bar into "Solution A" bottle. Add 9.4 g of Eagle's MEM containing kanamycin into "Solution B" bottle and dissolve the reagent in 1 l of ultrapure water (MilliQ water). Autoclave "Solution A" and "Solution B" separately. Cool both solution to about 60 °C and add "Solution B" to "Solution A". Place the bottle on ice and mix the solution using magnetic stirrer until the solution turns from turbid yellow to clear yellow (*see* **Note 3**). Add 10 ml of 200 mM L-glutamine 100×, 20 ml of Hi-FBS, and 31.5 ml of 7.5 % (w/v) sodium carbonate into the solution. Mix until all components are dissolved. Addition of all solutions will result in final clear orange-red mixture.
8. 10 % (v/v) formaldehyde. Dilute formaldehyde in distilled water. Store reagent at room temperature (*see* **Note 4**).
9. Methylene blue (6×) solution. Weight 2.25 g of methylene blue tetrahydrate powder and diluted in 500 ml of distilled water. Add 0.375 ml of 1 N sodium hydroxide to dissolve the methylene blue (*see* **Note 5**). Store the solution at room temperature.

10. Incubator with constant carbon dioxide (CO_2) supply and high humidity levels of at least 95 %.
11. Biosafety cabinet, Class II.
12. Light box and colony counter for plaque enumeration.

3 Methods

All equipment and reagent used in the cell culture must be sterile and proper sterilization technique must be used in the experiment. Perform cell culture incubation in a humidified incubator at 37 °C supplied with 5 % CO_2. All experiments must be conducted on a clean bench unless specified otherwise. Heat-inactivate all serum samples at 56 °C for 30 min and allow to cool to room temperature prior to use.

3.1 Preparation of Cell Monolayer in 12-Well Plates

1. Prepare a confluent monolayer of Vero cells (in 75 cm^2 flask).
2. Discard the old medium.
3. Wash cells with 10 ml of DPBS twice.
4. Trypsinize the cells with 1× trypsin–EDTA solution for 5–10 min and tap the flask corners gently to detach the cells from the flask's wall.
5. Resuspend the cells in 10 ml of complete medium by repeated pipetting.
6. Seed the cells at concentrations between 1.0×10^5 and 2.0×10^5 cells/well (12-well plate). For any cell lines used in this assay, it is recommended that cells to be propagated and stored at a low passage level to maintain susceptibility to CPE formation. Optimal seeding concentration for infection assay should also be predetermined before performing the PRNT assay.
7. Add 1 ml of complete medium in each well.
8. Tilt the plate horizontally about 3–5 times gently, to ensure that the cells form a uniform monolayer before incubating the cells in a humidified incubator (37 °C with 5 % CO_2) overnight. Allow the cells to reach confluency of approximately 70–90 %.
9. If cell confluency do not reach 70–90 %, incubate for another day.

3.2 Preparation of CHIKV-Antibody Immune Complex

1. Serially dilute sample twofold starting from 1:5 to 1:2560 with EMEM supplemented with 10 % Hi-FBS. If possible, use new pipette tips for preparation of each dilution.
2. Add 25 μl of diluted samples to 25 μl of CHIKV at a titer of 2.5×10^3 pfu/ml. The final concentration results in approximately 62.5 pfu for each well.

3. Incubate the virus–antibody mixture at 37 °C for 60 min (*see* **Note 6**).

3.3 Infection Assay

1. The assay can be formatted to 2-replicate sets or 3-replicate sets based on experimental requirements. The layout of each set is as indicated in Fig. 1.

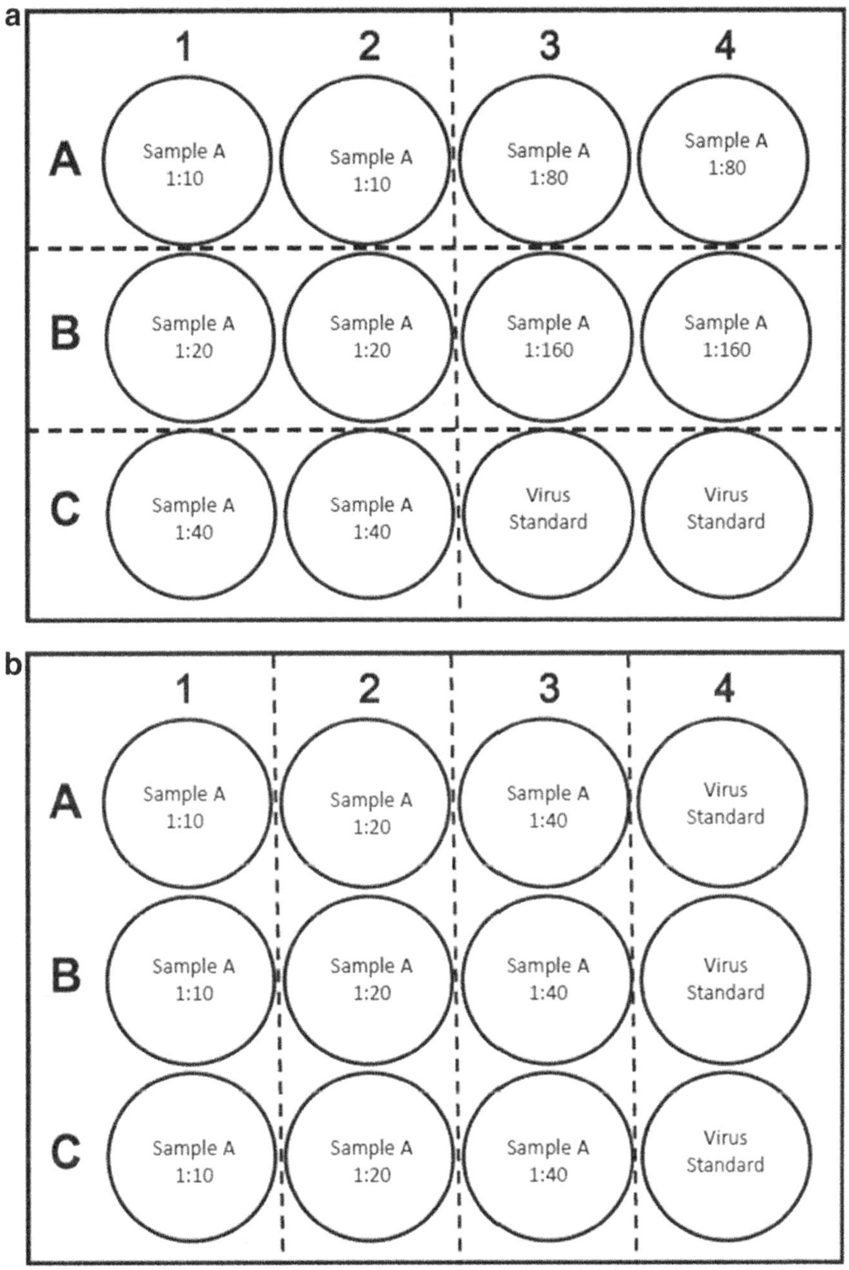

Fig. 1 Layout of 12-well plate for plaque reduction neutralization test (PRNT) assay of CHIKV. Layout of (**a**) two experimental replicates and (**b**) three experimental replicates per sample. Virus standard (negative control) indicates samples in the absence of antibodies

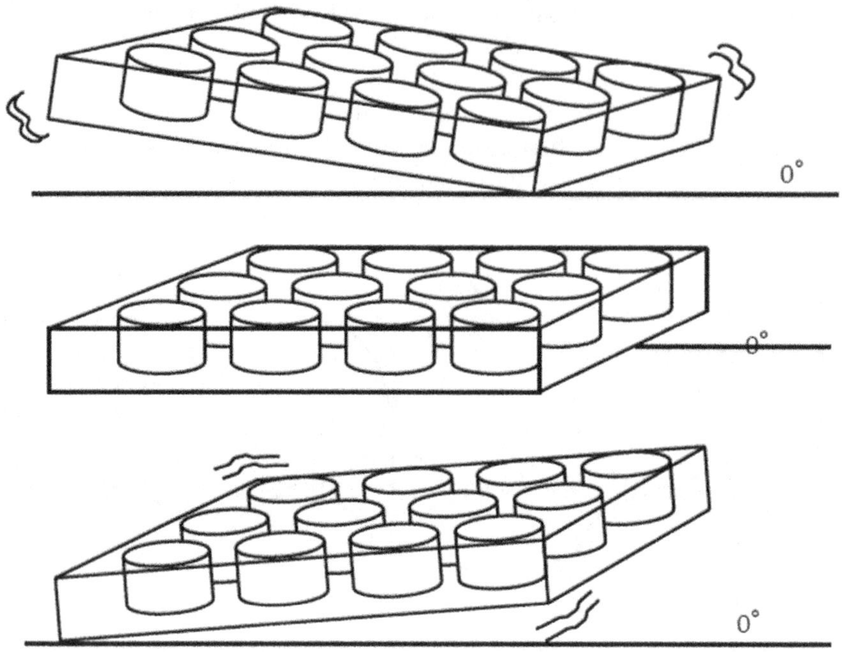

Fig. 2 Motion of direction during plate tilting. Gently tilt the plate for every 10 min for six times (total 60 min incubation time) to prevent the cells from drying up

2. Aspirate the old medium from cells in 12-well plate (*see* **Note 7**).
3. Add 100 μl of virus–antibody mixture to each well. Pipette the virus–antibody complexes mixture onto the wall of the wells. Avoid pipetting the solution directly onto the cells.
4. Incubate the plate at 37 °C for 60 min. During the 60 min incubation, tilt the plates gently every 10 min for a total of six times (total time: 60 min) as indicated in Fig. 2 (*see* **Note 8**).
5. Add 1.5 ml of maintenance medium and incubate the cells in humidified incubator at 37 °C, 5 % CO_2 for 3–5 days.

3.4 Plaque Visualization

1. Examine for the presence of plaques (cytopathic effect due to viral infection) under microscope or by naked eye. Plaque or cytopathic effect typically appears between day-2 and day-4 after CHIKV infection. The incubation period and plaque appearance varies depending on viral strains.
2. Once plaques are developed and confirmed by naked eye, fix the cells with 10 % formaldehyde for 60 min at room temperature (*see* **Note 9**). After this step, the following procedures can be done outside the biosafety cabinet.
3. Rinse the plates with running tap water but avoid placing plates directly under running tap water as this may damage the cell monolayer. Remove excess water by gentle tapping and by using paper towels.

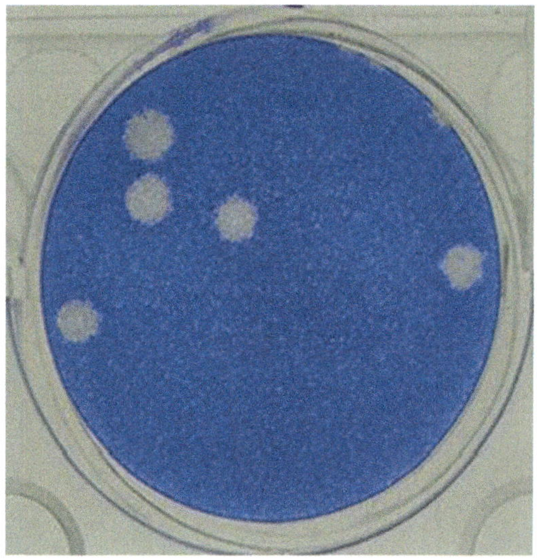

Fig. 3 Plaques of CHIKVS27 strain in Vero cell lines (photo courtesy of Dr. Chang-Kweng Lim)

4. Stain cells with 6× methylene blue solution for more than 1 h.
5. Repeat **step 3**.
6. Dry the plate by either placing the plate at room temperature (air-drying method) or in drying cabinet (incubation temperatures below 45 °C).
7. After drying the plates, count the plaques under a light box using a colony counter (Fig. 3).
8. Stained plates can be stored for long term at room temperature.

3.5 Determination of Plaque Reduction Neutralization Value

Calculate the number of plaque present in each well, at each serum dilution and control well. Neutralization titer is defined as the highest serum dilution with 50 % reduction in the number of plaques, 70 or 90 % as compared to the number of plaques in control monolayer wells in the absence of CHIKV antibody (Table 1). Higher percentage of reduction (>70 %) is speculated to possess higher specificity but the 50 % reduction is generally used to determine neutralizing titer.

3.6 Microneutralization Assay

Although plaque reduction neutralization test is the gold standard in measuring virus-neutralization and protective antibodies, the assay is time consuming and laborious. Recently, the development of a rapid micro-scale neutralization test for CHIKV or micro neutralization assay (Mnt assay) that utilizes the 96-well microplate format allows high sample throughput with shorter test duration. In Mnt assay, serial twofold dilutions are made in a 96-well microplate for samples to be incubated with challenge virus. Cells will

Table 1
An example of neutralization titer of a CHIKV patient serum sample

Sample	Plaques count	Average number of plaques
Virus standard	18, 29	23.5
1:10	7, 11	9
1:20	18, 9	13.5
1:40	14, 17	15.5
1:80	19, 21	20
1:160	13, 21	17.5

Neutralization titer is expressed as the maximum dilution of serum sample the yielded a >50 % plaque reduction in the virus inoculum as compared with standard virus sample. The CHIKV neutralizing antibody titer of the test sample (PRNT50) in the example given here is 1:10

then be added to serum–virus mixture in the microplate and then incubated for 1–4 days before fixing. Results can be determined by either by the PAP (Peroxidase-antiperoxidase) staining technique, cell viability or by observing CPE formation, depending of experiment parameters [2, 7, 8]. In an automated assay, a plate reader can be used to determine the optical density of the solubilized stain of live cells [7]. However, further studies are needed to validate these assays against the standard PRNT, to determine the utility and equivalence of each test to the PRNT.

4 Notes

1. Simian kidney cell lines (Vero and LLC-MK2) are suitable for virus plaquing because the cells are susceptible to CPE formation upon CHIKV infection [8, 9]. Vero cell is an adherent cell line derived from the kidney of African green monkey (*Cercopithecus aethiops*) and is widely used in virological studies. Mosquitoes cell lines (C6/36) are commonly used for CHIKV inoculation, isolation, and propagation.
2. Virus stocks are best used without repeated thawing and freezing cycles as this will decrease the virus titer and affect titration results. Ensure that the viruses used in the experiments have not been repeatedly thawed for more than three times.
3. Make sure that the solution is below 40 °C before placing the bottle on ice to prevent the glass bottle from breaking.
4. Use personal protective equipment (PPE) such as mask and gloves when diluting formaldehyde. Perform this step in a fume hood. Exposure to formaldehyde via inhalation or ingestion can be hazardous (US Environmental Protection Agency).

5. Use gloves when handling sodium hydroxide as the reagent is an irritant, and is corrosive to the skin, eyes, respiratory tract and gastrointestinal tract upon contact.
6. Extending the virus–antibody complex incubation period could result in partial virus inactivation [3]. To postpone the experiment, store the virus–antibody complex at 4 °C after the 1 h incubation period at 37 °C. The complexes can be stored up to 24 h. For optimum results, perform the assay immediately upon mixing.
7. Switch off the biosafety cabinet fan during infection assay to prevent the cells from drying up. Aspirate old medium in all the wells before adding the virus–antibody complexes into the wells. To ensure that the cells do not dry up during infection assay, start the experiment with only two 12-well plates at a time.
8. Confirm that plates are placed on a level rack in the incubator. Confirm that the medium is spread evenly across the cell monolayer. The plates could be incubated up to 90 min in this step. The 10 min incubation time of infection assay should preferably start immediately after the infection assay.
9. Check the plaque formation on the third day post-infection, and stop the experiment when the plaque can be observed by naked eye.

Acknowledgements

We would like to acknowledge Dr. Chang-Kweng Lim, Department of Virology I, National Institute of Infectious Diseases, Japan, for generously providing us with a figure on CHIKV plaques (Fig. 3). This work was supported in part by the research grant, Research on Emerging and Re-emerging Infectious Diseases (H26-shinkou-jitsuyouka-007), from the Ministry of Health, Labour and Welfare, Japan, the Environment Research and Technology Development Fund (S-8) of the Ministry of the Environment, and a Grant-in-Aid for Young Scientists (B) from JSPS (26870872).

References

1. Staples JE, Breiman RF, Powers AM (2009) Chikungunya fever: an epidemiological review of a re-emerging infectious disease. Clin Infect Dis 49(6):942–948
2. Mallilankaraman K, Shedlock DJ, Bao H, Kawalekar OU, Fagone P, Ramanathan AA, Ferraro B, Stabenow J, Vijayachari P, Sundaram SG, Muruganandam N, Sarangan G, Srikanth P, Khan AS, Lewis MG, Kim JJ, Sardesai NY, Muthumani K, Weiner DB (2011) A DNA vaccine against Chikungunya virus is protective in mice and induces neutralizing antibodies in mice and nonhuman primates. PLoS Negl Trop Dis 5(1):e928. doi:10.1371/journal.pntd.0000928
3. Roehrig J, Hombach J, Barrett A (2008) Guidelines for plaque-reduction neutralization testing of human antibodies to dengue viruses. Viral Immunol 21:123–132
4. Thomas SJ, Nisalak A, Anderson KB, Libraty DH, Kalayanarooj S, Vaughn DW, Putnak R,

Gibbons RV, Jarman R, Endy TP (2009) Dengue Plaque Reduction Neutralization Test (PRNT) in primary and secondary dengue virus infections: how alterations in assay conditions impact performance. Am J Trop Med Hyg 81(5):825–833. doi:10.4269/ajtmh.2009.08-0625

5. Dulbecco R, Vogt M, Strickland AGR (1956) A study of the basic aspects of neutralization of two animal viruses, Western equine encephalitis virus and poliomyelitis virus. Virology 2(2): 162–205
6. Russell PK, Nisalak A, Sukhavachana P, Vivona S (1967) A Plaque Reduction Test for dengue virus neutralizing antibodies. J Immunol 99(2): 285–290
7. Hobson-Peters J (2012) Approaches for the development of rapid serological assays for surveillance and diagnosis of infections caused by zoonotic flaviviruses of the Japanese encephalitis virus serocomplex. BioMed Res Int 2012:379738
8. Walker T, Jeffries CL, Mansfield KL, Johnson N (2014) Mosquito cell lines: history, isolation, availability and application to assess the threat of arboviral transmission in the United Kingdom. Parasit Vectors 7(1):382. doi:10.1186/1756-3305-7-382
9. Stim TB (1969) Arbovirus plaquing in two Simian kidney cell lines. J Gen Virol 5(3):329–338. doi:10.1099/0022-1317-5-3-329

Chapter 26

Reverse Genetics Approaches for Chikungunya Virus

Patchara Phuektes and Justin Jang Hann Chu

Abstract

Reverse genetic systems based on an infectious cDNA clone, a double-stranded copy of the viral genome carried on a plasmid vector, have greatly enhanced the understanding of RNA virus biology by facilitating genetic manipulation of viral RNA genomes. To date, infectious cDNA clones of Chikungunya virus (CHIKV) have been constructed using different combinations of plasmid vectors and/or bacterial host strains. Here, we describe our approaches for the construction of infectious cDNA clones of CHIKV and the protocol for genetic manipulation of the clones by site-directed mutagenesis.

Key words Reverse genetics, Chikungunya virus, Infectious cDNA clone, Site-directed mutagenesis

1 Introduction

Chikungunya virus (CHIKV) is a mosquito-borne alphavirus recently recognized as a major emerging epidemic-prone pathogen [1]. CHIKV is composed of a single-stranded, positive-sense RNA of approximately 12,000 nucleotides in length. The viral genome contains a 5′ cap and 3′ poly-A tail, and encodes two open reading frames (ORFs) flanked by a 5′ untranslated region (5′ UTR) and a 3′ untranslated region (3′ UTR; [2]). The viral RNA genome has an infectious nature in permissive host cells.

A reverse genetic system has been successfully developed to advance knowledge of various aspects of CHIKV including basic research into virus biology and applied research related to the development of diagnostics, therapeutics and vaccines [3–11]. This system is based on construction of an infectious cDNA clone, a double-stranded copy of the viral genome carried on a plasmid vector, and generation of a synthetic virus from the clone. Such a clone has enabled genetic manipulation of the CHIKV genome for investigation of the molecular mechanisms of CHIKV replication, and identification of the genetic determinants of virulence, pathogenesis, and transmission [3–11]. In general, infectious cDNA clones of CHIKV are constructed in plasmid vectors. A eukaryotic

Justin Jang Hann Chu and Swee Kim Ang (eds.), *Chikungunya Virus: Methods and Protocols*, Methods in Molecular Biology, vol. 1426, DOI 10.1007/978-1-4939-3618-2_26, © Springer Science+Business Media New York 2016

promoter sequence, such as the cytomegalovirus (CMV) immediate early promoter, or a promoter sequence recognized by a phage DNA-dependent RNA polymerase, such as bacteriophages T7 or SP6, is incorporated upstream of the viral genome for an in vivo and in vitro transcription of viral cDNA template, respectively. Depending on the incorporated promoter, clone-derived CHIKV is generated by transfecting either plasmid cDNA clone or RNA transcripts derived by in vitro transcription into permissive cells.

Viral cDNA instability arising by mutation, deletion, or rearrangement of the viral genome, resulting in a noninfectious clone, is the most common problem encountered in the development of reverse genetic systems for many RNA virus families [12–18]. This problem arises from the viral cDNA encoding products that are toxic to the bacterial host during propagation of the plasmid vector in bacterial host cells [18]. One of the approaches to overcome this problem is the use of medium or low-copy number plasmid vectors and appropriate bacterial host strains [15–17].

We have produced infectious cDNA clones of CHIKV clinical isolates, SGEHIDSD67Y2008 and SGEHICHD122508 using a low copy number plasmid vector, pSMART-LCKan and *E. coli* strain XL10-Gold as a bacterial host. These clones have been used for genetic manipulation by introduction of mutations into the viral genome in order to investigate genetic determinants of CHIKV replication and pathogenesis. In this chapter, we will describe technical details of our strategy for the construction of infectious cDNA clones of CHIKV and the protocol for modification of the viral genome by site-directed mutagenesis. The strategy and methods described in this chapter could be applicable for the generation of infectious cDNA clones and mutant viruses of other CHIKV strains as well as other RNA viruses.

The construction and modification of a CHIKV infectious cDNA clone is divided into the following steps (Fig. 1): (1) extraction of viral RNA and reverse transcription of the viral RNA genome into a single-stranded cDNA, (2) PCR amplification to generate two overlapping cDNA fragments spanning the entire viral genome using a single-stranded cDNA as a template, (3) cloning of the double-stranded cDNA fragments into a low-copy number plasmid vector followed by transformation into a bacterial host to multiply; two cDNA fragments are then assembled into a full-length cDNA clone, (4) in vitro transcription of the full-length cDNA clone into viral RNA, (5) transfection of the CHIKV RNA transcript into permissive cell lines to recover the clone-derived virus, (6) propagation of the recovered virus to increase viral titer for genetic and phenotypic characterization of the clone-derived virus compared to the parental virus, and (7) generation of mutant viruses using PCR-based mutagenesis and recombination-based cloning.

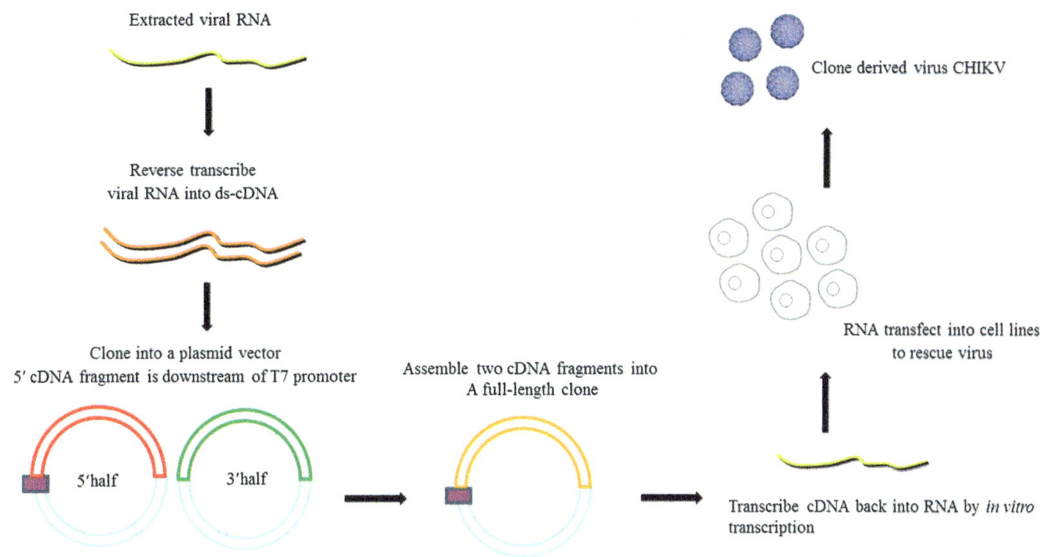

Fig. 1 Strategy for the construction of CHIKV full-length infectious cDNA clone. Viral RNA is extracted from virus infected cells and reverse-transcribed into cDNA. Two overlapping cDNA fragments are generated by PCR and cloned into a low-copy number plasmid vector. Full-length cDNA clone is constructed by joining the two cDNA fragments at a unique restriction site within the viral genome. A full length clone is linearized at the 3′ terminus of the viral genome prior to in vitro transcription into viral RNA. The CHIKV RNA transcripts are transfected into cells permissive for CHIKV replication to rescue clone derived-virus

2 Materials

2.1 Viral RNA Extraction

1. Cell culture supernatant containing CHIKV.
2. QIAamp® Viral RNA Mini kit (*see* **Note 1**).

2.2 Reverse Transcription

1. Reverse transcription: 200 U SuperScript™ III reverse transcriptase enzyme (*see* **Note 1**), 1× RT buffer, 0.5 mM dNTPs, 40 U ribonuclease Inhibitor (RI), 2 U ribonuclease H (RNase II), 10 mM DTT, 5 mM $MgCl_2$, and 30 pmol oligonucleotide primer.

2.3 PCR Amplification

1. PCR reaction: 2.5 U Q5® high-fidelity DNA polymerase (*see* **Note 1**), 1× PCR buffer, 0.2 mM dNTPs, 10 pmol each of oligonucleotide primer.

2.4 Agarose Gel Electrophoresis and Gel Extraction

1. 1 % TAE-agarose gel (100 ml): 1 g agarose, 100 ml 1× TAE buffer: 40 mM tris acetate, 1 mM EDTA.
2. 1 kb DNA ladder.
3. SYBR Green for DNA visualization.
4. Blue light transilluminator.
5. QIAquick Gel Extraction kit (*see* **Note 1**).

2.5 DNA Cloning

1. pSMART-LCKan plasmid in CloneSmart® Blunt Cloning Kit.
2. XL10-Gold ultracompetent *E. coli* cells.
3. Restriction endonuclease digestion: 1× buffer, 10 U of *Sac*I, *Not*I, or *Age*I.
4. DNA Ligation: 1 U T4 DNA ligase, 1× T4 DNA ligase buffer.
5. DNA phosphorylation: 10 U T4 polynucleotide kinase, 1× T4 polynucleotide kinase buffer, 20 μM dATP.
6. 2YT medium with kanamycin: 1.6 % tryptone (w/v), 0.5 % yeast extract (w/v), 85 mM NaCl, 30 μg/ml kanamycin.
7. 2YT agar with kanamycin: 2YT medium, 1.5 % agar, 30 μg/ml kanamycin.
8. SOC medium: 2 % tryptone (w/v), 1 % yeast extract (w/v), 8.5 mM NaCl, 20 mM glucose, 2.5 mM KCl, 10 mM $MgCl_2$.
9. Plasmid preparation kits: PureYield™ Plasmid Midiprep system (Promega), PureYield™ Plasmid Miniprep system (*see* **Note 1**).
10. QIAquick PCR purification kit (*see* **Note 1**).

2.6 RNA Transcription and Capping

1. Restriction endonuclease digestion: 1× buffer, 10 U *Not*I.
2. In vitro RNA transcription: mMESSAGE mMACHINE® T7 kit (*see* **Note 1**).
3. RNeasy Mini Kit (*see* **Note 1**).
4. 1 % TAE-agarose gel (*see* Subheading 2.4).
5. RiboRuler High Range RNA Ladder or 0.5–10 kb RNA Ladder.
6. 2× RNA loading buffer: 95 % formamide, 0.025 % SDS, 0.025 % bromophenol blue, 0.025 % xylene cyanol FF, 0.025 % ethidium bromide, 0.5 mM EDTA.

2.7 Recovery and Propagation of Clone Derived-CHIKV

1. Transfection reagents: Lipofectamine™ 2000 (*see* **Note 1**), Opti-MEM.
2. Cell lines: Baby hamster kidney cells (BHK-21), *Aedes albopictus* C6/36 cells.
3. Cell culture media: RPMI1640 or Leibovitz (L15) medium supplemented with 10 % FBS or 5 % FBS.
4. 24-well tissue culture trays.
5. T75 tissue culture flasks.
6. 1× PBS.

2.8 Site-Directed Mutagenesis

1. PCR reaction: CloneAmp HiFi PCR premix, 200 nM each of oligonucleotide primer.
2. DNA Cloning: In fusion HD cloning kit, XL10-Gold ultracompetent *E. coli* cells.

3 Methods

3.1 Viral RNA Extraction and Viral cDNA Synthesis

1. Propagate virus in cell culture until required, then harvest cell culture supernatant from CHIKV infected cells. Extract viral RNA using QIAamp® Viral RNA mini kit according to the manufacturer's instructions, and elute in 60 μl of AVE buffer (*see* **Note 2**).
2. Synthesize a first stranded cDNA covering the full-length genome of CHIKV using SuperScript™ III reverse transcriptase (*see* **Note 3**). Set up a reverse transcription reaction in a 0.2 ml tube in a reaction volume of 20 μl. Mix 8 μl of viral RNA and 3 μl (30 pmol) of the reverse primer specific to the 3′ end of CHIKV genome (Fig. 2; *see* **Note 4**) and heat for 5 min at 65 °C before chilling on ice for 5 min. Add 0.5 mM of dNTPs, 1× RT buffer, 10 mM DTT, 40 U of ribonuclease inhibitor, and 200 U of SuperScript™ III reverse transcriptase. Incubate the mixture at 42 °C for 1 h before inactivation of the enzyme by incubation at 70 °C for 15 min. Add 2 U of RNaseH and incubate the mixture at 37 °C for 20 min. Store the reverse transcription product at −20 °C until required.

3.2 PCR Amplification of Viral cDNA Fragments

1. Amplify two overlapping cDNA fragments of about 6 kb in length, the 5′ and 3′ half, by PCR using Q5® High-Fidelity DNA polymerase (*see* **Note 5**). For amplification of the 5′ half

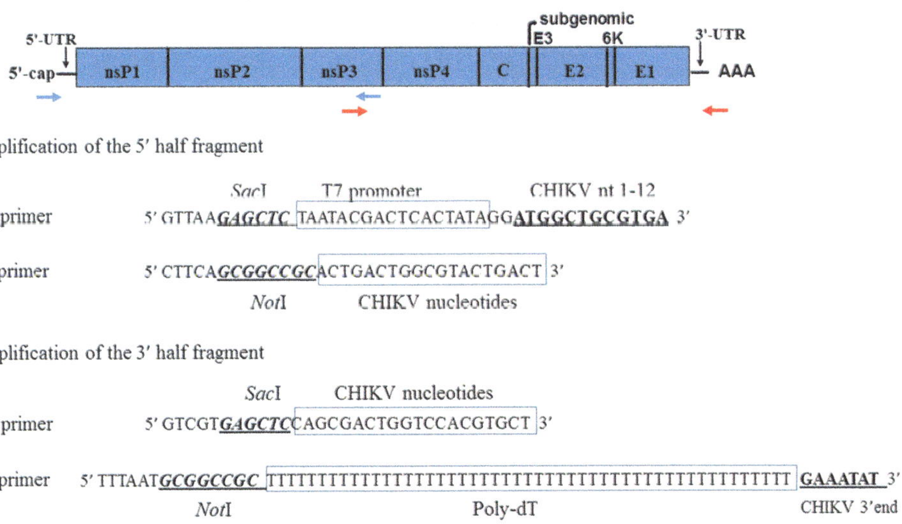

Fig. 2 Primer designs for the amplification of the 5′ half and 3′ half cDNA fragments. The genome structure of CHIKV is shown on the *top*. *Arrows* show position of primers within the genome. Forward primer for amplification of the 5′ half fragment contains a unique *Sac*I site, a T7 RNA polymerase promoter sequence and GG residues and 12 viral nucleotide sequence at the 5′ terminus of the viral genome. Reverse primer for amplification of the 3′ half fragment has seven viral nucleotide sequence at the 3′ terminus of the viral genome, a 48 poly A tail, and a unique *Not*I restriction site

cDNA fragment, a unique *Sac*I restriction site, a T7 RNA polymerase promoter sequence and GG residues are incorporated into the forward primer upstream of the 5′ terminus of the viral genome, and a unique *Not*I restriction site is incorporated into the reverse primer downstream of the viral sequence to facilitate cloning (Fig. 2; *see* **Note 6**). For amplification of the 3′ terminus cDNA fragment, a unique *Sac*I restriction site is incorporated into the forward primer upstream of the viral sequence, and a 48 poly A tail followed by a unique *Not*I restriction site are incorporated into the reverse primer downstream of the viral genome sequence (Fig. 2; *see* **Note 6**).

2. Perform the PCR reactions in 0.2 ml tube in a reaction volume of 50 μl following Q5® high-fidelity DNA polymerase's protocol. A typical PCR reaction contains 25 pmol each of forward and reverse primer, 0.2 mM dNTPs, 2.5 U of Q5 high-fidelity DNA polymerase, 2 μl of cDNA (from Subheading 3.1), and 1× PCR buffer. Perform the amplification with an initial denaturation at 98 °C 30 s, followed by 35 cycles of 98 °C 10 s, 70–72 °C 30 s and 72 °C 30 s/kb, then a final incubation at 72 °C for 2 min (*see* **Note 5**).
3. Perform agarose gel electrophoresis to confirm that the resulting PCR products are of the expected sizes. Run 7 μl of each PCR product through 1 % TAE-agarose gel at approximately 6 V/cm in 1× TAE buffer.
4. Purify the overlapping cDNA fragments for subsequent cloning by gel electrophoresis followed by gel extraction using a commercial kit. Mix the whole PCR products with dye loading buffer and SYBR Green before electrophoresis on 1 % TAE-agarose gel. Visualize DNA on a blue light transilluminator and excise the expected DNA band from the gel (*see* **Note 7**).

3.3 Cloning of Viral cDNAs into a Plasmid Vector

1. Generate phosphorylated ends of the purified cDNA fragments for subsequent blunt-end cloning using T4 polynucleotide kinase (*see* **Note 8**). Prepare a phosphorylation reaction in 0.2 ml tube in a reaction volume of 50 μl. Mix 250 μg of purified cDNA fragment, 1× T4 polynucleotide kinase buffer, 10 μl of 0.1 mM dATP, and 10 U of T4 polynucleotide kinase. Incubate the mixture at 37 °C for 30 min.
2. Purify the end-repaired cDNA fragments using QIAquick PCR purification kit to remove the kinase enzyme. Measure the concentration of cDNA fragments using a spectrophotometer.
3. Clone each purified blunt-end cDNA fragment into a pre-cut, blunt-end pSMART-LCKan vector using CloneSmart® Blunt Cloning Kit (Fig. 3). Prepare a ligation reaction in 0.2 ml tube in a reaction volume of 10 μl. Mix 500 ng of phosphorylated cDNA fragment with 2.5 μl of 4× CloneSmart® Vector Premix before adding 1 μl of ligase enzyme. Incubate the mixture at

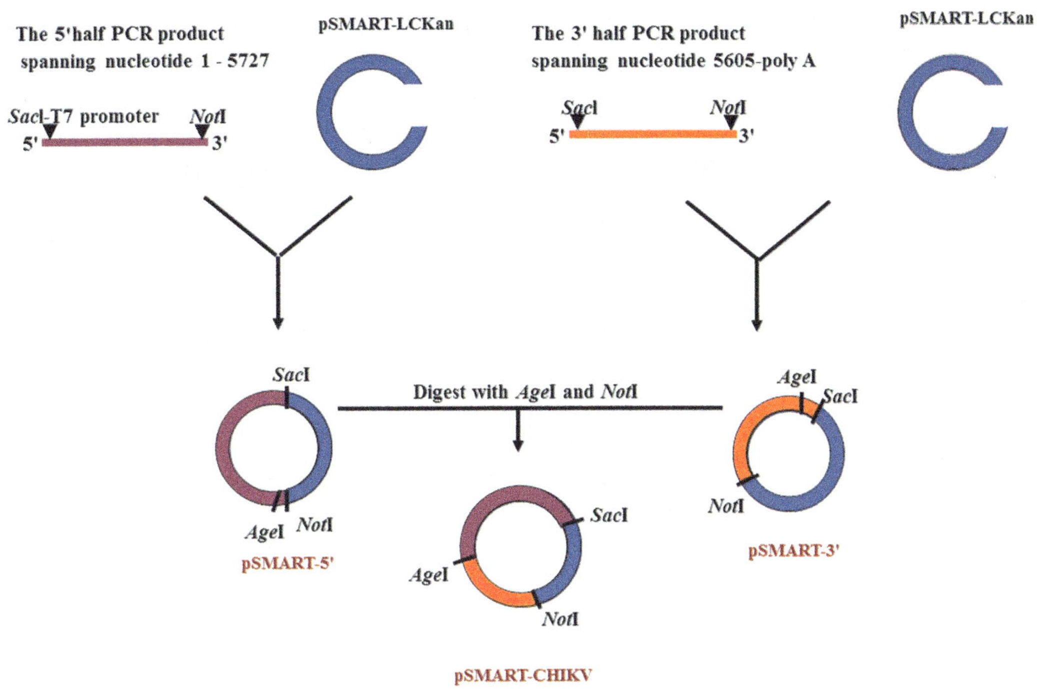

Fig. 3 Schematic diagram of the construction of 5′ and 3′ half clones using blunt-end pSMART-LCKan vector. The two half clones containing overlapping fragments are then assembled using unique restriction site, *NotI*, incorporated into the primers and a native *AgeI* site

21–25 °C for 2 h before heat-denaturation of the enzyme at 70 °C for 15 min.

4. Perform transformation by adding 2 μl of the DNA ligation mix into 100 μl of XL10-Gold cells, flick the tube gently to mix, then incubate the tube on ice for 30 min. Heat-pulse the tube in a 42 °C water bath for 30 s; incubate the tube on ice for 2 min before adding 1 ml of pre warmed SOC medium. Incubate the tube at 37 °C for 1 h with continuous shaking at 220 rpm. Plate the transformation mixture onto 2YT agar containing kanamycin 30 μg/ml and incubate the agar plate at 30 °C for 18–24 h (*see* **Note 9**).
5. Pick individual *E. coli* colonies and propagate in 2YT medium containing kanamycin 30 μg/ml at 30 °C for 18–24 h with continuous shaking at 220 rpm (*see* **Note 10**). Screen the clones by isolation of the plasmid DNA using the plasmid miniprep kit and characterization of plasmid DNA using restriction enzyme analysis. Sequence the cDNA clones containing the fragment of interest to ensure that no mutations have been introduced during the RT-PCR and cloning steps.
6. Assemble the two overlapping fragments into a full-length cDNA clone using a unique restriction site that is incorporated into the PCR primers and another site that is naturally existing

in the viral genome (Fig. 3). Set up the first digestion reaction using *Sac*I for both the 5′ half clone and the 3′ half clone. Incubate the reaction at 37 °C for 4 h, then run agarose gel electrophoresis to ensure that the digestion is complete. Purify the linearized plasmid DNA using QIAquick PCR purification kit before setting up the second digestion using *Age*I for both the 5′ half clone and the 3′ half clone (*see* **Note 11**). After confirmation of the complete digestion, separate the double-digested fragments by gel electrophoresis on 1 % TAE-agarose gel, excise the chosen fragments and gel purify using QIAquick gel extraction kit. Measure the quantity and quality of DNA by spectrophotometer.

7. Set up a ligation reaction in a 0.2 ml tube in a reaction volume of 10 μl. Mix 100 ng of vector DNA with an appropriate amount of insert, 1× T4 DNA ligase buffer, and 1 U of T4 DNA ligase. Generally, use insert-to-vector molar ratios of 1:1 or 3:1 (*see* **Note 12**). Perform the ligation reaction overnight at 4 °C. Use 2 μl of the DNA ligation mix for transformation as described above (**step 4**).
8. Carry out propagation of bacterial colonies and screening for the full-length cDNA clones as described above (**step 5**). Sequence the full-length cDNA clone to ensure that no mutations are introduced during the cloning steps.

3.4 CHIKV RNA Transcript Synthesis

1. Linearize 3–6 μg of the plasmid containing full-length cDNA of CHIKV with *Not*I. Analyze a small aliquot of the digestion reaction by gel electrophoresis to confirm that the plasmid is completely digested. Purify the linearized plasmid using QIAquick PCR purification kit and elute in nuclease-free water (*see* **Note 13**). Determine the quality and quantity of the purified plasmid by gel electrophoresis and spectrophotometer.
2. Transcribe the purified plasmid DNA into viral RNA in vitro using a T7 RNA transcription kit such as mMESSAGE mMACHINE® T7 kit. For mMESSAGE mMACHINE® T7 kit, set up the in vitro transcription reaction in a 1.5 ml tube in a reaction volume of 20 μl. The reaction mixture contains 1 μg of linearized plasmid DNA, 10 μl of 2× NTP/CAP, 2 μl of 10× reaction buffer, 2 μl of enzyme mix, and 1 μl of 30 mM GTP. Incubate the transcription reaction at 37 °C for 2 h. Add 1 μl of DNase to remove the template DNA followed by incubation at 37 °C for 15 min.
3. Purify the in vitro RNA transcripts using RNeasy Mini Kit following the protocol for RNA clean up and elute with 30 μl of RNase free water (*see* **Note 14**).
4. Analyze the quality and quantity of the purified RNA transcripts by running the RNA sample on a denaturing formaldehyde-agarose gel or non-denaturing agarose gel electrophoresis and a

spectrophotometer. We usually use non-denaturing 1 % TAE-agarose gel. Mix RNA ladder or RNA sample with RNA loading buffer; heat at 65 °C for 5 min before cooling on ice for about 2 min. Then load the RNA ladder or RNA sample into a well, run at 6 V/cm in 1× TAE buffer (*see* **Note 15**).

3.5 Recovery of Infectious Clone-Derived CHIKV by RNA Transfection

1. Seed BHK-21 cells into 24-well tissue culture trays at a seeding density of 8×10^4 cells per well, incubate for 18–24 h before transfection (*see* **Note 16**). Grow cells in RPMI supplemented with 10 % FBS to about 80 % confluency at the time of transfection.
2. Prior to transfection, replace growth media with 400 μl of fresh RPMI supplemented with 2 % FBS.
3. Transfect in vitro transcribed RNA into cells using Lipofectamine™ 2000 reagent. Dilute 3 μl of Lipofectamine in 47 μl of Opti-MEM and mix gently, then incubate for 5 min at room temperature. Dilute 1 μg of RNA in Opti-MEM to a final volume of 50 μl and mix gently. Combine diluted RNA with diluted Lipofectamine, mix gently and then incubate at room temperature for 20 min. Add 100 μl of RNA–Lipofectamine complex to each well, incubate for 2–3 days before harvesting virus (*see* **Note 17**).

3.6 Propagation and Characterization of Clone-Derived CHIKV

1. Propagate the clone-derived CHIKV virus by infecting 1 ml of the culture supernatant from the transfected cells into a sub-confluent T75 flask of C6/36 cells; incubate for 90 min, then wash infected cells with warm 1× PBS to remove unbound virus. Add 10 ml of L15 with 2 % FBS and incubate at 28 °C until cytopathic effect (CPE) is visible. If there is no visible CPE, harvest virus at about 4–6 days after infection depending on cell health (*see* **Note 18**).
2. Genetically characterize the clone-derived virus by RT-PCR and sequencing. Use 280 μl of virus suspension for viral RNA extraction. Perform RT-PCR as described above (Subheadings 3.1 and 3.2). Sequence purified PCR product by primer walking using Big-dye-terminator technology. Compare the sequence of cloned-derived virus to the parental viral.
3. Phenotypically characterize the clone-derived virus by comparing plaque morphology and growth kinetics to the parental virus following previously described protocols [19].

3.7 Site-Directed Mutagenesis to Introduce Mutations into the Full-Length cDNA Clone

1. Construct a sub-genomic clone containing the genomic region where mutation (s) will be introduced using unique restriction sites surrounding this region. Restriction enzyme digestion and cloning procedures are as described above (Subheading 3.3; Fig. 4).
2. Introduce mutations into sub-genomic clones by PCR based mutagenesis using In-Fusion HD cloning system following the

manufacturer's protocol. Briefly, design forward and reverse primers to have 15 bp overlaps with each other at their 5′ ends and to contain specific mutation(s) within the overlapping region (Fig. 4). Use CloneAmp HiFi PCR master mix for PCR amplification. Set up the reaction in a 0.2 ml tube in a final volume of 25 μl. Mix 12.5 μl of CloneAmp HiFi PCR premix, 200 nM of forward primer and reverse primer, and 100 ng of template DNA. Perform PCR for 35 cycles of 98 °C 10 s, 55 °C 5 s and 72 °C 5 s/kb. Analyze PCR product by gel electrophoresis and extract the expected PCR product from gel using gel extraction kit (Subheading 3.2, **step 4**). Set up the In-fusion reaction in a 0.2 ml tube in a final volume of 10 μl. Mix 100 ng of PCR product containing mutation(s) with 2 μl of In-fusion enzyme and incubate the mixture at 50 °C for 15 min. Transform 2 μl of purified product into XL10-gold as described above (Subheading 3.3, **step 4**).

3. Propagate and screen the plasmid clones as described above (Subheading 3.3, **step 5**). Sequence clones to test for the presence of a specific mutation(s).
4. Exchange a sub-genomic fragment containing mutation(s) with a corresponding region in the full-length cDNA clone using unique restriction sites and standard cloning procedure as previously described (Subheading 3.3; Fig. 4).

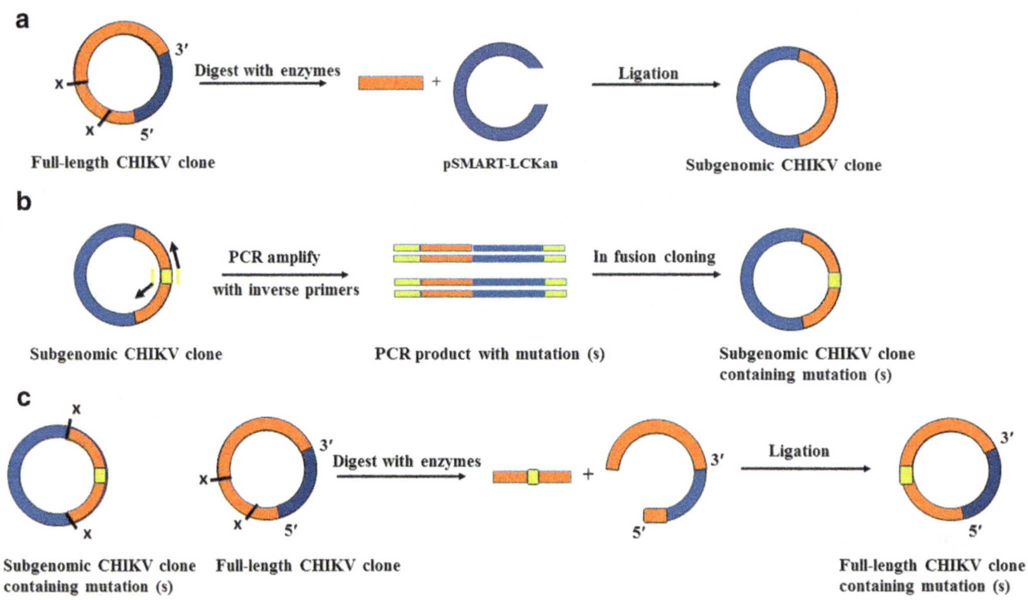

Fig. 4 Site-directed mutagenesis to introduce mutation(s) into the full-length cDNA clone. (**a**) Construction of a subgenomic cloning of CHIKV using two unique restriction sites, labeled X. (**b**) PCR amplification using inverse primers containing specific mutation(s) within the 15 bp overlapping region, indicated in *yellow*. (**c**) A sub-genomic fragment containing mutation(s) is cloned back into the full-length clone using two unique restriction sites, labeled X

4 Notes

1. Although we use the specified kits and reagents for our work, other commercial kits and reagents may also be suitable.
2. The viral titer of cell culture supernatant should be at least 10^6 pfu/ml. For QIAamp® Viral RNA mini kit, 280 μl of virus suspension is used for each RNA extraction in a mini column.
3. Although we have had good yields by synthesizing a single-stranded cDNA of the entire CHIKV genome using SuperScript™ III reverse transcriptase, other reverse transcriptase enzymes that can synthesize up to 12 kb of cDNA from RNA targets are also suitable. Alternatively, synthesize two half-length cDNA fragments with reverse primers that will be used for subsequent PCR amplification of 5′ and 3′ half fragments.
4. With the viral titer of $\geq 10^6$ pfu/ml and the described extraction protocol, we add the maximum volume of 8 μl (about 1– 2 μg) of the extracted viral RNA in the reverse-transcription reaction and obtained good results. Primer used for the reaction contains about 7 nucleotide sequences specific to the 3′ terminus of viral genome, followed by 48 poly T. If the 3′ terminus of the viral genome is not known, oligo-dT primers can be used.
5. Other high-fidelity DNA polymerases that can produce long amplicons can be used. If using Q5® High-Fidelity DNA polymerase, the online tool (NEB Tm Calculator) to determine the optimal annealing temperature for the primers used is recommended.
6. The GG residues are incorporated into the forward primer to improve transcription efficiency, without altering the infectivity of clone-derived virus. Since the designed forward primer for amplification of the 5′ fragment and the reverse primer for amplification of the 3′ fragment contain 12–7 nucleotides specific to the 5′ and 3′ terminus of the viral genome, respectively, the viral sequences at both termini should be confirmed by rapid amplification of cDNA ends using 5′ and 3′-RACE PCR reactions.
7. To avoid damage to DNA by UV light and enhance cloning efficiency, agarose gel is exposed only to visible light while excising the expected DNA band.
8. Q5® High-Fidelity DNA polymerase generates blunt ended PCR products. The purpose of phosphorylating the PCR product is so that ligation reaction can occur, since a pre-cut, blunt-end plasmid vector, pSMART-LCKan is in a de-phosphorylated form and that PCR-generated DNA is not phosphorylated.
9. To minimize the chance of viral cDNA instability in bacterial host, we grow the transformed bacteria at low temperature.

This approach is reported in the construction of infectious human enterovirus 71 [15] and dengue virus type 2 [18].

10. For small scale preparation of the low-copy number pSMART-LCKan, we grow bacteria in 3–5 ml of 2YT medium containing 30 μg/ml kanamycin. Use at least 3 ml of overnight culture for plasmid preparation to ensure that the amount of purified plasmid is enough for further analysis. For medium scale preparation of the plasmid, we grow bacteria in 100 ml of 2YT medium containing 30 μg/ml kanamycin and use all the overnight culture for plasmid purification. Growing a bacterial clone from glycerol stock culture may result in a low yield of plasmid DNA. Retransformation of plasmid DNA and growing bacterial culture from freshly grown colony are recommended to get a higher yield.
11. For sequential digestion, we set up a first digestion reaction using about 3–6 μg of DNA to ensure that the concentration of plasmid DNA is high enough for setting up a ligation reaction after a second digestion. Double-digestion is usually performed if the two restriction enzymes require the same reaction buffers and incubation conditions.
12. Perform a control ligation containing only plasmid vector in order to estimate background colonies.
13. The linearized plasmid can also be purified using phenol–chloroform extraction. After the digestion reaction, incubate the linearized plasmid at 50 °C with proteinase K (200 μg/ml) and SDS (0.5 %) for 30 min followed by purification using phenol–chloroform extraction and ethanol precipitation.
14. The RNA transcripts can be purified using 2.5 M lithium chloride according to mMESSAGE mMACHINE® T7 kit's protocol. We obtain similar results using both methods.
15. If the in vitro transcription reaction works well, the RNA transcript should be predominantly full-length. We usually dilute the RNA transcript at 1:10 and 1:100 with nuclease-free water before running a gel.
16. BHK-21 cells are used for transfection of viral RNA because the cells are permissive for CHIKV replication and that they have a high transfection efficiency.
17. Transfected cells should be observed daily. If high cellular toxicity occurs, lower the amount of RNA transcript to 0.5 μg or lower Lipofectamine to 2 μl.
18. Determine virus titer by plaque assay. Further passage virus in C6/36 cells for one more time to increase viral titer if required. We use clone-derived virus from the first or second passage for characterization. It is necessary to perform phenotypic characterization of clone-derived virus and compare to the parental virus. This is to ensure that the nonviral nucleotides added at the 5′ and 3′ terminus do not affect virus replication.

References

1. Nasci RS (2014) Movement of chikungunya virus into the Western hemisphere. Emerg Infect Dis 20:1394–1395
2. Weaver SC, Winegar R, Manger ID, Forrester NL (2012) Alphaviruses: population genetics and determinants of emergence. Antiviral Res 94:242–257
3. Delang L, Segura Guerrero N, Tas A et al (2014) Mutations in the chikungunya virus non-structural proteins cause resistance to favipiravir (T-705), a broad-spectrum antiviral. J Antimicrob Chemother 69:2770–2784
4. Tretyakova I, Hearn J, Wang E et al (2014) DNA vaccine initiates replication of live attenuated Chikungunya virus in vitro and elicits protective immune response in mice. J Infect Dis 209:1882–1890
5. Scholte FE, Tas A, Martina BE et al (2013) Characterization of synthetic Chikungunya viruses based on the consensus sequence of recent E1-226V isolates. PLoS One 8(8):e71047
6. Delogu I, Pastorino B, Baronti C et al (2011) In vitro antiviral activity of arbidol against Chikungunya virus and characteristics of a selected resistant mutant. Antiviral Res 90:99–107
7. Kümmerer BM, Grywna K, Gläsker S et al (2012) Construction of an infectious Chikungunya virus cDNA clone and stable insertion of mCherry reporter genes at two different sites. J Gen Virol 93:1991–1995
8. Gorchakov R, Wang E, Leal G et al (2012) Attenuation of Chikungunya virus vaccine strain 181/clone 25 is determined by two amino acid substitutions in the E2 envelope glycoprotein. J Virol 86:6084–6096
9. Tsetsarkin KA, McGee CE, Volk SM et al (2009) Epistatic roles of E2 glycoprotein mutations in adaption of chikungunya virus to Aedes albopictus and Ae. aegypti mosquitoes. PLoS One 4:e6835
10. Tsetsarkin K, Higgs S, McGee CE et al (2006) Infectious clones of Chikungunya virus (La Réunion isolate) for vector competence studies. Vector Borne Zoonotic Dis 6:325–337
11. Vanlandingham DL, Tsetsarkin K, Hong C et al (2005) Development and characterization of a double subgenomic chikungunya virus infectious clone to express heterologous genes in Aedes aegypti mosquitoes. Insect Biochem Mol Biol 35:1162–1170
12. Masters PS (1999) Reverse genetics of the largest RNA viruses. Adv Virus Res 53:245–264
13. Mishin VP, Cominelli F, Yamshchikov VF (2001) A 'minimal' approach in design of flavivirus infectious DNA. Virus Res 81:113–123
14. Sumiyoshi H, Hoke CH, Trent DW (1992) Infectious Japanese encephalitis virus RNA can be synthesized from in vitro-ligated cDNA templates. J Virol 66:5425–5431
15. Phuektes P (2009) Development of a reverse genetic system for human enterovirus 71 (HEV71) and the molecular basis of its growth phenotype and adaptation to mice. PhD thesis, Murdoch University. http://researchrepository.murdoch.edu.au/view/author/Phuektes,Patchara.html
16. Hurrelbrink RJ, Nestorowicz A, McMinn PC (1999) Characterization of infectious Murray Valley encephalitis virus derived from a stably cloned genome-length cDNA. J Gen Virol 80:3115–3125
17. Lai CJ, Zhao BT, Hori H, Bray M (1991) Infectious RNA transcribed from stably cloned full-length cDNA of dengue type 4 virus. Proc Natl Acad Sci U S A 88:5139–5143
18. Sriburi R, Keelapang P, Duangchinda T et al (2001) Construction of infectious dengue 2 virus cDNA clones using high copy number plasmid. J Virol Methods 92:71–82
19. Chen KC, Kam YW, Lin RT, Ng MM, Ng LF, Chu JJ (2013) Comparative analysis of the genome sequences and replication profiles of chikungunya virus isolates within the East, Central and South African (ECSA) lineage. Virol J 10:169

Chapter 27

Production of Chikungunya Virus-Like Particles and Subunit Vaccines in Insect Cells

Stefan W. Metz and Gorben P. Pijlman

Abstract

Chikungunya virus is a reemerging human pathogen that causes debilitating arthritic disease in humans. Like dengue and Zika virus, CHIKV is transmitted by *Aedes* mosquitoes in an epidemic urban cycle, and is now rapidly spreading through the Americas since its introduction in the Caribbean in late 2013. There are no licensed vaccines or antiviral drugs available, and only a few vaccine candidates have passed Phase I human clinical trials. Using recombinant baculovirus expression technology, we have generated CHIKV glycoprotein subunit and virus-like particle (VLP) vaccines that are amenable to large scale production in insect cells. These vaccines, in particular the VLPs, have shown high immunogenicity and protection against CHIKV infection in different animal models of CHIKV-induced disease. Here, we describe the production, purification, and characterization of these potent CHIKV vaccine candidates.

Key words Chikungunya virus, Baculovirus, Insect cells, Secreted E1 and E2, Virus-like particles, Production and purification

1 Introduction

Chikungunya virus (CHIKV) is an arthropod-borne (arbo)virus, which is transmitted by *Aedes* mosquito species. After its re-emergence in 2004 in Kenya, CHIKV has caused large-scale epidemics in Africa, Indian Ocean, and (South-East) Asia [1]. More recently, CHIKV has found its way to Europe (Italy 2007 [2]; France 2010 [3]) and emerged in the Caribbean in late 2013 [4]. The virus is expected to continue spreading to Central and South America, where it will then co-circulate with the most important arbovirus, dengue [5]. CHIKV causes significant morbidity such as the sudden onset of fever, myalgia, rash and in some cases severe, chronic arthralgia [6, 7]. CHIKV (genus *Alphavirus*; family *Togaviridae*) has a positive sense, single-stranded RNA genome of ~11 kb long and encodes two open reading frames (ORFs), the non-structural ORF and the structural ORF. The viral genome is encapsidated in a nucleocapsid which is tightly surrounded by a host-derived lipid

Justin Jang Hann Chu and Swee Kim Ang (eds.), *Chikungunya Virus: Methods and Protocols*, Methods in Molecular Biology, vol. 1426,DOI 10.1007/978-1-4939-3618-2_27, © Springer Science+Business Media New York 2016

envelope displaying trimeric E1/E2 glycoprotein spikes that mediate cell binding and entry [8].

The global threat of CHIKV is affecting millions and the unprecedented rapid spread of the virus demands effective countermeasures, including an efficacious vaccine. In the past decade, many studies have focused on generating CHIKV vaccine candidates, using a large variety of vaccine platforms, e.g., live-attenuated viruses, DNA vaccines, chimeric vector vaccines, glycoprotein subunits, and virus-like particle (VLP) vaccines [9–16]. Different expression platforms, i.e., mammalian cell transfection and baculovirus expression in insect cells, have been used for the production of CHIKV-E1 and E2 subunits and VLPs [14, 15]. The baculovirus insect cell expression technology is a well-established eukaryotic protein production platform with applications in veterinary and human vaccinology [17] and is well suited to produce essentially authentic arbovirus proteins [18, 19]. Indeed, the system has been used for the expression of secreted (s)E1 and secreted (s)E2 glycoprotein subunits [20] and CHIKV-VLPs [15]. Both modalities were immunogenic in animal models and conferred protection against CHIKV-induced disease with VLPs being better immunogens than subunits [15, 16]. In this chapter, we describe the production, purification, and analysis of the highly immunogenic CHIKV-VLPs [15, 16] and the glycoprotein subunits sE1 and sE1 [20], which were expressed in insect cells using recombinant baculovirus vectors. The generation of vector constructs and recombinant baculoviruses expressing the CHIKV-subunits and VLPs is explained in detail. Next, we detail the procedures of the production and purification of the subunits and VLPs, which are based on affinity chromatography and sucrose density gradient purification, respectively. Finally, a step-by-step description of subunit and VLP characterization based on glycosylation, furin-cleavage, and morphology is presented.

2 Materials

2.1 Cell Culture

1. *Spodoptera frugiperda* (*Sf*)21 insect cells (Invitrogen).
2. *Sf*9-easy titration (ET) insect cells [21].
3. Grace's insect-cell medium (Invitrogen) supplemented with 10 % fetal bovine serum (FBS).
4. *Sf*900II insect-cell medium (Invitrogen) supplemented with 5 % FBS and 200 μg/ml geneticin.
5. 75 cm^2 culture flasks.
6. 6-well culture plates.
7. Cell scrapers.

2.2 Generation of Recombinant Baculovirus Ac-sE1, Ac-sE2, and Ac-S27

1. Synthetically generated CHIKV S27 structural polyprotein DNA (GeneArt®).
2. Phusion High Fidelity DNA polymerase.
3. 10 mM dNTPs mix.
4. Forward primers (*see* **Note 1**):
   - sE1-Fw (ggggacaagtttgtacaaaaaagcaggcttaggatccaccatggc cacataccaagaggctgc).
   - sE2-Fw (ggggacaagtttgtacaaaaaagcaggcttaggatccaccatgagt cttgccatcccagttatg).
5. Reverse primers (*see* **Note 1**):
   - sE1-Rv (ggggaccactttgtacaagaaagctgggtaaagcttctaatgatg atgatgatgatgcatccatgacatcgccgtagcgg).
   - sE2-Rv (ggggaccactttgtacaagaaagctgggtaaagcttctaatgat gatgatgatgatgctgcagcagctataataatacagaa).
6. MilliQ water.
7. Silica Bead DNA Gel Extraction Kit (Thermo Scientific).
8. pDONR207 donor plasmid (Invitrogen).
9. pDEST8 expression plasmid (Invitrogen).
10. BP Clonase™ II enzyme mix (Invitrogen).
11. LR Clonase™ II enzyme mix (Invitrogen).
12. Proteinase K solution (100 μg/ml).
13. Electrocompetent DH5α-*E. coli* bacteria.
14. Luria broth (LB) liquid culture medium.
15. LB-agar plates containing 7 μg/ml gentamycin.
16. LB-agar plates containing 7 μg/ml gentamycin and 100 μg/ml ampicillin.
17. Tris–EDTA (TE) buffer: 10 mM EDTA, 25 mM Tris–HCl pH 8.0.
18. DH10Bac electrocompetent *E. coli* bacteria (*see* **Note 2**).
19. LB-agar plates containing 7 μg/ml gentamycin, 50 μg/ml kanamycin, 10 μg/ml tetracycline, 100 μg/ml X-gal, and 40 μg/ml IPTG.
20. Primers for analyzing recombined bacmids:
    - M13-Fw (taaagcacggccag).
    - M13-Rv (caggaaacagatatgac).
    - Genta-Rv (agccacctactcccaacatc).
21. *Sf*21 insect-cells.
22. Serum-free Grace's insect-cell medium.

23. Grace's insect-cell medium supplemented with 10 % FBS and 50 μg/ml gentamycin.
24. 6-well cell culture plate.
25. FectoFly™ I (Polyplus transfection).
26. Rocking plateau.
27. 75 cm^2 culture flasks.
28. Cell scrapers.
29. *Sf*9-ET insect-cells.
30. *Sf*900II insect-cell medium supplemented with 5 % FBS and 200 μg/ml geneticin (G418).
31. 60-well microtiter plate.
32. Fluorescent microscope.

2.3 Production of CHIKV-sE1 and -sE2 Subunits and VLPs

1. *Sf*21 insect-cells.
2. Serum-free Grace's insect-cell medium.
3. Phosphate buffered saline (PBS).
4. Rocking plateau.

2.4 Purification of sE1 and sE2 Subunits

1. Talon® spin columns 0.5 ml (Clontech).
2. Talon washing buffer (TWB): 20 mM Tris–HCl, 100 mM NaCl, pH 7.9.
3. Talon elution buffer (TEB): 20 mM Tris–HCl, 100 mM NaCl, 300 mM imidazole, pH 7.9.
4. Roller plateau.

2.5 Purification of CHIKV-VLPs

1. 7 % (w/v) polyethylene glycol (PEG)-6000.
2. 0.5 M NaCl.
3. Glycine-tris-sodium chloride-EDTA (GTNE) buffer: 200 mM Glycine, 50 mM Tris–HCl, 100 mM NaCl, 1 mM EDTA, pH 7.3.
4. Discontinuous 70 % (w/v), 40 % (w/v) sucrose in GTNE gradient at 4 °C.
5. SW55 soft ultracentrifuge tubes 5 ml (Beckman).
6. SW55 rotor (Beckman).
7. Pasteur pipettes.
8. Roller plateau.

2.6 CHIKV Protein Analysis

1. Glycoprotein denaturing buffer (New England Biolabs).
2. G7 reaction buffer (New England Biolabs).
3. 10 % NP40 buffer (New England Biolabs).
4. 500 units/μl PNGase F (New England Biolabs).

5. MilliQ water.
6. SDS-PAGE and Western blot equipment.
7. Rabbit-polyclonal anti-E1 and anti-E2 [20].
8. *Sf*21 insect-cells.
9. Grace's insect-cell medium (Invitrogen) serum free.
10. 50 μM furin inhibitor I (Calbiochem).
11. 6-well culture plate.
12. PBS.
13. Rocking plateau.
14. PBS–Tween 0.05 %.
15. *Sf*9-ET insect-cells.
16. *Sf*900II insect-cell medium supplemented with 5 % FBS and 200 μg/ml geneticin.
17. *Sf*900II insect-cell medium supplemented with 0.2 mg/ml cholesterol.
18. HCl-acidified *Sf*900II medium with pH 5.8, 5.5, and 5.0.
19. Fluorescent microscope.
20. Copper 400 square mesh grids (Veco).
21. Argon gas discharger.
22. Filter paper.
23. 2 % uranyl acetate.
24. Transmission electron microscope.

3 Methods

3.1 Cell Culture

1. Maintain adherent *Sf*21-cells as monolayer cell cultures in a closed 75 cm^2 culture flask at 27 °C (without additional CO_2) in Grace's insect cell medium, supplemented with 10 % FBS. Passage cells when the culture reaches ~80 % confluency, by dislodging them from the bottom of the flask using a Pasteur pipette and resuspending them in the culture medium by pipetting up and down. Split the cells 1:5 in fresh supplemented medium for culture maintenance. Leave the cells for 1 h at 27 °C for attachment to the bottom of the culture flask or plate.
2. Maintain *Sf*9-easy titration (ET) cells as a monolayer cell culture in *Sf*900II insect cell medium supplemented with 5 % FBS and 200 μg/ml geneticin at 27 °C. Passage cells when the culture reaches ~80 % confluency, by scraping them from the flask bottom using a cell scraper and resuspending them in the culture medium by pipetting up and down. Split the cells 1:5 in fresh supplemented medium for culture maintenance.

3.2 Generation of Recombinant Baculovirus *Ac*-sE1, *Ac*-sE2, and *Ac*-S27

Generation of recombinant baculovirus expressing the sE1 and sE2 subunits and the complete CHIKV S27 structural cassette is based on the Bac-to-Bac baculovirus expression system, using an adapted *Autographa californica* nucleopolyhedrovirus (*Ac*MNPV) backbone [22]. The synthetically generated S27 structural polyprotein cloning fragment (Genbank accession # AF369024) containing AttB1/2 recombination sites that enable Gateway ® cloning will be used as a template for generation of the sE1 and sE2 cloning fragments.

1. To PCR amplify the sE1 and sE2 coding fragments, set up a 20 μl PCR mix containing:
   - 20 ng synthetic S27 DNA.
   - 4 μl 5× Phusion HF buffer.
   - 0.25 μl 10 mM dNTPs.
   - 0.5 μl 10 mM Fw-primer (*see* **Note 1**).
   - 0.5 μl 10 mM Rv-primer (*see* **Note 1**).
   - 0.5 μl Phusion DNA polymerase.
   - Fill to 20 μl with MilliQ water.
2. To amplify the target sequences, start with initial denaturation at 95 °C for 3 min, followed by 35 cycles of denaturation at 95 °C for 30 s, annealing at 60 °C for 30 s, elongation at 72 °C for 3 min. Finish with a final elongation at 72 °C of 7 min. Purify the amplicons from agarose gel by using Silica Bead DNA Gel Extraction Kit following the manufacturer's protocol.
3. Add 150 ng of synthetic S27 DNA or sE1 and sE2 amplicons to 150 ng of pDONR207 donor plasmid in 8 μl TE-buffer.
4. Add 2 μl BP Clonase™ II enzyme mix and incubate for 1 h at room temperature (RT).
5. Stop the reaction by adding 1 μl Proteinase K and incubate for 10 min at 37 °C.
6. Transform 1 μl of the BP-reaction mix into electrocompetent DH5α *E. coli* bacteria. After transformation, recover the bacteria for 1 h in 1 ml LB-medium.
7. Plate 100 μl of transformed bacteria on LB-agar plates supplemented with 7 μg/ml gentamycin and incubate overnight at 37 °C.
8. Verify correctly recombined pDONR-sE1, pDONR-sE2, and pDONR-S27 colonies by sequencing.
9. Add 150 ng of pDONR-sE1, pDONR-sE2, or pDONR-S27 to 150 ng of the pFastBacI (pFB) analog pDEST8 destination vector in 8 μl TE-buffer.
10. Add 2 μl LR Clonase™ II enzyme mix and incubate for 1 h at RT.

11. Stop the reaction by adding 1 μl Proteinase K and incubate for 10 min at 37 °C.
12. Transform 1 μl of the LR-reaction mix into electrocompetent DH5α *E. coli* bacteria. After transformation, recover the bacteria for 1 h in 1 ml LB-medium.
13. Plate 100 μl of transformed bacteria on LB-agar plates supplemented with 7 μg/ml gentamycin and 100 μg/ml ampicillin and incubate overnight at 37 °C.
14. Verify correctly recombined pFB-sE1, pFB-sE2, and pFB-S27 by PCR using any available internal primers and restriction/digestion analysis.
15. Transform 1 μl of pFB-sE1, pFB-sE2, and pFB-S27 into electrocompetent DH10Bac *E. coli* bacteria and recover for 4 h in 1 ml LB medium at 37 °C.
16. Plate 100 μl of 10^{-1} and 10^{-2} dilutions of the transformed DH10Bac cells on LB-agar plates supplemented with 7 μg/ml gentamycin, 50 μg/ml kanamycin, 10 μg/ml tetracycline, 100 μg/ml X-gal, and 40 μg/ml IPTG. Incubate overnight at 37 °C.
17. White colonies indicate correct recombination of sE1, sE2 or the S27 structural cassette into the bacmid. Analyze correctly recombined bacmids by PCR using GentaFw, M13Fw, and M13Rv primers (*see* **Notes 3** and **4**).
18. Store the bacmids at 4 °C.
19. Seed 8×10^5 *Sf*21-cells per well in a 6-well plate in 2 ml of Grace's medium without antibiotics and let cells attach to the bottom of the well (*see* **Notes 5** and **6**).
20. Add 10 μl bacmid DNA to 90 μl serum-free Grace's medium.
21. Add 5 μl FectoFly I to 95 μl serum-free Grace's medium.
22. Mix the bacmid DNA and the FectoFly solution and incubate for 30 min at RT.
23. After incubation, add the 200 μl FectoFly I-DNA mix dropwise to the cells and incubate for 4 h at 27 °C on a rocking plateau.
24. Replace the medium with Grace's medium supplemented with 10 % FBS and 50 μg/ml gentamycin and incubate for ~4–5 days until clear baculovirus cytopathic effect (CPE) is visible (*see* **Notes 7** and **8**).
25. Harvest cells by pipetting up and down and separate the supernatant from the cell fraction by centrifugation of 5 min at $1500 \times g$.
26. Seed 8×10^6 *Sf*21-cells in a 75 cm^2 culture flask in 12 ml of Grace's medium without antibiotics and let cells attach to the bottom of the flask.

27. Remove the medium from the cells and add the 2 ml of *Ac*-sE1, *Ac*-sE2, or *Ac*-S27 containing supernatant retrieved after bacmid transfection.
28. Add an additional 2 ml of SFM to the cells and incubate for 4 h on a rocking plateau at 27 °C.
29. Remove medium from the cells and add 12 ml of Grace's medium supplemented with 10 % FBS and 50 μg/ml gentamycin and incubate at 27 °C for ~2–3 days until clear baculovirus CPE is visible.
30. Harvest the cells by scraping them from the bottom of the flask and resuspend them in the supernatant. Spin down the supernatant for 5 min at 1500 × *g* and store the medium at 4 °C.
31. Prepare a 90 μl dilution series of 10^{-1} to 10^{-9} of the baculovirus (BV) suspension in 1.5 ml Eppendorf tubes (*see* **Note 9**).
32. Dilute *Sf*9-ET cells to a final concentration of 1.5×10^6 cells/ml in Sf900II medium supplemented with 5 % FBS and 200 μg/ml geneticin.
33. Add 90 μl of the *Sf*9-ET cell suspension to each virus dilution and mix well.
34. Add 10 μl of cell/virus suspension to each well of a 60-well microtiter plate. Fill six wells per dilution and start with six wells of uninfected *Sf*9-ET cells at the bottom of the plate. Fill the remaining rows by the different virus dilution, starting with the lowest dilution.
35. Incubate cells for ~5 days at 27 °C.
36. Observe infected wells using an inverted fluorescence microscope.
37. Read out by using GFP expression as a sign of infection.
38. Accumulate the number of infected and uninfected wells per dilution, starting with the lowest concentration for infected wells and with the highest concentration for uninfected cells.
39. Calculate the percentage of accumulated infected wells (AIW) per dilution and calculate the virus titer using the following formula (*see* **Note 10**):

    $TCID_{50}/ml = 10^{(a+x)} \times 200/ml$.

    $a = -(\log n)$.

    n = the highest dilution of which the percentage of AIW is higher than 50 %.

    b = percentage of AIW of dilution n.

    c = percentage of AIW of a ten times dilution of dilution n.

    x = relative percentage of AIW = $(b - 50)/(b - c)$.

3.3 Production of CHIKV-sE1 and -sE2 subunits and VLPs

1. Seed 8×10^6 *Sf*21-cells in a 75 cm^2 culture flask in 12 ml of Grace's medium without antibiotics and let cells attach to the bottom of the flask.
2. Infect the cells with a multiplicity of infection (MOI) = 10. Based on the titer of *Ac*-sE1, *Ac*-sE2, and *Ac*-S27, calculate the correct amount of virus needed and add this to 4 ml of serum-free Grace's medium.
3. Replace the medium by the virus solution and incubate for 4 h at 27 °C on a rocking plateau.
4. Replace the medium for 10 ml SFM and incubate for ~72 h until clear baculovirus CPE is visible.
5. Harvest the subunit or VLP-containing medium fraction by separating it from the cell fraction by centrifugation. Wash the cell fraction with PBS and finally store it at −20 °C for subsequent protein expression analysis.

3.4 Purification of sE1 and sE2 Subunits

The CHIKV-sE1 and -sE2 subunits contain a C-terminal poly-histidine tail to enable efficient purification using Talon® spin columns.

1. Equilibrate the Talon® spin columns with 8 ml TWB.
2. Load the subunit-containing supernatant on the column and collect the flowthrough.
3. Reload the flowthrough onto the same column.
4. When all medium has passed the column, wash three times with 5 ml TWB.
5. To elute the bound subunits from the Talon-resin, add 0.2 ml TEB to the column and vortex vigorously (*see* **Notes 11** and **12**). Leave the column on a roller bench for 10 min at 4 °C.
6. Centrifuge the column at $100 \times g$ for 2 min at 4 °C and collect the eluted fraction. This contains the sE1 and sE2 subunits.
7. Store the subunits at −80 °C for further analysis and use.

3.5 Purification of CHIKV-VLPs

1. Precipitate the secreted VLPs and other proteins fractions from the medium by adding 7 % (w/v) polyethylene glycol (PEG)-6000 and 0.5 M NaCl to the medium fraction and incubate for 2 h on a roller bench at 4 °C.
2. Centrifuge the precipitates for 15 min at $4000 \times g$ at 4 °C and dissolve the pellet in 1 ml GTNE-buffer of 4 °C.
3. Prepare a discontinuous 70 % (w/v), 40 % (w/v) sucrose in GTNE gradient at 4 °C (*see* **Notes 13–15**).
4. Load the 1 ml VLP solution carefully on top of the 40 % sucrose fraction without disrupting the gradient.
5. Centrifuge the loaded gradient for 2 h at $70{,}000 \times g$ in an ultracentrifuge at 4 °C.

6. Carefully isolate the 70–40 % VLP-containing interphase band and resuspend it in 5 ml GTNE buffer.
7. Pellet the VLPs by centrifugation for 30 min at 85,000 × *g* in an ultracentrifuge at 4 °C and resuspend in 50 μl GTNE before storage at −80 °C for subsequent analysis and use.

3.6 CHIKV Protein Analysis

Characterization of the produced CHIKV subunits and VLPs are based on glycosylation states, furin-dependent cleavage maturity, fusogenic activity, and morphology.

1. Treat protein samples with 1 μl denaturing buffer in 9 μl MilliQ water for 10 min at 95 °C.
2. Add 2 μl G7 reaction buffer, 2 μl 10 % NP40 buffer, 0.5 μl PNGase F in 4.5 μl MilliQ water and incubate for 1 h at 37 °C.
3. Analyze both treated and non-treated protein samples by SDS-PAGE and Western blot (WB) using rabbit-polyclonal anti-E1 and anti-E2 [20], 1:15,000 and 1:20,000 diluted in PBS-Tween.
4. Seed 8 × 10^5 *Sf*21-cells per well in a 6-well plate in 2 ml of Grace's medium without antibiotics and let cells attach to the bottom of the well.
5. Infect the cells with an MOI of 10 with *Ac*-sE2 and *Ac*-S27 and add this to 1 ml of serum-free Grace's medium.
6. Replace the medium by the virus solution and incubate for 4 h at 27 °C on a rocking plateau.
7. Remove the 1 ml infection medium and add 2 ml Grace's medium containing 50 μM of furin inhibitor I and incubate for 72 h at 27 °C.
8. Harvest the cell and medium fraction by pipetting the cells loose. Separate the cells from the medium by centrifugation.
9. Wash the cells once with PBS and finally resuspend in 100 μl PBS.
10. Store the cell and medium fraction at −20 °C.
11. Analyze both treated and non-treated protein samples by SDS-PAGE and WB using rabbit-polyclonal anti-E1 and anti-E2 [20], 1:15,000 and 1:20,000 diluted in PBS-Tween.
12. Seed 8 × 10^5 *Sf*9-ET cells per well in a 6-well plate in 2 ml of *Sf*900II medium supplemented with 5 % FBS and 200 μg/ml geneticin and let cells attach to the bottom of the well.
13. Infect the cells with an MOI of 10 with *Ac*-GFP (*see* **Note 16**), *Ac*-sE1, *Ac*-sE2, and *Ac*-S27 and add this to 1 ml of *Sf*900II medium supplemented with 5 % FBS and 200 μg/ml geneticin. Incubate for 4 h at 27 °C on a rocking plateau.
14. Replace the infection medium with *Sf*900II medium supplemented with 0.2 mg/ml cholesterol and incubate for 72 h at 27 °C.

15. Subject infected cells for 2 min to acidified medium with pH 5.8, 5.5, and 5.0.
16. Score syncytia formation 4 h post induction with an inverted fluorescence microscope.
17. Treat copper 400 square mesh grids with argon gas discharge. Once discharged, use the grids within 1 h. After 1 h, discharge again.
18. Load 5 μl purified VLP sample to the grid and incubate for 2 min at RT.
19. Remove sample carefully using filter paper.
20. Wash the grid 5 × 2 min in MilliQ water by placing the grid upside down on 5 μl MilliQ water droplets placed on Parafilm.
21. Treat the grids with 2 % uranyl acetate for 15 s and remove the excess uranyl acetate carefully using filter paper.
22. Air-dry the grids and analyze the samples with a transmission electron microscope.

4 Notes

1. The primers used to amplify the sE1 and sE2 coding fragments are based on the CHIKV-S27 sequence (Genbank accession # AF369024). The primers contain AttB1/2 recombination sequences to enable Gateway® cloning (underlined sequences). In addition, sE1-Rv and sE2-Rv have a poly-His tail for efficient purification of the secreted glycoproteins.
2. The electrocompetent DH10Bac *E. coli* cells are stably transformed with a bacmid encoding the complete AcMNPV genome, a mini-F replicon to enable single-copy replication in bacteria and the attTn7 transposition site needed for insertion of de CHIKV coding fragments. In addition to the bacmid, the cells maintain at tetracycline-resistant helper plasmid encoding the enzymes required for transposition.
3. To verify successful Tn7 transposition of sE1, sE2, and S27 structural cassette into the bacmid, perform a PCR using M13-Fw and M13-Rv. The annealing sites of these primers flank the Tn7 transposition site (attTn7) and the size of the amplicon indicates successful incorporation of the insert. An additional PCR using M13-Fw and Genta-Rv is performed. The gentamycin resistance marker is co-recombined into the bacmid together with the CHIKV coding sequences. Thus, by using Genta-Rv as an internal primer, one can be sure that the recombination was successful.
4. PCR analysis of white colonies and thus correctly recombined bacmids can show a mixed phenotype of correct transposition

and so-called "empty bacmids." As long as amplicons indicate at least partial recombination, one can retransform the isolated bacmid into DH10β *E. coli* bacteria, which results in the loss of the empty bacmids.

5. During transfection of the bacmid DNA, it is important that cells are in their log-phase. If cell densities are too high or cells have reached their growth plateau, transfection efficiency declines rapidly.
6. The presence of FBS in the medium does not affect transfection efficiency.
7. Transfected cell will develop CPE caused by both the FectoFly I transfection reagent as well as baculovirus replication, which is characterized by enlarged nuclei and thus enlarged cells, cell fusion, and stalling of cell division.
8. When cells are transfected or infected with the S27 structural cassette, clear CPE caused by CHIKV capsid can be observed, which is a good indication of the level of infection. Dense nuclear bodies are formed due to the auto-assembly of nucleocapsids in the nucleus of infected cells.
9. During the preparation of the virus dilution series and the loading of the wells, it is important to refresh pipet tips at every dilution. Using the same tips results in false virus titers.
10. Average baculovirus titers range from of 5×10^7 to 5×10^8 $TCID_{50}$/ml. It is advised to redo the EPDA when viral titers are $>2 \times 10^9$.
11. When the subunits are eluted from the Talon-resin, it is important to completely resuspend the resin in the elution buffer to maximize elution efficiency.
12. To ensure protein stability, all purification steps should preferably be executed at 4 °C.
13. To ensure VLP stability, all purification steps should preferably be executed at 4 °C.
14. Before the soft SW55-tubes are used to prepare the sucrose gradient, leave them in water for 1 h at RT. This will decrease the risk of crack being formed during ultracentrifugation.
15. When preparing the 70 %, 40 % sucrose gradient, start with pipetting 3 ml of 40 % sucrose into the tube. Next, place a glass Pasteur pipette in the tube and pipet 1 ml 70 % sucrose into the Pasteur pipette. This will place the 70 % sucrose underneath the 3 ml 40 % sucrose by gravity and will ensure a clear and tight gradient boundary.
16. To ensure that the formation of syncytia is caused by CHIKV-E1 and not by the GP64 fusion protein of the baculovirus itself, one should include a GFP control (*Ac*-GFP). This control will only show syncytia formation when the pH of the medium drops to the range in which GP64 is fusogenically active ($pH < 5.0$).

References

1. Powers A, Logue C (2007) Changing patterns of chikungunya virus: re-emergence of a zoonotic arbovirus. J Gen Virol 88:2363–2377
2. Enserink M (2007) Epidemiology. Tropical disease follows mosquitoes to Europe. Science 317:1485–1487
3. Gould E, Gallian P, De Lamballerie X, Charrel R (2010) First cases of autochthonous dengue fever and chikungunya fever in France: from bad dream to reality! Clin Microbiol Infect 16:1702–1704
4. Enserink M (2014) Infectious diseases. Crippling virus set to conquer Western Hemisphere. Science 344:678–679
5. Guzman M, Harris E (2014) Dengue. Lancet 385:453–465
6. Borgherini G, Poubeau P, Jossaume A, Gouix A, Cotte L et al (2008) Persistent arthralgia associated with chikungunya virus: a study of 88 adult patients on reunion island. Clin Infect Dis 47:469–475
7. Suhrbier A, Jaffar-Bandjee M, Gasque P (2012) Arthritogenic alphaviruses—an overview. Nat Rev Rheumatol 8:42–429
8. Strauss J, Strauss E (1994) The alphaviruses: gene expression, replication, and evolution. Microbiol Mol Biol Rev 58:491–562
9. Levitt N, Ramsburg H, Hasty S, Repik P, Cole F et al (1986) Development of an attenuated strain of chikungunya virus for use in vaccine production. Vaccine 4:157–162
10. Mallilankaraman K, Shedlock D, Bao H, Kawalekar O, Fagone P et al (2011) A DNA vaccine against chikungunya virus is protective in mice and induces neutralizing antibodies in mice and nonhuman primates. PLoS Negl Trop Dis 5:e928
11. Muthumani K, Lankaraman K, Laddy D, Sundaram S, Chung C et al (2008) Immunogenicity of novel consensus-based DNA vaccines against chikungunya virus. Vaccine 26:5128–5134
12. Wang D, Suhrbier A, Penn-Nicholson A, Woraratanadharm J, Gardner J et al (2011) A complex adenovirus vaccine against chikungunya virus provides complete protection against viraemia and arthritis. Vaccine 29:2803–2809
13. Wang E, Volkova E, Adams A, Forrester N, Xiao S et al (2008) Chimeric alphavirus vaccine candidates for chikungunya. Vaccine 26: 5030–5039
14. Akahata W, Yang ZY, Andersen H, Sun S, Holdaway H et al (2010) A virus-like particle vaccine for epidemic Chikungunya virus protects nonhuman primates against infection. Nat Med 16:334–338
15. Metz S, Gardner I, Geertsema C, Thuy L, Goh L et al (2013) Effective chikungunya virus-like particle vaccine produced in insect cells. PLoS Negl Trop Dis 7(3):e2124
16. Metz S, Martina B, van den Doel P, Geertsema C, Osterhaus A et al (2013) Chikungunya virus-like particles are more immunogenic in a lethal AG129 mouse model compared to glycoprotein E1 or E2 subunits. Vaccine 31: 6092–6096
17. van Oers MM, Pijlman GP, Vlak JM (2015) Thirty years of baculovirus-insect cell protein expression: from dark horse to mainstream technology. J Gen Virol 96:6–23
18. Metz S, Pijlman G (2011) Arbovirus vaccines; opportunities for the baculovirus-insect cell expression system. J Invertebr Pathol 107(Suppl):S16–S30
19. Pijlman G (2015) Enveloped virus-like particles as vaccines against pathogenic arboviruses. Biotechnol J 10:1–12
20. Metz S, Geertsema C, Martina B, Andrade P, Heldens J et al (2011) Functional processing and secretion of Chikungunya virus E1 and E2 glycoproteins in insect cells. Virol J 8: 353–361
21. Hopkins R, Esposito D (2009) A rapid method for titrating baculovirus stocks using the Sf-9 Easy Titer cell line. Biotechniques 47:785–788
22. Kaba S, Salcedo A, Wafula P, Vlak J, van Oers M (2004) Development of a chitinase and v-cathepsin negative bacmid for improved integrity of secreted recombinant proteins. J Virol Methods 122:113–118

Chapter 28

Protocols for Developing Novel Chikungunya Virus DNA Vaccines

Christopher Chung, Kenneth E. Ugen, Niranjan Y. Sardesai, David B. Weiner, and Kar Muthumani

Abstract

To date, there have been several million infections by the Chikungunya virus (CHIKV), a mosquito-transmitted emerging pathogen that is considered to be taxonomically an Old World RNA virus. Although original CHIKV outbreaks were restricted to India, East Asian countries, Northern Italy, and France, a recent sharp rise had been identified in 41 countries or territories in the Caribbean, Central America, South America, and North America. A total of 1,012,347 suspected and 22,579 laboratory-confirmed CHIKV cases have been reported from these areas, which signals an increasing risk to the US mainland. Unlike past epidemics that were usually associated with *Ae. aegypti* transmission, the Caribbean outbreak was associated with *Ae. albopictus* transmission as the principal mosquito vector. In addition, the substantial increase in the number of deaths during this epidemic, as well as incidence of neurologic disease, suggests that CHIKV may have become more virulent. Currently, there are no licensed vaccines or therapeutics available for CHIKV or its associated disease pathologies. Therefore, development of new vaccines and therapies that could confer immunity and/or treat clinical symptoms of CHIKV is greatly desired. This chapter describes the use of entirely cutting edge technologies/methodologies developed by our group for the development and evaluation of novel DNA vaccines against CHIKV.

Key words Chikungunya virus, Envelope protein, DNA vaccine, Cellular immunity, Humoral responses, Neutralization assay

1 Introduction

Chikungunya virus (CHIKV) is a mosquito-borne virus of the Togaviridae family, belonging to the genus Alphaviridae, which has recently gained renewed interest as an emerging pathogen and potential biological weapon [1, 2]. CHIKV was first recognized in epidemic form in East Africa between 1952 and 1953, during which time it was isolated from the blood of a febrile patient in Tanzania [3, 4]. Since then, additional outbreaks have been reported, with the largest thus far affecting mainly Asia and Africa

Justin Jang Hann Chu and Swee Kim Ang (eds.), *Chikungunya Virus: Methods and Protocols*, Methods in Molecular Biology, vol. 1426,DOI 10.1007/978-1-4939-3618-2_28, © Springer Science+Business Media New York 2016

in 2004. In 2005–2006, epidemics were noted in India and La Reunion Islands in the Indian Ocean. As of now, there are an estimated 1.3 million people across 13 states infected in India [5]. More recently, in December 2013, two laboratory-confirmed cases of CHIKV were discovered in Saint Martin in patients without a travel history. Since then, there has been an increased observed incidence rate of CHIKV infection, with thousands of cases identified in New Guinea and the Caribbean [6]. Increased global travel could pose a potential risk by spreading the CHIKV disease to non-endemic regions. Alarmingly, Europe, the USA, Canada, and Australia have recently had documented cases of CHIKV. Overall, CHIKV has been identified in nearly 40 countries with continued spread to new regions [7, 8].

Chikungunya virus is an enveloped, single stranded, positive sense RNA virus with a genome size of 11–12 kb. This genome includes two open reading frames encoding several nonstructural (nsP1 to nsP4; 9, 10) and structural proteins (Capsid, E3, E2, 6K, and E1; 5). CHIKV has a complex replication cycle and, consequently, its genome is easily susceptible to mutation. Due to this mutation rate, there are currently three distinct genotypes that have been identified based on the gene sequence of the envelope protein (E1): East/Central/South African, West African, and Asian [11]. CHIKV has been listed by the National Institute of Allergy and Infectious Diseases (NIAID) as a Category C Priority Pathogen, due to its potential to be engineered for mass dissemination.

CHIKV is transmitted to humans via the *Aedes* species mosquito. In Africa, transmission follows a sylvatic cycle comprising nonhuman primates and various species of forest-dwelling mosquitoes. However, in Asia, an urban cycle is observed in which mosquitoes spread CHIKV from infected to uninfected humans [11]. Traditionally, *Aedes aegypti* has been the primary vector but a mutation in the CHIKV envelope (E1-A226V) during the La Reunion outbreak has resulted in increased infectivity in *Aedes albopictus* [12, 13]. Now, *Aedes albopictus* is the main vector in the Indian Ocean Islands [14]. Of note, these two species of mosquito are both found in North America [15–17]. In addition to mosquito vector-based spread, blood borne transmission is possible and mother-to-child transmission has been reported in newborns of viremic women who developed disease within a week prior to delivery [18, 19].

Chikungunya fever is not typically associated with mortality, but substantial morbidity has been observed. Once transmitted, CHIKV has an incubation period of about 2–6 days. There is no prodromal phase and a sudden onset of symptoms typically appears 4–7 days post-infection [7]. Clinically, Chikungunya fever can be divided into an acute phase and a chronic phase. The acute phase lasts from a few days to a couple weeks and is characterized by high fevers, rigors, headaches, maculopapular or petechial rash, nausea, vomiting, myalgia, polyarthralgia, joint stiffness, and photophobia [3, 5, 7]. Moreover, some of these patients who experience the acute phase will later progress to the chronic phase. The chronic phase consists of persistent, disabling polyarthritis, which can last anywhere in the range of a few weeks to several years. Chronic joint pain occurs in 30–40 % of those infected and can cause sufferers to adopt bent or stooping postures [7]. In fact, Chikungunya is derived from the Swahili or Makonde word meaning "to become contorted" or "that which bends up." Evidence suggests that persistence of the alpha virus in synovial tissues, along with ongoing stimulation of innate and adaptive immune responses, is likely responsible for chronic arthritic disease [20–23]. Reported Chikungunya fever-related complications include respiratory failure, cardiovascular decompensation, meningoenchepalitis, severe acute hepatitis, severe cutaneous effects, other CNS problems, and kidney failure [24–28]. On occasion, fatalities are observed at a rate of 1 in 1000 [7]. CHIKV infections resulting in death disproportionately affect the elderly, children, and those with underlying medical conditions [11, 25, 27].

CHIKV infection is treated symptomatically with nonsteroidal anti-inflammatory drugs, steroids, bed rest, fluid replacement and hydration, and mild exercise [3, 11, 29]. At this time, there are no approved antiviral treatments for Chikungunya fever, although potential treatment options being assessed include CHIKV antibodies, ribavirin, interferon-alpha, and chloroquine [3, 30–36]. From a prophylactic standpoint, there are no licensed vaccines that are currently available. Notably, one attenuated CHIKV vaccine candidate developed by the US Army did reach Phase II clinical trials in 2000, involving a total of 58 study subjects, all of whom developed neutralizing antibodies [37]. The vaccine, however, is no longer being pursued because of concerns related to virus passaging in uncertified cell cultures during development. It was also later discovered that the vaccine candidate was only attenuated by two point mutations, thus raising the risk for reversion [38].

Given the recent Chikungunya outbreaks, there has been a renewed effort for vaccine development. In addition to the formalin inactivated and live attenuated vaccines developed and evaluated by the US Army, other approaches that have been explored include chimeric alpha virus vaccines, purified immunoglobulin isolated from convalescent CHIKV patients, and virus-like particles [7, 37, 39–45]. In addition to these various approaches, DNA vaccines are a promising vaccination method, which our group has great familiarity with and has studied in the context of CHIKV [9, 46, 47].

Initial DNA vaccine reports surfaced almost 20 years ago. DNA, as a vaccine platform, has been studied in numerous clinical protocols. The technology is attractive for many reasons including its excellent safety profile and the simplicity of engineering and production compared to the development of live attenuated and killed viral vaccines. In addition, antivector immunity is not a concern for this platform. In theory this platform has a great deal to offer to the field of vaccinology. However, performance issues have plagued this approach in the past limiting its utility. Early studies of DNA vaccines in small animals were promising, but a major issue arose regarding immune potency in nonhuman primates (NHPs) and in the clinic. A number of clinical studies have evaluated DNA versus alternate delivery platforms and formulations. In these studies the overall response rates and magnitude of the induced responses in humans were poor. There are numerous strategies being investigated to enhance the immunogenicity of DNA vaccines. These include plasmid sequence optimization, adjuvants, improved strategies such as prime boosting protocols, and improved plasmid delivery. Plasmid optimization includes codon optimization for expression in human cells, RNA optimization to improve mRNA stability and more efficient translation on the ribosome, the addition of leader sequences to enhance translation efficiency, and the creation of consensus immunogens to induce more cross reactive responses. We have reported improved in vivo expression using such plasmids; however, these approaches alone do not engender immunity in the range that was induced by viral vectors [48–51].

With DNA vaccines, it is possible to construct a consensus vaccine that allows induction of a cross-reactive immune response across distinct phylogenetic groups. By taking the most common sequence across many different isolates, one can develop responses to conserved regions, allowing for an increase in cross-reactive responses. This approach was demonstrated in a study by Muthumani et al. in which a consensus construct was designed against E1, E2, and the core capsid protein of CHIKV by aligning 21 sequences isolated between 1952 and 2006 [47]. The construct was then further modified with codon and RNA optimization, addition of a strong Kozak sequence, and substitution of a signal peptide with an immunoglobulin E leader sequence to improve vaccine efficacy. Mice immunized with this construct induced strong IFN-gamma response, T cell proliferation, and increased total IgG, along with elevated anti-E1 specific, anti-E2 specific, and anti-capsid specific IgG antibodies.

This concept was further explored in a study by Mallilankraman et al., which demonstrated the protective effect of a DNA vaccine against CHIKV in a murine model and the induction of CHIKV neutralizing antibodies in mice and nonhuman primates [9]. Specifically, a single consensus envelope vaccine expressing all three CHIKV envelope glycoproteins (E3 + E2 + E1) was constructed and designated CHIKV-Env. After confirming in vitro expression, CHIKV-Env was shown to be able to induce $CD8^+$ CTL responses in mice after three immunizations and induce higher levels of envelope specific serum IgG than mice immunized with DNA vaccine constructs containing E1, E2, or E3 alone. Additionally, neutralizing and hemagglutination inhibition antibodies were significantly greater than constructs with E1 or E2 alone. Upon challenge with a mouse adapted CHIKV, vaccination with CHIKV-Env resulted in decreased production of pro-inflammatory cytokines IL-6 and TNF-α, restoration and maintenance of body weight, lower viremia, and decreased pathological abnormalities in the brain, heart, liver, lungs, and kidneys compared to naïve controls. Rhesus macaque studies demonstrated that three of four CHIKV-Env DNA immunized monkeys have detectable envelope-specific functional CTL activity. Moreover, all four monkeys developed neutralizing antibody titers. Further work with patient samples revealed that significant neutralizing antibody and hemagglutination inhibition antibodies were found in convalescent human samples, suggesting that these antibody titers correlate with the ability of the host to clear CHIKV during the course of a natural infection.

Further work was performed to determine the potential adjuvant role of nonstructural protein-2 expressing DNA plasmid (pnsP2) on protective immunity elicited by a CHIKV-Env expressing DNA vaccine (CHIKV-Env) [46]. Co-immunization with pEnv and pnsP2 in mice resulted in higher total anti-CHIKV Env IgG binding titer in sera when compared to mice receiving pEnv only. Cellular response induced by pEnv+pnsP2 was also significantly higher than that induced by pEnv alone. Co-immunization led to decreased viremia, increased rate of survival, and significant reduction of TNF-α and IL-6 production upon challenge. The pEnv+pnsP2 group failed to demonstrate any clinical signs of infection during the study and 100 % survived until day 12 or beyond. Brain tissue sections demonstrated normal morphology for only the group that received both plasmids.

In this book chapter, the methodology involved in the development and assessment of Chikungunya virus DNA vaccines is described and summarized. Specifically, the chapter discusses the approach for constructing a DNA vaccine for Chikungunya virus, along with various experimental procedures for confirming expression and efficacy of the DNA construct in vitro and in vivo. Methods for detecting in vitro expression include transfection into a mammalian cell line followed by confirmation using Western blot and immunofluorescence assays. Methods to demonstrate in vivo functionality is based on DNA vaccination in murine and non human primate (macaque) models, followed by assays measuring either cellular or humoral responses. Protocols to demonstrate the efficacy of the DNA vaccine against CHIKV in vitro include neutralization assays and mouse challenge studies. Several of the techniques presented in this chapter are not necessarily specific to DNA vaccines and can be broadly generalized to other vaccine development approaches.

2 Materials

2.1 *In Vitro* Transfection

1. 293T human embryonic kidney cells (*see* **Note 1**).
2. DMEM (Dulbecco's Modified Eagle Medium).
3. D10 medium: DMEM, 10 % heat-inactivated fetal bovine serum (FBS), and penicillin–streptomycin mixture (100 IU/ml penicillin and 100 μg/ml streptomycin). Under sterile conditions combine 50 ml of heat-inactivated FBS, 5 ml of penicillin–streptomycin mixture, and 445 ml of DMEM. Filter–sterilize and store at 4 °C.
4. PBS.
5. CHIKV DNA vaccine and control pVax1 constructs.
6. TurboFectin 8.0 transfection reagent (OriGene Technologies).
7. Cell lysis buffer (Cell Signaling Technology) supplemented with protease inhibitor cocktail (Roche).

2.2 SDS-PAGE and Western Blot

1. NuPAGE LDS sample buffer.
2. NuPAGE reducing agent.
3. NuPAGE antioxidant.
4. NuPAGE transfer buffer.
5. 10 % Bis-Tris gel.
6. MOPS-SDS running buffer: 50 mM MOPS, 50 mM Tris-Base, 0.1 % SDS, 1 mM EDTA, pH 7.7.
7. Immobilon-FL membrane (EMD Millipore).
8. 10 % methanol in NuPAGE transfer buffer (*see* **Note 2**).
9. Odyssey blocking buffer (LI-COR Biosciences).
10. Odyssey blocking buffer with 0.1 % Tween 20.
11. Odyssey blocking buffer with 0.1 % Tween 20 and 0.01 % SDS.
12. PBS with 0.1 % Tween 20.
13. Anti-CHIKV antibody: Commercially available CHIKV antibody, Chikungunya patient serum, or CHIKV-immunized mouse sera.
14. IR Dye secondary antibody.
15. Odyssey CLx Infrared Imaging System (LI-COR Biosciences).

2.3 Immunofluorescence Assay

1. 293T human embryonic kidney cells.
2. D10 medium.
3. PBS.
4. CHIKV DNA and control pVax1 DNA vaccine construct.
5. 2-Chamber tissue culture treated glass slides (BD Falcon).

6. 2 % paraformaldehyde in water.
7. 0.1 % glycine, 1 % BSA in PBS.
8. Primary CHIKV antibody and fluorescently labeled secondary antibody.
9. Nuclear counterstains DAPI (diamidino-2-phenylindole).
10. Fluoromount G (Electron Microscopy Sciences).
11. Laser scanning confocal microscope.

2.4 DNA Vaccine Mice Immunization

1. Balb/C or C57BL/6 mice (*see* **Note 3**).
2. Isoflurane.
3. CHIKV DNA vaccine and control pVax1 construct.
4. In vivo electroporation machine (Inovio Pharmaceutics Inc.).

2.5 Isolation of Splenocytes from Immunized Mice

1. Stomacher and stomacher bags (Seward Ltd).
2. Ammonium chloride-potassium (ACK) lysis buffer (Cambrex BioScience).
3. R10 medium: RPMI 1640 with 2 mM/l L-glutamine, 10 % heat-inactivated fetal bovine serum, and penicillin–streptomycin mixture (100 IU/ml penicillin and 100 μg/ml streptomycin). Under sterile conditions combine 50 ml of heat-inactivated FBS, 5 ml of penicillin–streptomycin mixture, and 445 ml of RPMI 1640 with 2 mM/l L-glutamine. Filter-sterilize and store at 4 °C.
4. PBS.
5. 40-μm cell strainer.

2.6 IFN-γ Enzyme-Linked Immuno Spot Assay

1. 96-well plates for ELISpot assays (Millipore).
2. Anti-mouse IFN-γ capture antibody (R&D Systems).
3. 1 % BSA in PBS: Dissolve 1 g of BSA in 100 ml of PBS.
4. Isolated splenocytes from immunized mice.
5. R10 medium.
6. PBS.
7. Concanavalin A (Con A).
8. Peptide pools (*see* **Note 4**).
9. Biotinylated anti-mouse IFN-γ antibody.
10. Streptavidin-alkaline phosphatase.
11. BCIP/NBT-plus substrate (MabTech).
12. ELISpot Reader (CTL Limited).

2.7 Enzyme-Linked Immunosorbent Assay (ELISA)-binding

1. Vaccinated mice sera samples.
2. 96-well high binding polystyrene plates (Corning).
3. Recombinant CHIKV protein.
4. PBS-T: PBS, 0.2 % Tween 20. Add 1 ml of Tween 20 to 500 ml of PBS.
5. 1 % FBS in PBS-T. Add 1 ml of FBS to 99 ml of PBS-T. Filter-sterilize.
6. 10 % FBS in PBS. Add 50 ml of FBS to 450 ml of PBS. Filter-sterilize.
7. HRP-conjugated goat anti-mouse kappa light chain (Bethyl Laboratories).
8. SIGMAFAST OPD (Sigma-Aldrich).
9. Stop solution: 2 N H_2SO_4.
10. Biotek EL312e Bio-Kinetics reader.

2.8 Flow Cytometry and Intracellular Cytokine Staining (ICS) Assay

1. Isolated splenocytes from immunized mice.
2. U-bottom 96-well plate.
3. Vaccine specific peptide pools (5 μg/ml per peptide).
4. Protein transport inhibitor cocktail (brefeldin A and monensin; eBioscience).
5. Cell stimulation cocktail (with protein transport inhibitors, phorbol 12-myristate 13-acetate-PMA, ionomycin, brefeldin A, and monensin; eBioscience).
6. R10 medium.
7. PBS.
8. LIVE/DEAD FixableAqua Dead Cell Stain Kit or LIVE/DEAD FixableViolet Dead Cell Stain Kit (Life Technologies).
9. FACS buffer: PBS containing 0.1 % sodium azide and 1 % FBS.
10. BD Cytofix/Cytoperm (BD Biosciences).
11. BD Perm/Wash (BD Biosciences).
12. Surface staining fluorochrome-conjugated antibodies: CD19 (V450; clone 1D3; BD Biosciences), CD4 (FITC; clone RM4-5; eBioscience), CD8 (APC-Cy7; clone 53-6.7; BD Biosciences), and CD44 (A700; clone IM7; Biolegend).
13. Intracellular staining fluorochrome-conjugated antibodies: IFN-γ (APC; clone xMG1.2; Biolegend), TNF-α (PE; clone MP6-xT22; eBioscience), CD3 (PerCP/Cy5.5; clone 145-2C11; Biolegend), and IL-2 (PeCy7; clone JES6-SH4; eBioscience).
14. LSRII flow cytometer (BD Biosciences).
15. FlowJo software (Tree Star).

2.9 Viral Neutralization Assay

1. Vero cells.
2. M10 medium: MEM, 10 % heat-inactivated fetal bovine serum, and penicillin–streptomycin mixture (100 IU/ml penicillin and 100 μg/ml streptomycin). Under sterile conditions combine 50 ml of heat-inactivated FBS, 5 ml of penicillin–streptomycin mixture, and 445 ml of MEM. Filter-sterilize and store at 4 °C.
3. CHIKV-Pseudovirus (*see* **Note 5**).
4. CHIKV-Vaccinated mice sera samples.
5. Reporter Lysis Buffer (Promega).
6. BriteLite Plus Kit (PerkinElmer).
7. Flat-bottom 96-well plate.
8. GloMax Multi Detection System (Promega).

2.10 DNA Plasmid Based Immunization Studies in Rhesus Macaques

1. Rhesus macaques.
2. CHIKV DNA vaccine and control pVax1 construct.
3. In vivo electroporation machine.
4. EDTA tubes.
5. Ficoll-Paque PLUS (GE Healthcare).
6. R10 medium.
7. Ammonium chloride-potassium (ACK) lysis buffer.
8. 90 % FBS, 10 % DMSO mixture.

3 Methods

3.1 DNA Vaccine Design

When designing a DNA vaccine, it is helpful to create a construct that will not only be well expressed, but will provide a broad protective response (Fig. 1).

1. Obtain CHIKV capsid and envelope gene sequences, which are readily available in the NCBI Genbank database.
2. Use appropriate bioinformatics DNA software to perform multiple sequence alignment using a ClustalW algorithm. This will generate a consensus gene sequence, which can be used for DNA vaccine design (*see* **Note 6**).
3. Optimize consensus sequence for expression by codon and RNA optimization, addition of an immunoglobulin E leader sequence, and addition of a strong Kozak sequence. Clone the gene into a suitable mammalian expression vector and formulate in water for immunization (*see* **Note 7**).

3.2 In Vitro Transfection

1. Plate 0.6×10^6 293T cells or other suitable cell line in 2 ml of D10 medium per well in a six-well plate and allow cells to adhere overnight in a 37 °C incubator. Prepare enough wells

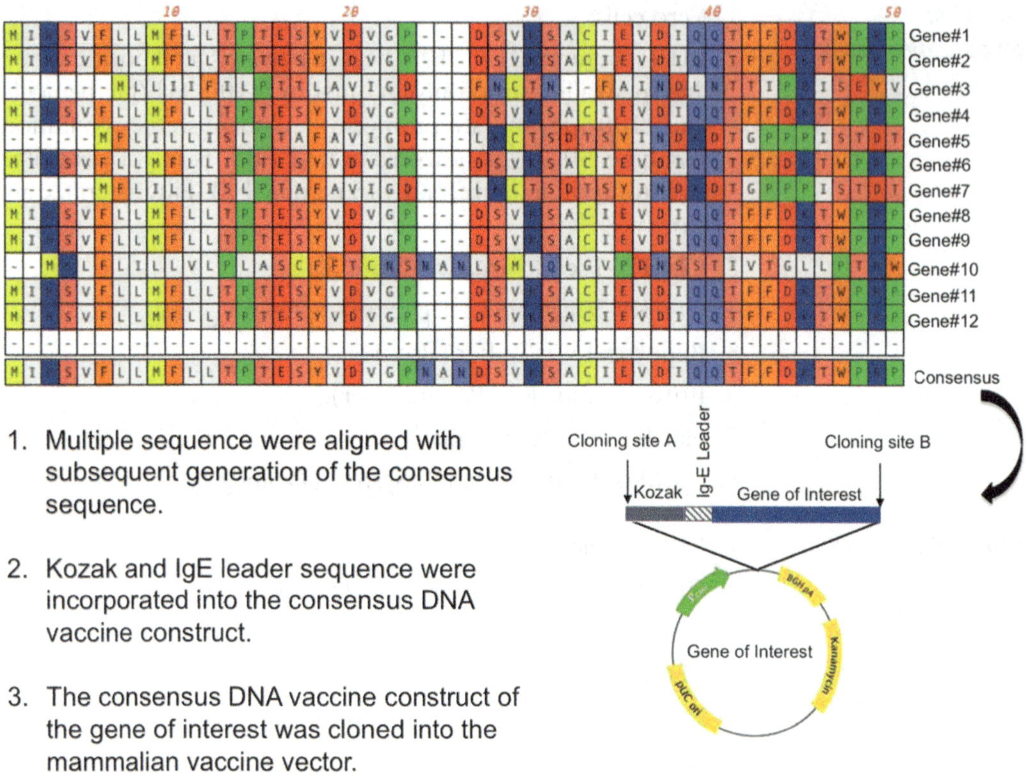

Fig. 1 Consensus DNA vaccine design. Gene sequences for the viral capsid or envelope protein from multiple strains are obtained through the NCBI Genbank database. Approprite software is used to align the multiple sequences using a ClustalW algorithm. A consensus gene sequence, which is the gene of interest, is obtained and then subjected to codon and RNA optimization, addition of an IgE leader and Kozak sequences. This resulting final DNA construct is then subsequently cloned into a suitable mammalian expression vector

for each DNA sample, as well as one additional for pVax1 negative control DNA.

2. The next day, check cells to ensure they are 55–75 % confluent.
3. For each sample prepare an Eppendorf tube containing 100 μl of serum-free media (DMEM) combined with 20 μl of TurboFectin 8.0 and incubate at room temperature for 5 min.
4. Add 5 μg of the DNA sample of interest to this mixture and allow incubation at room temperature for another 30 min.
5. Just prior to the end of this incubation period, replenish the cell media. Slowly add the TurboFectin/DNA mixture to the cells (*see* **Note 8**). Return cells to the incubator.
6. 48–72 h post-transfection, carefully remove and collect the supernatant.

7. Collect the cells by washing each well vigorously with 2 ml of cold PBS or by using a cell scraper. Transfer the cell solution into 15 ml tubes and spin at 4 °C for 5 min at 200×*g*.
8. Remove the supernatant and resuspend the cells with 100 μl of cell lysis buffer supplemented with a protease inhibitor cocktail.
9. Perform three cycles of freeze-thaws using dry ice and a 37 °C water bath.
10. Spin the samples at maximum speed for 5 min and collect the supernatant containing the expressed protein of interest.
11. Samples can be used immediately or stored at −20 °C for future analysis.

3.3 SDS-PAGE and Western Blot

1. Denature the isolated protein in NuPAGE LDS Sample Buffer and NuPAGE Reducing Agent and incubate for 10 min at 70 °C.
2. Separate the proteins using sodium dodecyl sulfate polyacrylamide gel electrophoresis (SDS-PAGE) on a 10 % Bis-Tris gel in MOPS SDS Running Buffer with 500 μl of NuPAGE Antioxidant supplemented to the upper (inner) chamber. Set the voltage to 150 V and run the gel for approximately 1 h.
3. Transfer the gel to an Immobilon-FL membrane at 30 V for 1 h in 10 % methanol NuPAGE transfer buffer with NuPAGE Antioxidant.
4. Following transfer, block the membrane in Odyssey Blocking Buffer overnight at 4 °C.
5. The following day, incubate the membrane for 1 h at room temperature with a primary CHIKV antibody, Chikungunya patient serum, or CHIKV-immunized mouse sera at the appropriate dilution (1:250 to 1:500) in Odyssey Blocking Buffer with 0.1 % Tween 20.
6. Wash the membrane four times for 5 min each at room temperature in PBS with 0.1 % Tween 20.
7. Add IR Dye secondary antibody, diluted at 1:15,000 in Odyssey Blocking Buffer with 0.1 % Tween 20 and 0.01 % SDS and incubate for 1 h.
8. Wash the membrane again four times, 5 min each, using PBS with 0.1 % Tween 20.
9. Finally, rinse the membrane with PBS and visualize using the Odyssey CLx Infrared Imaging System.

3.4 Immunofluorescence Assay (IFA)

1. Seed a suitable cell line (2×10^5 cells) in 2-chamber tissue culture treated glass slides and allow cells to adhere overnight.
2. Transfect with the DNA vaccine construct or control pVax1 vector as described before.

3. 48 h post-transfection, aspirate media from glass chambers and wash with PBS.
4. Fix cells with 2 % paraformaldehyde for 20 min at room temperature.
5. Aspirate the paraformaldehyde and wash twice with PBS. Add 0.3 % Triton-X in PBS and incubate at 4 °C for 10 min.
6. Incubate overnight at 4 °C with a primary antibody against CHIKV (*see* **Note 9**).
7. Wash off excess primary antibody with PBS.
8. Add a fluorescently labeled secondary antibody and incubate for 2 h at 37 °C (*see* **Note 10**).
9. Wash off excess secondary antibody with PBS.
10. Counterstain with DAPI diluted at 1:10,000 in PBS for 5 min.
11. Mount coverslip with about 20 μl of Fluoromount G (*see* **Note 11**).
12. Acquire images using a laser scanning confocal microscope.

3.5 DNA Vaccine Mice Immunization

1. Immunize isoflurane-anesthetized mice ($n = 5$ for each group) with the appropriate DNA construct (CHIKV-Capsid, CHIKV-Envelope, control pVax1, etc.) at 25 μg in a total volume of 25 to 50 μl in the quadriceps or tibialis anterior muscle.
2. Immediately follow with in-vivo electroporation (two pulses of 0.2 A, 52 ms/pulse, 4 s firing delay) at the site of injection (*see* **Note 12**).
3. Repeat immunizations 2 weeks apart for a total of three immunizations.
4. Following the final immunization, sacrifice mice for immunology assays or move into a BSL-3 facility for CHIKV challenge studies (*see* **Note 13**).

3.6 Isolation of Splenocytes from Immunized Mice

1. Harvest spleens 1–2 weeks after final vaccination.
2. Collect individual spleens in 10 ml of R10 (*see* **Note 14**).
3. Transfer spleen to stomacher bag and stomach at high speed for 60 s.
4. Pour through cell strainer into new 50 ml conical tube. Wash sides of bag with 10 ml sterile PBS and add through 40 μm strainer to 50 ml conical tube.
5. Centrifuge at 300 × *g* for 10 min. Pour off supernatant.
6. Resuspend cell pellet in 5 ml ACK Lysing Buffer and incubate for 5 min.
7. Fill tube completely with PBS to dilute.

8. Spin at 300×*g* for 5 min and pour off supernatant (*see* **Note 15**).
9. Resuspend pellet in 20 ml of R10.
10. Add cell suspension to new 50 ml conical tube through a 40 μm cell strainer.
11. Count cells and resuspend at a concentration of 1.0×10^7 cells/ml (*see* **Note 16**).

3.7 IFN-γ Enzyme-Linked Immuno Spot Assay (See Note 17)

1. Coat 96-well ELISpot plates with anti-mouse IFN-γ capture antibody in sterile PBS (1:100) and incubate for 24 h at 4 °C.
2. Wash plates 3–4× with 200 μl/well of PBS and then block for at least 2 h with 1 % BSA in PBS.
3. Remove the blocking media and in each well, add splenocytes from vaccinated mice (200,000 cells in 100 μl) along with either 100 μl of media alone (negative control), 100 μl of media with Con A diluted at 1:1500 (positive control), or 100 μl of media with peptide pools (10 μg/ml).
4. Incubate plates for 12–48 h at 37 °C in 5 % CO_2.
5. Dispose cells and wash 3–4× with 200 μl/well of PBS.
6. Dilute the biotinylated anti-mouse IFN-γ antibody at 1:100 in PBS and add 100 μl/well.
7. Wrap plate in saran wrap and incubate overnight at 4 °C.
8. Wash plates 3–4× with 200 μl/well of PBS (*see* **Note 18**).
9. Dilute streptavidin-alkaline phosphatase at 1:1000 in PBS and add 100 μl/well. Incubate for 2 h at room temperature in the dark.
10. Wash plates 3–4× with 200 μl/well of PBS.
11. Add 100 μl/well of BCIP/NBT solution and develop until spots emerge (*see* **Note 19**).
12. Stop reaction by washing extensively with running tap water (*see* **Note 20**).
13. Leave the plates face down on the bench and allow to dry overnight.
14. Quantify with an automated ELISpot reader (*see* **Note 21**).

3.8 ELISA-binding

1. Coat 96-well high binding polystyrene plates with recombinant CHIKV protein (100 μl/well at 0.25 to 1.0 μg/ml) and incubate overnight at 4 °C.
2. Wash plates 3–4× with 200 μl/well of PBS-T and block with 10 % FBS in PBS for 1 h at room temperature.
3. Wash plates 3–4× with 200 μl/well of PBS-T.

4. Prepare serial dilutions of samples at log or half-log steps in 1 % FBS in PBS-T (*see* **Note 23**).
5. Add diluted samples and incubate for 1 h at room temperature.
6. Wash plates 3–4× with 200 μl/well of PBS-T.
7. Add 100 μl of HRP-conjugated goat anti-mouse kappa light chain diluted at 1:20,000 in 1 % FBS in PBS-T and incubate for 1 h at room temperature.
8. Wash plates 3–4× with 200 μl/well of PBS-T.
9. Prepare SIGMAFAST OPD detection substrate and add 100 μl/well for 10 min at room temperature.
10. Stop the reaction by adding 100 μl/well of 2 N H_2SO_4 stop solution.
11. Read plate at 450 nm with the Biotek EL312e Bio-Kinetics reader.

3.9 Flow Cytometry (FACS) and Intracellular Cytokine Staining (ICS) Assay (See Note 25)

1. Determine the number of splenocytes required for staining. For each test sample, a final concentration of no less than 1×10^6 cells in 50 μl of staining buffer is suggested.
2. Based on number of samples necessary, aliquot cells into a U-bottom 96-well plate.
3. Stimulate splenocytes with pooled vaccine specific peptides (5 μg/ml per peptide) for 5–6 h at 37 °C in 5 % CO_2 in the presence of protein transport inhibitor cocktail. Use cell stimulation cocktail and R10 media as positive and negative controls, respectively.
4. Spin down at 300 × *g* for 5 min to pellet cells and decant the supernatant.
5. Wash cells with 200 μl of PBS.
6. Spin down at 300 × *g* for 5 min to pellet cells and decant the supernatant.
7. To stain for Live/Dead, resuspend cell pellets in 50 μl of either Live/Dead Aqua (1:500 in PBS) or Live/Dead Violet (1:20 in PBS) and incubate in the dark at room temperature for 10 min.
8. Spin down at 300 × *g* for 5 min to pellet cells and decant the supernatant.
9. Wash cells with 200 μl of PBS.
10. Spin down at 300 × *g* for 5 min to pellet cells and decant the supernatant.
11. For surface staining, determine the total volume require (50 μl per sample) and make up the staining cocktail by diluting each antibody to the desired concentration in FACS buffer.

12. Resuspend cells in the staining cocktail and incubate in the dark at room temperature for 20–30 min.
13. Spin down at 300 × *g* for 5 min to pellet cells and decant the supernatant (*see* **Note 26**).
14. Wash cells with 200 μl of FACS buffer and spin down at 300 × *g* for 5 min to pellet cells. Decant supernatant.
15. Fix cells in 100 μl BD fix/perm buffer for 20 min at 4 °C.
16. Centrifuge at 400 × *g* for 5 min and decant supernatant.
17. Prepare intracellular staining cocktail by diluting antibodies to desired concentration in BD perm/wash buffer. Add 50 μl of the staining cocktail to each sample and incubate in the dark at 4 °C for 1 h.
18. Centrifuge at 400 × *g* for 5 min and decant supernatant.
19. Wash cells with 200 μl of FACS buffer and spin down at 300 × *g* for 5 min. Decant supernatant. Repeat 1 ×.
20. Resuspend cells in 200 μl of FACS buffer for analysis (*see* **Note 27**).
21. Collect data using a LSRII flow cytometer and analyze using FlowJo software (Tree Star). Use SPICE v5. Boolean gating with FlowJo software to examine T cell polyfunctionality [53].

3.10 Viral Neutralization Assay (See Note 28)

1. Plate Vero cells in M10 media at a concentration of 3.0×10^6 cells/well in a 6-well plate with 2 ml of media/well.
2. Allow cells to grow at 37 °C, 5 % CO_2 for 24 h.
3. Carefully remove media from wells using a 1000 μl pipette of Pipette-Aid and add 2 ml of fresh M10 media to each well.
4. Add 50–100 pg of pseudovirus to appropriate wells.
5. Add 200 μl of serum to appropriate wells (*see* **Note 29**).
6. Incubate for 48 h at 37 °C, 5 % CO_2.
7. Carefully remove the media from cells. Add 2 ml of sterile PBS to each well and wash the cells off of the plate using a 1000 μl pipette.
8. Once the cells are in solution, transfer them to a 15 ml conical tube and spin down at 4 °C for 5 min at 200 × *g*. Carefully discard the supernatant (*see* **Note 30**).
9. Add 200 μl of Reporter Lysis Buffer to each tube.
10. Resuspend cells by pipetting and transfer to labeled Eppendorf tubes.
11. Working at the bench, transfer tubes to dry ice to freeze and then thaw with hands or in a 37 °C water bath. Repeat freeze/thaw for a total of 3×.

12. Spin samples in a microcentrifuge for 5 min at max speed then transfer the supernatant to new, labeled tubes (*see* **Note 31**).
13. In a flat-bottom 96-well plate, add 100 μl of each sample in duplicate.
14. Add 100 μl of the BriteLite Plus reagent to each sample. Mix well by pipetting.
15. Wait for at least 1 min, but no more than 15 min before reading the plate using the GloMax Multi Detection system.
16. Use the reciprocal of the highest serum dilution at which Vero cells remain intact with 100 % suppression of cytopathic effect to determine neutralization titers. Use GraphPad Prism 5 software for nonlinear regression fitting with sigmoidal dose–response (variable slope) and to determine the IC50 value.

3.11 DNA Plasmid Based Immunization Studies in Rhesus Macaques (See Note 32)

1. Immunize five to ten rhesus macaques at weeks 0, 4, and 8 with 1 mg (10 mg/ml) of each DNA vaccine via intramuscular delivery to the quadriceps muscle followed by electroporation. Immunize an equivalent number of control animals with the pVax1 vector.
2. Bleed rhesus macaques at 2-week intervals collecting 5 ml for serum studies and 10 ml in EDTA tubes for peripheral blood mononuclear cell (PBMC) isolation.
3. Isolate PBMCs using standard Ficoll-Paque centrifugation and resuspending in complete R10 culture medium.
4. Lyse red blood cells using ACK lysis buffer.
5. Resuspend cells in R10 medium, count, and then freeze down (10×10^6 cells/ml) using 90 % FBS 10 % DMSO mixture until further use.
6. Similar immune assays may be performed as has been described before for mice.

4 Notes

1. Alternatively, any suitable mammalian cell line such as RD, BHK-21, or Vero cells may be used.
2. Substitute 20 % methanol in place of 10 % methanol if simultaneously transferring two membranes instead of one.
3. House 6–8-week-old Balb/C or C57BL/6 mice in temperature-controlled, light-cycled facilities in accordance with the guidelines of the National Institutes of Health (Bethesda) and the host research Institutional Animal Care and Use Committee (IACUC). Mice are divided into experimental groups as appropriate for the animal study.

4. Peptide pools consist of 15-mers overlapped by 9 or 11 amino acids, which span the entire length of the appropriate protein.
5. Pseudoviruses are obtained by co-transfection with an envelope-expressing plasmid plus a lentiviral backbone plasmid that lacks envelope. Co-transfection generates pseudovirus particles able to infect cells but unable to produce infectious progeny due to an absence of a complete genome. The envelope protein solely determines viral receptor binding, membrane fusion, entry, and neutralizing antibody responses. A CHIKV pseudovirus incorporates CHIKV envelope glycoproteins into lentiviral pseudoparticles. For this assay, the pseudovirus must contain the luciferase gene in its backbone.
6. When designing a DNA vaccine, it is helpful to create a construct that will not only be well expressed, but will provide a broad protective response. Hence, the consensus vaccine approach is recommended.
7. In the case of multiple antigens encoded in a single DNA plasmid, as was constructed in the study by Mallilankaraman et al., furin cleavage sites may be added to facilitate proper processing [9].
8. Add the mixture dropwise. Gently swirl plates afterwards to allow even distribution.
9. An example of a primary and secondary antibody combination we have used successfully is mouse anti-Env IgG antibodies from CHIKV immunized mouse sera (1:500) with Alexa Fluor 488-Anti mouse IgG (Invitrogen, Molecular Probes).
10. All steps including and following the addition of the fluorescently labeled secondary antibody should be performed in the dark or slides should be covered so as to minimize contact with light.
11. Ideally, allow for mounted coverslips to dry overnight prior to imaging. However, slides may be imaged sooner with caution.
12. DNA vaccines delivery using intramuscular injection with an electroporation (EP) technique has been shown to increase plasmid uptake, enhance the immunogenic response of the vaccine and to demonstrate vaccine efficacy in human clinical trial recently [9, 52–56].
13. Perform injections for a CHIKV viral challenge similarly as outlined for DNA vaccine mice immunizations. Administer an injection of 1×10^7 pfu of CHIKV-viral strain in 25 μl volume subcutaneously into the dorsum of each hind foot. Fourteen days following the challenge, monitor mice daily for survival as well as signs of infection such as foot swelling, loss of body weight, and lethargy. Measure foot swelling (height by breadth)

by digital caliper. Euthanize animals with greater than 30 % body mass loss, collect serum samples for cytokine quantification and immune analysis. On post-challenge days 7 and 14, collect blood samples from the tail or via retro-orbital bleeding. Analyze viremia by a plaque assay (pfu/ml).

14. If pooling spleens, add all spleens to one tube in an appropriate volume of R10 (i.e., add three spleens to 30 ml R10).
15. If cell pellet still contains a large amount of erythrocytes, repeat ACK step.
16. Cells are now ready to be used for either intracellular cytokine staining or Enzyme-Linked Immuno Spot (ELISpot) assay.
17. Perform an ELISpot assay to measure antigen specific T cell immune response. IFN-γ is a type II interferon important for innate and adaptive immunity with multiple functions, mostly related to T helper and cytotoxic T cell response to infection. IFN-γ ELISpot assays can be performed to determine the ability of an immunogen to induce cellular responses and for the purposes of identifying a dominant epitope.
18. From this point onwards, plates may be washed by preparing a basin containing PBS and submerging the plates for each wash.
19. Time for color development may vary anywhere from 10 to 40 min depending on assay conditions and reagent providers. Refer to the wells containing Con A to determine when to stop color development.
20. Remove the back of plate and also rinse the back of membrane to ensure the reaction is stopped completely.
21. Normalize raw values to SFU per million splenocytes after subtraction of background (media only) wells.
22. Perform ELISA assays with sera from vaccinated mice (vaccine construct or control plasmid pVax1) to measure antibody expression kinetics and target antigen binding. Collect sera samples from vaccinated mice at various time points.
23. Flow cytometry and intracellular cytokine staining assays allow further characterization of immune responses and determination of T cell polyfunctionality.
24. If intracellular staining is not required, skip **steps 14–18**.
25. If cells will not be analyzed on the same day, resuspend cells in FACS buffer with 1 % paraformaldehyde. Store at 4 °C and run samples within 24 h of staining.
26. Viral neutralization assays are to determine the highest serum dilution at which cytopathic effects are suppressed.
27. Sera may be pooled from multiple animals.
28. Excess supernatant may be removed using a 100 or 200 μl pipette.

29. Cell lysate may be stored short-term at −20 °C, long-term at −80 °C, or used immediately for the luciferase assay.
30. Following successful mouse challenge studies, the next step in vaccine development is to test efficacy in a nonhuman primate model.

Acknowledgements

We acknowledge Seleeke Flingai and Emma Reuschel of the Weiner laboratory for significant contributions and/or critical reading and editing of these methods.

References

1. Teo TH, Lum FM, Claser C, Lulla V, Lulla A, Merits A, Renia L, Ng LF (2013) A pathogenic role for CD4+ T cells during Chikungunya virus infection in mice. J Immunol 190(1):259–269. doi:10.4049/jimmunol.1202177
2. Long KM, Whitmore AC, Ferris MT, Sempowski GD, McGee C, Trollinger B, Gunn B, Heise MT (2013) Dendritic cell immunoreceptor regulates Chikungunya virus pathogenesis in mice. J Virol 87(10):5697–5706. doi:10.1128/JVI.01611-12
3. Burt FJ, Rolph MS, Rulli NE, Mahalingam S, Heise MT (2012) Chikungunya: a re-emerging virus. Lancet 379(9816):662–671. doi:10.1016/S0140-6736(11)60281-X
4. Ross RW (1956) The Newala epidemic. III. The virus: isolation, pathogenic properties and relationship to the epidemic. J Hyg 54(2):177–191
5. Her Z, Kam YW, Lin RT, Ng LF (2009) Chikungunya: a bending reality. Microbes Infect 11(14–15):1165–1176. doi:10.1016/j.micinf.2009.09.004
6. Van Bortel W, Dorleans F, Rosine J, Blateau A, Rousset D, Matheus S, Leparc-Goffart I, Flusin O, Prat C, Cesaire R, Najioullah F, Ardillon V, Balleydier E, Carvalho L, Lemaitre A, Noel H, Servas V, Six C, Zurbaran M, Leon L, Guinard A, van den Kerkhof J, Henry M, Fanoy E, Braks M, Reimerink J, Swaan C, Georges R, Brooks L, Freedman J, Sudre B, Zeller H (2014) Chikungunya outbreak in the Caribbean region, December 2013 to March 2014, and the significance for Europe. Euro Surveill 19(13)
7. Schwartz O, Albert ML (2010) Biology and pathogenesis of chikungunya virus. Nat Rev Microbiol 8(7):491–500. doi:10.1038/nrmicro2368
8. Suhrbier A, Jaffar-Bandjee MC, Gasque P (2012) Arthritogenic alphaviruses—an overview. Nat Rev Rheumatol 8(7):420–429. doi:10.1038/nrrheum.2012.64
9. Mallilankaraman K, Shedlock DJ, Bao H, Kawalekar OU, Fagone P, Ramanathan AA, Ferraro B, Stabenow J, Vijayachari P, Sundaram SG, Muruganandam N, Sarangan G, Srikanth P, Khan AS, Lewis MG, Kim JJ, Sardesai NY, Muthumani K, Weiner DB (2011) A DNA vaccine against Chikungunya virus is protective in mice and induces neutralizing antibodies in mice and nonhuman primates. PLoS Negl Trop Dis 5(1):e928. doi:10.1371/journal.pntd.0000928
10. Silva LA, Khomandiak S, Ashbrook AW, Weller R, Heise MT, Morrison TE, Dermody TS (2014) A single-amino-acid polymorphism in Chikungunya virus E2 glycoprotein influences glycosaminoglycan utilization. J Virol 88(5):2385–2397. doi:10.1128/JVI.03116-13
11. Thiboutot MM, Kannan S, Kawalekar OU, Shedlock DJ, Khan AS, Sarangan G, Srikanth P, Weiner DB, Muthumani K (2010) Chikungunya: a potentially emerging epidemic? PLoS Negl Trop Dis 4(4):e623. doi:10.1371/journal.pntd.0000623
12. Dubrulle M, Mousson L, Moutailler S, Vazeille M, Failloux AB (2009) Chikungunya virus and Aedes mosquitoes: saliva is infectious as soon as two days after oral infection. PLoS One 4(6):e5895. doi:10.1371/journal.pone.0005895
13. Schuffenecker I, Iteman I, Michault A, Murri S, Frangeul L, Vaney MC, Lavenir R, Pardigon N, Reynes JM, Pettinelli F, Biscornet L, Diancourt L, Michel S, Duquerroy S, Guigon G, Frenkiel MP, Brehin AC, Cubito N, Despres P, Kunst F, Rey FA, Zeller H, Brisse S (2006) Genome microevolution of chikungunya

viruses causing the Indian Ocean outbreak. PLoS Med 3(7):e263

14. Vazeille M, Moutailler S, Coudrier D, Rousseaux C, Khun H, Huerre M, Thiria J, Dehecq JS, Fontenille D, Schuffenecker I, Despres P, Failloux AB (2007) Two Chikungunya isolates from the outbreak of La Reunion (Indian Ocean) exhibit different patterns of infection in the mosquito, Aedes albopictus. PLoS One 2(11):e1168
15. Eisen L, Moore CG (2013) Aedes (Stegomyia) aegypti in the continental United States: a vector at the cool margin of its geographic range. J Med Entomol 50(3):467–478
16. Pesko K, Westbrook CJ, Mores CN, Lounibos LP, Reiskind MH (2009) Effects of infectious virus dose and bloodmeal delivery method on susceptibility of Aedes aegypti and Aedes albopictus to chikungunya virus. J Med Entomol 46(2):395–399
17. Reiskind MH, Lounibos LP (2013) Spatial and temporal patterns of abundance of Aedes aegypti L. (Stegomyia aegypti) and Aedes albopictus (Skuse) [Stegomyia albopictus (Skuse)] in southern Florida. Med Vet Entomol 27(4):421–429. doi:10.1111/mve.12000
18. Gerardin P, Barau G, Michault A, Bintner M, Randrianaivo H, Choker G, Lenglet Y, Touret Y, Bouveret A, Grivard P, Le Roux K, Blanc S, Schuffenecker I, Couderc T, Arenzana-Seisdedos F, Lecuit M, Robillard PY (2008) Multidisciplinary prospective study of mother-to-child chikungunya virus infections on the island of La Reunion. PLoS Med 5(3):e60. doi:10.1371/journal.pmed.0050060, 07-PLME-RA-1274 [pii]
19. Ramful D, Carbonnier M, Pasquet M, Bouhmani B, Ghazouani J, Noormahomed T, Beullier G, Attali T, Samperiz S, Fourmaintraux A, Alessandri JL (2007) Mother-to-child transmission of Chikungunya virus infection. Pediatr Infect Dis J 26(9):811–815. doi:10.1097/INF.0b013e3180616d4f
20. Poo YS, Nakaya H, Gardner J, Larcher T, Schroder WA, Le TT, Major LD, Suhrbier A (2014) CCR2 deficiency promotes exacerbated chronic erosive neutrophil-dominated Chikungunya virus arthritis. J Virol 88(2):6862–6872. doi:10.1128/JVI.03364-13
21. Hoarau JJ, Gay F, Pelle O, Samri A, Jaffar-Bandjee MC, Gasque P, Autran B (2013) Identical strength of the T cell responses against E2, nsP1 and capsid CHIKV proteins in recovered and chronic patients after the epidemics of 2005–2006 in La Reunion Island. PLoS One 8(12):e84695. doi:10.1371/journal.pone.0084695
22. Labadie K, Larcher T, Joubert C, Mannioui A, Delache B, Brochard P, Guigand L, Dubreil L, Lebon P, Verrier B, de Lamballerie X, Suhrbier A, Cherel Y, Le Grand R, Roques P (2010) Chikungunya disease in nonhuman primates involves long-term viral persistence in macrophages. J Clin Invest 120(3):894–906. doi:10.1172/JCI40104, 40104 [pii]
23. Lidbury BA, Rulli NE, Suhrbier A, Smith PN, McColl SR, Cunningham AL, Tarkowski A, van Rooijen N, Fraser RJ, Mahalingam S (2008) Macrophage-derived proinflammatory factors contribute to the development of arthritis and myositis after infection with an arthrogenic alphavirus. J Infect Dis 197(11):1585–1593. doi:10.1086/587841
24. Farnon EC, Sejvar JJ, Staples JE (2008) Severe disease manifestations associated with acute chikungunya virus infection. Crit Care Med 36(9):2682–2683. doi:10.1097/CCM.0b013e3181843d94
25. Couderc T, Chretien F, Schilte C, Disson O, Brigitte M, Guivel-Benhassine F, Touret Y, Barau G, Cayet N, Schuffenecker I, Despres P, Arenzana-Seisdedos F, Michault A, Albert ML, Lecuit M (2008) A mouse model for Chikungunya: young age and inefficient type-I interferon signaling are risk factors for severe disease. PLoS Pathog 4(2):e29
26. Queyriaux B, Simon F, Grandadam M, Michel R, Tolou H, Boutin JP (2008) Clinical burden of chikungunya virus infection. Lancet Infect Dis 8(1):2–3. doi:10.1016/S1473-3099(07)70294-3
27. Robin S, Ramful D, Le Seach F, Jaffar-Bandjee MC, Rigou G, Alessandri JL (2008) Neurologic manifestations of pediatric chikungunya infection. J Child Neurol 23(9):1028–1035. doi:10.1177/0883073808314151, 0883073808314151 [pii]
28. Simon F, Paule P, Oliver M (2008) Chikungunya virus-induced myopericarditis: toward an increase of dilated cardiomyopathy in countries with epidemics? Am J Trop Med Hyg 78(2):212–213, doi:78/2/212 [pii]
29. Couderc T, Khandoudi N, Grandadam M, Visse C, Gangneux N, Bagot S, Prost JF, Lecuit M (2009) Prophylaxis and therapy for Chikungunya virus infection. J Infect Dis 200(4):516–523. doi:10.1086/600381
30. Brighton SW (1984) Chloroquine phosphate treatment of chronic Chikungunya arthritis. An open pilot study. S Afr Med J 66(6):217–218
31. Briolant S, Garin D, Scaramozzino N, Jouan A, Crance JM (2004) In vitro inhibition of Chikungunya and Semliki Forest viruses repli-

cation by antiviral compounds: synergistic effect of interferon-alpha and ribavirin combination. Antiviral Res 61(2):111–117

32. de Lamballerie X, Leroy E, Charrel RN, Ttsetsarkin K, Higgs S, Gould EA (2008) Chikungunya virus adapts to tiger mosquito via evolutionary convergence: a sign of things to come? Virol J 5:33. doi:10.1186/1743-422X-5-33, 1743-422X-5-33 [pii]
33. Khan M, Santhosh SR, Tiwari M, Lakshmana Rao PV, Parida M (2010) Assessment of in vitro prophylactic and therapeutic efficacy of chloroquine against Chikungunya virus in vero cells. J Med Virol 82(5):817–824. doi:10.1002/jmv.21663
34. Ravichandran R, Manian M (2008) Ribavirin therapy for Chikungunya arthritis. J Infect Dev Ctries 2(2):140–142
35. Schilte C, Couderc T, Chretien F, Sourisseau M, Gangneux N, Guivel-Benhassine F, Kraxner A, Tschopp J, Higgs S, Michault A, Arenzana-Seisdedos F, Colonna M, Peduto L, Schwartz O, Lecuit M, Albert ML (2010) Type I IFN controls chikungunya virus via its action on nonhematopoietic cells. J Exp Med 207(2): 429–442. doi:10.1084/jem.20090851
36. Smee DF, Alaghamandan HA, Kini GD, Robins RK (1988) Antiviral activity and mode of action of ribavirin 5′-sulfamate against Semliki Forest virus. Antiviral Res 10(6): 253–262
37. Edelman R, Tacket CO, Wasserman SS, Bodison SA, Perry JG, Mangiafico JA (2000) Phase II safety and immunogenicity study of live chikungunya virus vaccine TSI-GSD-218. Am J Trop Med Hyg 62(6):681–685
38. Gorchakov R, Wang E, Leal G, Forrester NL, Plante K, Rossi SL, Partidos CD, Adams AP, Seymour RL, Weger J, Borland EM, Sherman MB, Powers AM, Osorio JE, Weaver SC (2012) Attenuation of Chikungunya virus vaccine strain 181/clone 25 is determined by two amino acid substitutions in the E2 envelope glycoprotein. J Virol 86(11):6084–6096. doi:10.1128/JVI.06449-11
39. Hallengard D, Kakoulidou M, Lulla A, Kummerer BM, Johansson DX, Mutso M, Lulla V, Fazakerley JK, Roques P, Le Grand R, Merits A, Liljestrom P (2014) Novel attenuated Chikungunya vaccine candidates elicit protective immunity in C57BL/6 mice. J Virol 88(5):2858–2866. doi:10.1128/JVI.03453-13
40. Noranate N, Takeda N, Chetanachan P, Sittisaman P, A-Nuegoonpipat A, Anantapreecha S (2014) Characterization of chikungunya virus-like particles. PLoS One 9(9):e108169. doi:10.1371/journal.pone.0108169
41. Brandler S, Ruffie C, Combredet C, Brault JB, Najburg V, Prevost MC, Habel A, Tauber E, Despres P, Tangy F (2013) A recombinant measles vaccine expressing chikungunya virus-like particles is strongly immunogenic and protects mice from lethal challenge with chikungunya virus. Vaccine 31(36):3718–3725. doi:10.1016/j.vaccine.2013.05.086
42. Kumar M, Sudeep AB, Arankalle VA (2012) Evaluation of recombinant E2 protein-based and whole-virus inactivated candidate vaccines against chikungunya virus. Vaccine 30(43):6142–6149. doi:10.1016/j.vaccine.2012.07.072
43. McClain DJ, Pittman PR, Ramsburg HH, Nelson GO, Rossi CA, Mangiafico JA, Schmaljohn AL, Malinoski FJ (1998) Immunologic interference from sequential administration of live attenuated alphavirus vaccines. J Infect Dis 177(3):634–641
44. Tiwari M, Parida M, Santhosh SR, Khan M, Dash PK, Rao PV (2009) Assessment of immunogenic potential of Vero adapted formalin inactivated vaccine derived from novel ECSA genotype of Chikungunya virus. Vaccine 27(18): 2513–2522. doi:10.1016/j.vaccine.2009.02.062, S0264-410X(09)00289-8 [pii]
45. Wang E, Kim DY, Weaver SC, Frolov I (2011) Chimeric Chikungunya viruses are nonpathogenic in highly sensitive mouse models but efficiently induce a protective immune response. J Virol 85(17):9249–9252. doi:10.1128/JVI.00844-11
46. Bao H, Ramanathan AA, Kawalakar O, Sundaram SG, Tingey C, Bian CB, Muruganandam N, Vijayachari P, Sardesai NY, Weiner DB, Ugen KE, Muthumani K (2013) Nonstructural protein 2 (nsP2) of Chikungunya virus (CHIKV) enhances protective immunity mediated by a CHIKV envelope protein expressing DNA Vaccine. Viral Immunol 26(1):75–83. doi:10.1089/vim.2012.0061
47. Muthumani K, Lankaraman KM, Laddy DJ, Sundaram SG, Chung CW, Sako E, Wu L, Khan A, Sardesai N, Kim JJ, Vijayachari P, Weiner DB (2008) Immunogenicity of novel consensus-based DNA vaccines against Chikungunya virus. Vaccine 26(40):5128–5134. doi:10.1016/j.vaccine.2008.03.060
48. Hokey DA, Weiner DB (2006) DNA vaccines for HIV: challenges and opportunities. Springer Semin Immunopathol 28(3):267–279. doi:10.1007/s00281-006-0046-z
49. Kutzler MA, Weiner DB (2008) DNA vaccines: ready for prime time? Nat Rev Genet 9(10):776–788. doi:10.1038/nrg2432
50. Laddy DJ, Yan J, Khan AS, Andersen H, Cohn A, Greenhouse J, Lewis M, Manischewitz J, King LR, Golding H, Draghia-Akli R, Weiner

DB (2009) Electroporation of synthetic DNA antigens offers protection in nonhuman primates challenged with highly pathogenic avian influenza virus. J Virol 83(9): 4624–4630. doi:10.1128/JVI.02335-08, JVI.02335-08 [pii]

51. Muthumani G, Laddy DJ, Sundaram SG, Fagone P, Shedlock DJ, Kannan S, Wu L, Chung CW, Lankaraman KM, Burns J, Muthumani K, Weiner DB (2009) Co-immunization with an optimized plasmid-encoded immune stimulatory interleukin, high-mobility group box 1 protein, results in enhanced interferon-gamma secretion by antigen-specific CD8 T cells. Immunology 128(1 Suppl):e612–e620. doi:10.1111/j.1365-2567.2009.03044.x, IMM3044 [pii]

52. Bagarazzi ML, Yan J, Morrow MP, Shen X, Parker RL, Lee JC, Giffear M, Pankhong P, Khan AS, Broderick KE, Knott C, Lin F, Boyer JD, Draghia-Akli R, White CJ, Kim JJ, Weiner DB, Sardesai NY (2012) Immunotherapy against HPV16/18 generates potent TH1 and cytotoxic cellular immune responses. Sci Transl Med 4(155):155ra138. doi:10.1126/scitranslmed.3004414

53. Muthumani K, Falzarano D, Reuschel EL, Tingey C, Flingai S, Villarreal DO, Wise M, Patel A, Izmirly A, Aljuaid A, Seliga AM, Soule G, Morrow M, Kraynyak KA, Khan AS, Scott DP, Feldmann F, LaCasse R, Meade-White K, Okumura A, Ugen KE, Sardesai NY, Kim JJ, Kobinger G, Feldmann H, Weiner DB (2015) A synthetic consensus anti-spike protein DNA vaccine induces protective immunity against Middle East respiratory syndrome coronavirus in nonhuman primates. Sci Transl Med 7(301):301ra132. doi:10.1126/scitranslmed.aac7462

54. Flingai S, Plummer EM, Patel A, Shresta S, Mendoza JM, Broderick KE, Sardesai NY, Muthumani K, Weiner DB (2015) Protection against dengue disease by synthetic nucleic acid antibody prophylaxis/immunotherapy. Sci Rep 5:12616. doi:10.1038/srep12616

55. Muthumani K, Wise MC, Broderick KE, Hutnick N, Goodman J, Flingai S, Yan J, Bian CB, Mendoza J, Tingey C, Wilson C, Wojtak K, Sardesai NY, Weiner DB (2013) HIV-1 Env DNA vaccine plus protein boost delivered by EP expands B- and T-cell responses and neutralizing phenotype in vivo. PLoS One 8(12):e84234. doi:10.1371/journal.pone.0084234

56. Trimble CL, Morrow MP, Kraynyak KA, Shen X, Dallas M, Yan J, Edwards L, Parker RL, Denny L, Giffear M, Brown AS, Marcozzi-Pierce K, Shah D, Slager AM, Sylvester AJ, Khan A, Broderick KE, Juba RJ, Herring TA, Boyer J, Lee J, Sardesai NY, Weiner DB, Bagarazzi ML (2015) Safety, efficacy, and immunogenicity of VGX-3100, a therapeutic synthetic DNA vaccine targeting human papillomavirus 16 and 18 E6 and E7 proteins for cervical intraepithelial neoplasia 2/3: a randomised, double-blind, placebo-controlled phase 2b trial. Lancet 386(10008):2078–2088. doi:10.1016/S0140-6736(15)00239-1, pii, S0140-6736(15)00239-1

INDEX

A

Alphavirus ... 3, 51, 60, 63, 68, 75, 87, 93, 114, 119, 137, 154, 158, 179, 196, 198, 211, 243, 244, 250, 251, 255, 263, 273, 283, 297
Antibody
  monoclonal ... 66, 80, 159–161, 225, 274
  primary ... 145–147, 149, 150, 172, 235, 236, 238, 239, 322
  secondary ... 49, 66, 145, 146, 148–150, 172, 230, 236, 238, 316, 317, 321, 322, 327
Antigen ... 3, 9, 39, 40, 51, 53, 68, 79, 81, 143, 144, 146, 157, 231, 235–240, 327, 328
Antivirals ... 201, 243–251, 255–261, 264, 313
Aquacide ... 95, 99
Aquacide II ... 95, 99
Arbidol ... 244, 251
Arbovirus ... 64, 129, 144, 273, 297, 298
Arthralgia ... 4, 63, 64, 75, 151, 243, 264, 273, 297
Artificial blood meal ... 129–141

B

Baculovirus ... 52, 263–272, 298–300, 302–305, 308
Bioinformatics ... 47, 191, 195–199, 201, 319

C

C6/36 ... 65, 77, 80, 88–91, 144, 153, 218, 275, 280, 286, 291, 294
Carboxymethyl cellulose (CMC) ... 94, 99
Cellular immunity ... 88, 315
Chikungunya virus ... 3–8, 11–19, 21–36, 51, 54, 58–59, 63–64, 67, 69, 75–82, 87–91, 93–102, 105–116, 119–127, 130, 132, 143–151, 153–161, 179–193, 195–199, 201–207, 211–223, 225 232, 235–240, 243–251, 255–261, 263–281, 283–294, 297, 312–329
Chloroquine ... 244, 248, 313
Common ancestor ... 24, 28–30
Crystal violet ... 72, 95–99, 102
Cytopathic effect (CPE) ... 71, 72, 80, 88, 91, 93, 94, 130, 136–139, 141, 147, 153, 156–158, 161, 239, 244, 248, 256, 260, 274, 276, 280, 291, 303, 304, 308, 326, 328

D

Detection ... 39–50, 65, 66, 68, 76–79, 93, 94, 105–116, 120, 132–134, 136, 139, 143–151, 235–240, 264, 273, 319, 324, 326
DNA
  alignment (*see* Multiple alignment)
  cloning ... 286
  extraction ... 290, 299
  quantifiaction ... 106
  vaccine ... 298, 311–329

E

Electrophoresis
  one dimensional ... 40–41, 44–45
  two dimensional ... 41–42, 45–46
Envelope protein
  E1 glycoprotein ... 51, 164, 165
  E2 glycoprotein ... 51–60, 164, 165, 298
  secreted E1 and E2 ... 165
Enzyme-linked immunosorbent assay (ELISA) ... 39–41, 43, 44, 48, 49, 51, 54, 60, 65, 66, 68, 76, 77, 79–80, 93, 94, 105, 328
Epitopes ... 40, 52, 201–207, 226, 328
  prediction ... 201–207
Evolution ... 3–8, 21–36, 94
Evolutionary network ... 22, 26–28

F

Favipiravir ... 247
Flow cytometry ... 93, 144, 146, 149, 192, 318, 324–325, 328
Fluorescence microscopy ... 94, 264, 266–270, 304, 307
Formalin-fixed ... 235–240
Fusion ... 165, 168, 170, 170, 172, 176, 177, 177, 225, 225, 226, 227, 227, 228, 229, 229, 231, 232, 264, 264, 266, 268–269, 269, 271, 286, 308, 327, 244, 292, 164, 165, 170, 175
Human cDNA library ... 165
Humoral responses ... 315

Justin Jang Hann Chu and Swee Kim Ang (eds.), *Chikungunya Virus: Methods and Protocols*, Methods in Molecular Biology, vol. 1426, DOI 10.1007/978-1-4939-3618-2, © Springer Science+Business Media New York 2016

I

Immune modulators 249–250
Immunodominant antigen 51
Immunofluorescence (IFA) assay 65, 68, 93, 94, 105, 143–151, 226, 236, 270, 273, 315–317, 321, 322
Immunogold labeling 155, 156, 159, 160
Immunohistochemistry 235–237
Infection
  intracerebral 212–217
  intranasal 212–213, 215, 217
  subcutaneous 218–221, 235
Infectious cDNA clone 283–285
Insect cells 51–60, 88, 264, 266–267, 271, 297–308
Internal ribosome entry site (IRES) 264, 265, 271

L

Label free 256

M

Macaques 315, 319, 326
Machine learning 201, 202, 204, 206, 207
Major histocompatibility complex (MHC) 201, 202, 206
Mass spectrometry
  GeLC-MS/MS 179–193
  LC-MS/MS 39, 40, 42–43, 47, 180
Mice 77, 80, 82, 153, 212–222, 226, 227, 231, 235, 236, 239, 314, 315, 317–319, 322, 323, 327
Microneutralization. *See* Plaque, reduction neutralization test
Molecular surveillance 195
Mosquito
  Aedes aegypti 4, 85, 87, 119, 134–135, 263, 312
  Aedes albopictus 4, 22, 75, 80, 87, 88, 119, 134, 135, 144, 211, 263, 286, 312
  cells 87–91, 275, 280
  infection 120, 122–123
  processing 132, 136–139
Multiple alignment 24–25
Mus musculus (*see* Mice)

N

Negative staining 154, 155, 160, 238–240
Neutralizing/neutralization
  antibody 68, 274, 280, 315, 327
  assay 72, 273–281, 315, 319, 325–326, 328

O

Oral infection 120

P

Paraffin-embedded 235–240
Peptide 40, 43, 45, 47–49, 87, 179–199, 202, 203, 206, 260, 314, 317, 318, 323, 324, 327
Phylogenetics 4, 5, 22, 24, 26, 27, 34, 40, 43, 45, 47–49, 87, 179–199, 202, 203, 206, 314, 327
Phylogenetic tree 12, 18–19, 26, 28–30, 79
Phylogeography 22, 28, 31–32, 36
Plaque
  assays 21, 90, 93–102, 116, 144, 147, 153, 156, 158, 161, 218, 271, 274, 294, 328
  reduction neutralization test 273–281
Polymerase chain reaction (PCR) 14–19, 23, 32, 53, 55, 59, 65, 76–78, 105–116, 121, 124, 141, 143, 165, 169, 173–175, 284–307
Production 51, 52, 88, 100, 264, 271, 297–308
Proliferation 195, 256, 257, 314
Protein
  interactions 163–178, 195–198
  purification 53, 158, 294, 298
  structure similarity 196
Protein–protein interactions 195, 198
Proteome 39, 180, 197, 199

R

Real-time
  cell analysis (RTCA) 256–258, 260
  reverse transcription PCR (RT-PCR) 14–16, 18, 65, 76, 78, 80, 81, 94, 105–116, 131, 133, 134, 139, 143, 150, 289, 291
Recombinant protein 51–53, 58, 60, 264
Reverse genetics 283–294
Reverse transcription PCR (RT-PCR) 14–16, 18, 78, 80, 81, 94, 105–116, 133, 134, 139, 143, 150, 289, 291
Ribavirin 246, 249, 251, 256–260, 313

S

Saliva collection 127, 132, 137, 138, 141
Screening 65, 163–168, 195, 201, 226, 229–232, 246–248, 256, 264, 290
Sequencing 5, 11, 12, 14–18, 23, 33, 56, 78, 141, 165, 169, 175, 184, 185, 291, 302
Serological methods 65, 105
Serology 51–60
Serum-free 90, 121, 123, 239, 265, 267, 299–301, 303, 305, 306, 320
Sf9 ....51–60, 271
Site-directed mutagenesis 284, 286, 291–292
Structural bioinformatics 198
Sucrose gradient 153, 158, 159, 308
SYBR Green I 14–16, 106, 107, 110–112, 115
Synthetic peptide 39–49

T

TaqMan®....106, 107, 111, 113, 115
Throat Swab....75–82

U

Urine....75–82, 314, 315

V

Vero cell line....256, 257, 260, 261, 274, 279
Viral RNA
  extraction....14, 53, 76, 107, 108, 111, 285, 287, 291
  quantification....93
Viral transmission dynamics....137
Virus
  propagation....87–97, 286
  titration....94, 143
Virus–host interactions....164, 195–199
Virus-like particle (VLP)....52, 297–308

Y

Yeast two-hybrid (Y2H)....163–178, 195

Made in the USA
Monee, IL
13 September 2026

62976350R00195